TODAY'S TECHNICIAN ™

CLASSROOM MANUAL
FOR MANUAL TRANSMISSIONS & TRANSAXLES

SIXTH EDITION

TODAY'S TECHNICIAN ™

CLASSROOM MANUAL
FOR MANUAL TRANSMISSIONS & TRANSAXLES

JACK ERJAVEC

MICHAEL RONAN

JACK ERJAVEC,
SERIES EDITOR

SIXTH EDITION

CENGAGE
Learning™

Australia • Canada • Mexico • Singapore • Spain • United Kingdom • United States

Today's Technician™: Classroom Manual for Manual Transmissions and Transaxles, 6th Edition

Jack Erjavec
Michael Ronan

SVP, GM Skills & Global Product Management: Dawn Gerrain

Product Team Manager: Erin Brennan

Senior Director, Development: Marah Bellegarde

Senior Product Development Manager: Larry Main

Content Developer: Mary Clyne

Product Assistant: Maria Garguilo

Vice President, Marketing Services: Jennifer Ann Baker

Senior Production Director: Wendy Troeger

Production Director: Andrew Crouth

Senior Content Project Manager: Cheri Plasse

Senior Art Director: Bethany Casey

Cover Image(s): © cla78/Shutterstock

For product information and technology assistance, contact us at
Cengage Learning Customer & Sales Support, 1-800-354-9706

For permission to use material from this text or product,
submit all requests online at **www.cengage.com/permissions.**

Further permissions questions can be e-mailed to
permissionrequest@cengage.com

Library of Congress Control Number: 2014940367

Book-only ISBN: 978-1-3052-6176-1
Package ISBN: 978-1-3052-6178-5

Cengage Learning
20 Channel Center Street
Boston MA 02210
USA

Cengage Learning is a leading provider of customized learning solutions with office locations around the globe, including Singapore, the United Kingdom, Australia, Mexico, Brazil, and Japan. Locate your local office at: **www.cengage.com/global**

Cengage Learning products are represented in Canada by Nelson Education, Ltd.

To learn more about Cengage Learning, visit **www.cengage.com**

Purchase any of our products at your local college store or at our preferred online store **www.cengagebrain.com**

Notice to the Reader

Printed in the United States of America
Print Number: 02   Print Year: 2017

CONTENTS

CONTENTS

Thanks to the support the *Today's Technician* series has received from those who teach automotive technology, Cengage Learning, the leader in automotive-related textbooks, is able to live up to its promise to provide new editions of the series every few years. We have listened and responded to our critics and our fans and present this updated and revised sixth edition. By revising this series on a regular basis, we can respond to changes in the industry, changes in technology, changes in the certification process, and to the ever-changing needs of those who teach automotive technology.

The *Today's Technician* series by Cengage features textbooks that cover all mechanical and electrical systems of automobiles and light trucks (while the heavy-duty trucks portion of the series does the same for heavy-duty vehicles). Principally, the individual titles correspond to the main areas of ASE (National Institute for Automotive Service Excellence) certification.

This new edition, like the last, was designed to give students a chance to develop the same skills and gain the same knowledge that today's successful technician has. This edition also reflects the changes in the guidelines established by the National Automotive Technicians Education Foundation (NATEF) in 2013.

The purpose of NATEF is to evaluate technician training programs against standards developed by the automotive industry and recommend qualifying programs for certification (accreditation) by ASE (National Institute for Automotive Service Excellence). Programs can earn ASE certification upon the recommendation of NATEF. NATEF's national standards reflect the skills that students must master. ASE certification through NATEF evaluation ensures that certified training programs meet or exceed industry-recognized, uniform standards of excellence.

Additional titles include remedial skills and theories common to all of the certification areas and advanced or specific subject areas that reflect the latest technological trends. Each text is divided into two volumes: a Classroom Manual and a Shop Manual.

The technician of today and for the future must know the underlying theory of all automotive systems and be able to service and maintain those systems. Dividing the material into two volumes provides the reader with the information needed to begin a successful career as an automotive technician without interrupting the learning process by mixing cognitive and performance learning objectives into one volume.

The design of Cengage's *Today's Technician* series was based on features that are known to promote improved student learning. The design was further enhanced by a careful study of survey results, in which the respondents were asked to value particular features. Some of these features can be found in other textbooks, while others are unique to this series.

Each Classroom Manual contains the principles of operation for each system and subsystem. The Classroom Manual also contains discussions on design variations of key components used by the different vehicle manufacturers. It also looks into emerging technologies that will be standard or optional features in the near future. This volume is organized to build upon basic facts and theories. The primary objective of this volume is to allow the reader to gain an understanding of how each system and subsystem operates. This understanding is necessary to diagnose the complex automobiles of today and tomorrow. Although the basics contained in the Classroom Manual provide the knowledge needed for diagnostics, diagnostic procedures appear only in the Shop Manual. An understanding of the underlying theories is also a requirement for competence in the skill areas covered in the Shop Manual.

A spiral-bound Shop Manual covers the "how-to's." This volume includes step-by-step instructions for diagnostic and repair procedures. Photo Sequences are used to illustrate some of the common service procedures. Other common procedures are listed and are accompanied with fine line drawings and photos that allow the reader to visualize and conceptualize the finest details of the procedure. This volume also contains the reasons for performing the procedures, as well as when that particular service is appropriate.

The two volumes are designed to be used together and are arranged in corresponding chapters. Not only are the chapters in the volumes linked together, the contents of the chapters are also linked. This linking of content is evidenced by marginal callouts that refer the reader to the chapter and page that the same topic is addressed in the other volume. This feature is valuable to instructors. Without this feature, users of other two-volume textbooks must search the index or table of contents to locate supporting information in the other volume. This is not only cumbersome but also creates additional work for an instructor when planning the presentation of material and when making reading assignments. It is also valuable to the students; with the page references they also know exactly where to look for supportive information.

Both volumes contain clear and thoughtfully selected illustrations, many of which are original drawings or photos specially prepared for inclusion in this series. This means that the art is a vital part of each textbook and not merely inserted to increase the numbers of illustrations.

The page layout, used in the series, is designed to include information that would otherwise break up the flow of information presented to the reader. The main body of the text includes all of the "need-to-know" information and illustrations. The wide side margins of each page contain many of the special features of the series. These features present truly "nice-to-know" information such as: simple examples of concepts just introduced in the text, explanations or definitions of terms that are not defined in the text, examples of common trade jargon used to describe a part or operation, and exceptions to the norm explained in the text. This type of information is placed in the margin, out of the normal flow of information. Many textbooks attempt to include this type of information and insert it in the main body of text; this tends to interrupt the thought process and cannot be pedagogically justified. By placing this information off to the side of the main text, the reader can select when to refer to it.

Jack Erjavec

Series Advisor

HIGHLIGHTS OF THIS EDITION–CLASSROOM MANUAL

The text was updated throughout to include the latest developments. Some of these new topics include dual-clutch systems, various limited-slip differential designs, six-speed transmissions, constantly variable transmissions, and self-shifting manual transmissions. New coverage of high performance topics is integrated throughout the classroom manual to reflect recent advances in the fast-growing automotive performance industry. Chapter 11, Electronically Controlled and Automated Transmissions, covers one of the latest trends in manual transmissions. These transmissions are manual transmissions without a driver-operated clutch. They rely on electronics and hydraulics and provide increased performance and a reduction in fuel consumption and emissions while providing the driver with the conveniences of an automatic transmission.

Chapter 1 introduces the purpose of the main system and how it links to the rest of the vehicle. The chapter also describes the purpose and location of the subsystems, as well as the major components of the system and subsystems. The goal of this chapter is to establish a basic understanding that students can base their learning on. All systems and subsystems that will be discussed in detail later in the text are introduced and their primary purpose described. The second chapter covers the underlying basic theories of operation for the topic of the text. This is valuable to the student and the instructor because it covers the theories that other textbooks assume the reader knows. All related basic physical, chemical, and thermodynamic theories are covered in this chapter.

The order of the topics reflects the most common reviewer suggestions. Many of this edition's updates include current electronic applications. Current model transmissions are used as examples throughout the text. Some are discussed in detail. This includes six- and seven-speed and constantly variable transmissions. This new edition also has more information on nearly all manual transmission-related topics. Finally, the art has been updated throughout the text to enhance comprehension and improve visual interest.

HIGHLIGHTS OF THIS EDITION–SHOP MANUAL

Along with the Classroom Manual, the Shop Manual was updated to match current trends. Service information related to the new topics covered in the Classroom Manual is included in this manual. New coverage of high performance topics is integrated throughout the shop manual to reflect recent advances in the fast-growing automotive performance industry. Twenty-seven detailed photo sequences show students what to expect when they perform the same procedure. They can also provide a student with familiarity of a system or type of equipment they may not be able to perform at their school. Although the main purpose of the textbook is not to prepare someone to successfully pass an ASE exam, all of the information required to do so is included in the textbook.

To stress the importance of safe work habits, Chapter 1 is dedicated to safety. Included in this chapter are common shop hazards, safe shop practices, safety equipment, and legislation concerning the safe handling of hazardous materials and wastes. This chapter also includes precautions that must be adhered to when servicing a hybrid vehicle. Chapter 2 covers the basic skills a transmission technician uses to earn a living, including basic diagnostics. Also included in this chapter are those tools and procedures that are commonly used to diagnose and service manual transmissions and drivelines. In summary, this chapter describes what it takes to be a successful technician, typical pay plans for technicians, service information sources, preparing repair orders, ASE certification, and the laws and regulations a technician should be aware of.

The rest of the chapters have been thoroughly updated. Much of the updating focuses on the diagnosis and service to new systems, as well as those systems instructors have said they need more help in. Currently accepted service procedures are used as examples throughout the text. These procedures also served as the basis for new job sheets that are included in the text. Finally, the art has been updated throughout the text to enhance comprehension and improve visual interest.

CLASSROOM MANUAL

Features of this manual include:

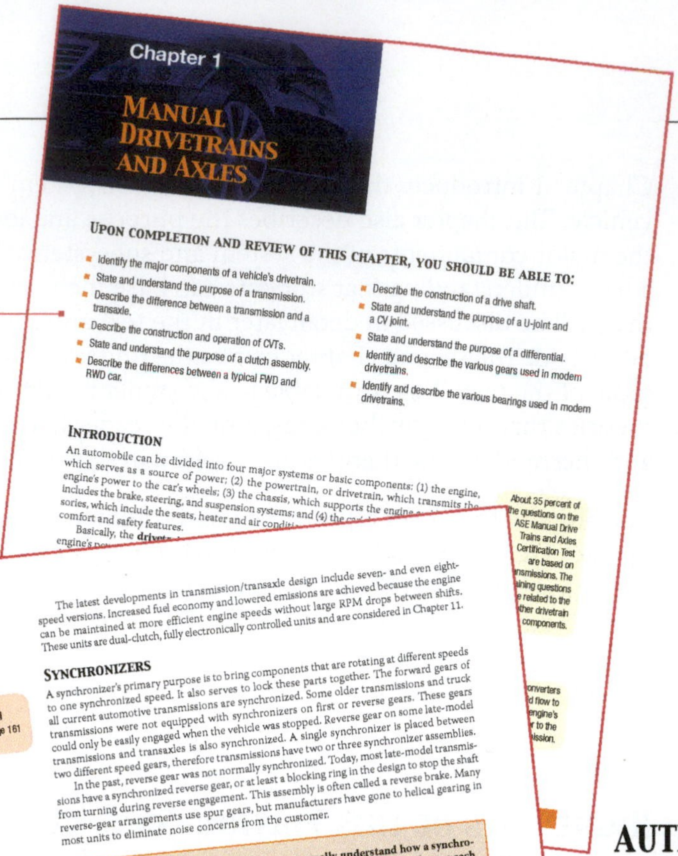

COGNITIVE OBJECTIVES

These objectives define the contents of the chapter and define what the student should have learned upon completion of the chapter.

Each topic is divided into small units to promote easier understanding and learning.

CROSS-REFERENCES TO THE SHOP MANUAL

Reference to the appropriate page in the Shop Manual is given whenever necessary. Although the chapters of the two manuals are synchronized, material covered in other chapters of the Shop Manual may be fundamental to the topic discussed in the Classroom Manual.

MARGINAL NOTES

These notes add "nice-to-know" information to the discussion. They may include examples or exceptions, or may give the common trade jargon for a component.

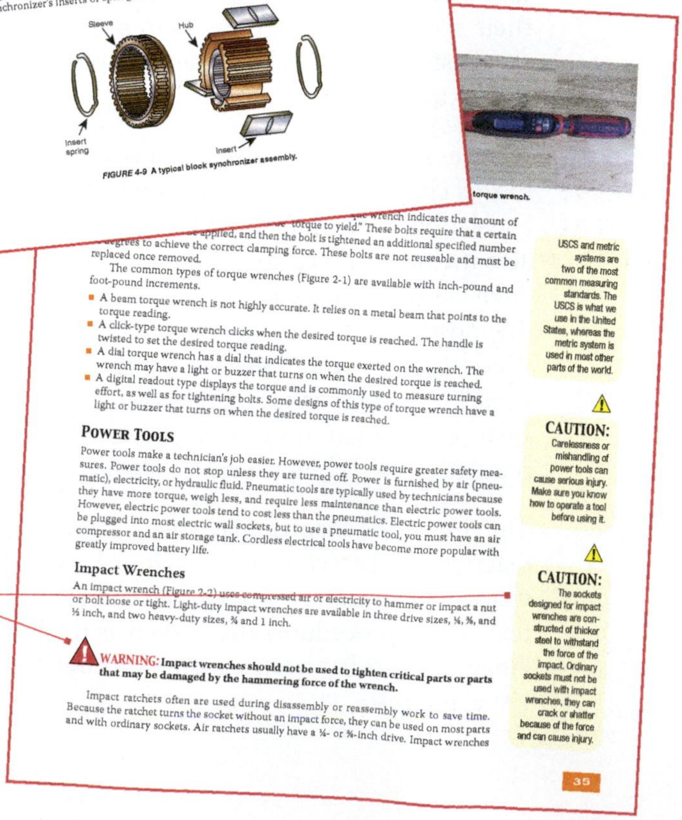

CAUTIONS AND WARNINGS

Throughout the text, warnings are given to alert the reader to potentially hazardous materials or unsafe conditions. Cautions are given to advise the student of things that can go wrong if instructions are not followed or if a nonacceptable part or tool is used.

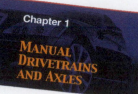

Chapter 1

MANUAL DRIVETRAINS AND AXLES

UPON COMPLETION AND REVIEW OF THIS CHAPTER, YOU SHOULD BE ABLE TO:

- Identify the major components of a vehicle's drivetrain.
- State and understand the purpose of a transmission.
- Describe the difference between a transmission and a transaxle.
- Describe the construction and operation of CVTs.
- State and understand the purpose of a clutch assembly.
- Describe the differences between a typical FWD and RWD car.
- Describe the construction of a drive shaft.
- State and understand the purpose of a U-joint and a CV joint.
- State and understand the purpose of a differential.
- Identify and describe the various gears used in modern drivetrains.
- Identify and describe the various bearings used in modern drivetrains.

INTRODUCTION

An automobile can be divided into four major systems or basic components: (1) the engine, which serves as a source of power; (2) the powertrain, or drivetrain, which transmits the engine's power to the car's wheels; (3) the chassis, which supports the engine and includes the brake, steering, and suspension systems; and (4) the accessories, which include the seats, heater and air conditioning, comfort and safety features.

Basically, the **drivetrain** ...

engine's power ...

About 35 percent of the questions on the ASE Manual Drive Trains and Axles Certification Test are based on transmissions. The remaining questions are related to the other drivetrain components.

The latest developments in transmission/transaxle design include seven- and even eight-speed versions. Increased fuel economy and lowered emissions are achieved because the engine can be maintained at more efficient engine speeds without large RPM drops between shifts. These units are dual-clutch, fully electronically controlled units and are considered in Chapter 11.

SYNCHRONIZERS

A synchronizer's primary purpose is to bring components that are rotating at different speeds to one synchronized speed. It also serves to lock these parts together. The forward gears of all current automotive transmissions are synchronized. Some older transmissions and truck transmissions were not equipped with synchronizers on first or reverse gears. These gears could only be easily engaged when the vehicle was stopped. Reverse gear on some late-model transmissions and transaxles is also synchronized. A single synchronizer is placed between two different speed gears, therefore transmissions have two or three synchronizer assemblies.

In the past, reverse gear was not normally synchronized. Today, most late-model transmissions have a synchronized reverse gear, or at least a blocking ring in the design to stop the shaft from turning during reverse engagement. This assembly is often called a reverse brake. Many reverse-gear arrangements use spur gears, but manufacturers have gone to helical gearing in most units to eliminate noise concerns from the customer.

AUTHOR'S NOTE: Sometimes the best way to really understand how a synchronizer works is to hold a complete synchro assembly in your hands and move each part around and back and forth. Notice what else moves when the parts are in their various positions, then think about the rest of the transmission parts and their movements.

Synchronizer Designs

There are four types of synchronizers used in synchromesh transmissions: block, disc and plate, plain, and pin. The most commonly used type on current transmissions is the block type. All synchronizers use friction to synchronize the speed of the gear and shaft before the connection is made.

Block synchronizers consist of a hub (called a **clutch hub**), sleeve, blocking ring, and inserts or spring-and-ball detent devices (Figure 4-9). The synchronizer sleeve surrounds the synchronizer assembly and meshes with the external splines of the hub. The hub is internally splined to the transmission's main shaft. The outside of the sleeve is grooved to accept the shifting fork. Three slots are equally spaced around the outside of the hub and are fitted with the synchronizer's inserts or spring-and-ball detent assemblies.

FIGURE 4-9 A typical block synchronizer assembly.

AUTHOR'S NOTES

This feature includes simple explanations, stories, or examples of complex topics. These are included to help students understand difficult concepts.

---

wrench indicates the amount of "torque to yield." These bolts require that a certain ... applied, and then the bolt is tightened an additional specified number of degrees to achieve the correct clamping force. These bolts are not reuseable and must be replaced once removed.

The common types of torque wrenches (Figure 2-1) are available with inch-pound and foot-pound increments.

- A beam torque wrench is not highly accurate. It relies on a metal beam that points to the torque reading.
- A click-type torque wrench clicks when the desired torque is reached. The handle is twisted to set the desired torque reading.
- A dial torque wrench has a dial that indicates the torque exerted on the wrench. The wrench may have a light or buzzer that turns on when the desired torque is reached.
- A digital readout type displays the torque and is commonly used to measure turning effort, as well as for tightening bolts. Some designs of this type of torque wrench have a light or buzzer that turns on when the desired torque is reached.

Power Tools

Power tools make a technician's job easier. However, power tools require greater safety measures. Power tools do not stop unless they are turned off. Power is furnished by air (pneumatic), electricity, or hydraulic fluid. Pneumatic tools are typically used by technicians because they have more torque, weigh less, and require less maintenance than electric power tools. However, electric power tools tend to cost less than the pneumatics. Electric power tools can be plugged into most electric wall sockets, but to use a pneumatic tool, you must have an air compressor and an air storage tank. Cordless electrical tools have become more popular with greatly improved battery life.

Impact Wrenches

An impact wrench (Figure 2-2) uses compressed air or electricity to hammer or impact a nut or bolt loose or tight. Light-duty impact wrenches are available in three drive sizes, ¼, ⅜, and ½ inch, and two heavy-duty sizes, ¾ and 1 inch.

WARNING: Impact wrenches should not be used to tighten critical parts or parts that may be damaged by the hammering force of the wrench.

Impact ratchets often are used during disassembly or reassembly work to save time. Because the ratchet turns the socket without an impact force, they can be used on most parts and with ordinary sockets. Air ratchets usually have a ¼- or ⅜-inch drive. Impact wrenches

torque wrench.

USCS and metric systems are two of the most common measuring standards. The USCS is what we use in the United States, whereas the metric system is used in most other parts of the world.

⚠ **CAUTION:**
Carelessness or mishandling of power tools can cause serious injury. Make sure you know how to operate a tool before using it.

⚠ **CAUTION:**
The sockets designed for impact wrenches are constructed of thicker steel to withstand the force of the impact. Ordinary sockets must not be used with impact wrenches, they can crack or shatter because of the force and can cause injury.

A BIT OF HISTORY

This feature gives the student a sense of the evolution of the automobile. This feature not only contains nice-to-know information, but also should spark some interest in the subject matter.

TERMS TO KNOW LIST

A list of new terms appears next to the Summary.

REVIEW QUESTIONS

Short answer essay, fill-in-the-blank, and ASE-style review questions are found at the end of each chapter. These questions are designed to accurately assess the student's competence in the stated objectives at the beginning of the chapter.

SUMMARIES

Each chapter concludes with a summary of key points from the chapter. These are designed to help the reader review the chapter contents.

Energy Conversion

Energy conversion occurs when one form of energy is changed to another form. Because energy is not always in the desired form, it must be converted to a form we can use. Some of the most common automotive energy conversions are discussed here.

Chemical to Thermal Energy. Chemical energy in gasoline or diesel fuel is converted to thermal energy when the fuel burns in the engine cylinders.

Chemical to Electrical Energy. The chemical energy in a battery (Figure 2-4) is converted to electrical energy to power many of the accessories on an automobile. In some hybrid and all electric vehicles, the battery is used to power the drive wheels.

Electrical to Mechanical Energy. In an automobile, the battery supplies electrical energy to the starting motor, and this motor converts the electrical energy to mechanical energy to crank the engine.

Thermal to Mechanical Energy. The thermal energy that results from the burning of the fuel in the engine is converted to mechanical energy. The firing impulses on the pistons create rotational motion of the crankshaft. That motion is transferred to the drive train and is used to move the vehicle.

Mechanical to Electrical Energy. The generator is driven by mechanical energy from the engine. The generator converts this energy to electrical energy, which powers the electrical accessories on the vehicle and recharges the battery.

Electrical to Radiant Energy. Radiant energy is light energy. In the automobile, electrical energy is converted to thermal energy, which heats up the inside of light bulbs so that they illuminate and release radiant energy.

Solar to Electrical Energy. Solar energy can be converted to electrical energy through the use of photovoltaic panels. Automotive applications include roof panels that power cooling fans while the vehicle is parked.

Kinetic to Thermal Energy. To stop a vehicle, the brake system must change the kinetic energy of the moving vehicle to kinetic and static thermal and heat energy. This is the result of friction, which is discussed later in this chapter.

Kinetic to Mechanical to Electrical Energy. Hybrid vehicles have a system tive braking, that uses the energy of the moving vehicle mechanical energy used to operate the the batteries (Figure 2-4)

TERMS TO KNOW

(continued)

Helical gear
Horsepower
Hypoid gear
Idler gear
Journal
Lug nut
Mild hybrid
Overdrive
Parallel hybrid
Planetary carrier
Planetary gear
Planetary pinions
Radial load
Rear-wheel drive (RWD)
Ring gear
Roller bearing
Series hybrid
Spur gear
Stud
Sun gear
Thrust bearing
Torque
Transaxle
Transfer case
Underdrive
Universal joint (U-joint)
Worm gear

SUMMARY

- The rotating or turning effort of the engine's crankshaft is called engine torque.
- Gears are used to apply torque to other rotating parts of the drive train and to multiply torque.
- Transmissions offer various gear ratios through the meshing of various-sized gears.
- Reverse gear is accomplished by adding a third gear to a two-gear set. This gear, the reverse idler gear, causes the driven gear to rotate in reverse.
- The operation of a CVT is based on a steel belt linking two variable pulleys.
- Self-shifting manual transmissions are available. They are manual transmissions that use electronic or hydraulic actuators to shift the gears and work the clutch.
- Connected to the rear of the crankshaft is the flywheel, which serves many functions, including acting as the driving member of the clutch assembly.
- The clutch assembly is comprised of another driving disc, the pressure plate, and a driven disc, the clutch disc.
- The clutch disc is mounted to the input shaft of the transmission and carries the engine's torque to the transmission when the clutch assembly is engaged.
- In FWD cars, the transmission and drive axle is located in a single assembly called a transaxle. In RWD cars, the drive axle is connected to the transmission through a drive shaft.
- The drive shaft and its joints are called the drive line of the car.
- Universal joints allow the drive shaft to change angles in response to movements of the car's suspension and rear axle assembly.
- The rear axle housing encloses the entire rear-wheel driving axle assembly.
- The primary purpose of the differential is to allow a difference in driving wheel speed when the vehicle is rounding a corner or curve. The ring and pinion in the drive axle also multiples the torque it receives from the transmission.
- On FWD cars, the differential is part of the transaxle assembly.
- The drive axles on FWD cars extend from the sides of the transaxle to the drive wheels. CV joints are fitted to the axles to allow the axles to move with the car's suspension.
- 4WD vehicles typically use a transfer case, which relays engine torque to both a front and rear driving axle.
- Bearings are used to reduce the friction caused by something rotating within something else.

REVIEW QUESTIONS

Short Answer Essays

1. What are the primary purposes of a vehicle's drive train?
2. What is the basic advantage of a limited-slip differential?
3. What is the purpose of an idler gear?
4. Why are transmissions equipped with many different forward gear ratios?
5. What is the primary difference between a transaxle and a transmission?
6. Why are U-joints and CV joints used in the drive line?
7. What does a differential (final drive) unit do to the torque it receives?
8. What is the purpose of the clutch assembly? How does it work?
9. What kind of gears are commonly used in today's automotive drive trains?
10. When are ball- or roller-type bearings used?

24

To stress the importance of safe work habits, the Shop Manual dedicates one full chapter to safety. Other important features of this manual include:

PERFORMANCE-BASED OBJECTIVES

These objectives define the contents of the chapter and define what the student should have learned upon completion of the chapter. These objectives also correspond with the list of required tasks for NATEF certification. *Each NATEF task is addressed*.

Although this textbook is not designed to simply prepare someone for the certification exams, it is organized around the NATEF task list. These tasks are defined generically when the procedure is commonly followed and specifically when the procedure is unique for specific vehicle models. Imported and domestic model automobiles and light trucks are included in the procedures.

CAUTIONS AND WARNINGS

Throughout the text, warnings are given to alert the reader to potentially hazardous materials or unsafe conditions. Cautions are given to advise the student of things that can go wrong if instructions are not followed or if a non-acceptable part or tool is used.

BASIC TOOLS LISTS

Each chapter begins with a list of the basic tools needed to perform the tasks included in the chapter.

CUSTOMER CARE

This feature highlights those little things a technician can do or say to enhance customer relations.

CROSS-REFERENCES TO THE CLASSROOM MANUAL

Reference to the appropriate page in the Classroom Manual is given whenever necessary. Although the chapters of the two manuals are synchronized, material covered in other chapters of the Classroom Manual may be fundamental to the topic discussed in the Shop Manual.

SPECIAL TOOLS LISTS

Whenever a special tool is required to complete a task, it is listed in the margin next to the procedure.

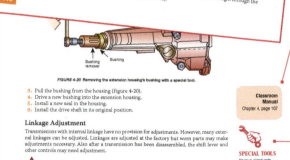

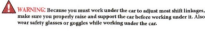

PHOTO SEQUENCES

Many procedures are illustrated in detailed Photo Sequences. These detailed photographs show the students what to expect when they perform particular procedures. They also can provide the student a familiarity with a system or type of equipment, which the school may not have.

SERVICE TIPS

Whenever a short-cut or special procedure is appropriate, it is described in the text. These tips are generally those things commonly done by experienced technicians.

MARGINAL NOTES

These notes add "nice-to-know" information to the discussion. They may include examples or exceptions, or may give the common trade jargon for a component.

PHOTO SEQUENCE 5

DISASSEMBLY OF A TYPICAL TRANSMISSION

P5-1 Place the transmission in neutral, then remove the bolts that attach the shift-lever cover.

P5-2 Lift the cover and shifter out of the extension housing. As you remove parts, thoroughly inspect them. Replace all worn or damaged parts.

P5-3 Remove the retaining pin for the shift rail and linkage.

SERVICE TIP: There are many different types of transmission fluids specified by manufacturers. The only way you will know that you are refilling a unit with the correct lubricant is to refer to the appropriate service manual. It is very important that transmissions and differentials are lubed with the correct lubricant.

Multiviscosity lubricants maintain their viscosity over a wide range of temperatures.

If the transmission and final drive use the same lubricant and have a common reservoir, check the level at the transaxle fill plug. If the transmission and final drive have separate fluid reservoirs, check each individually and refill them with the proper lubricant.

Always refer to your shop manual for refilling the axle housing with lubricant. The required types of lubricants and their capacities vary with the type of axle and with the manufacturer. The common transmission/differential lubricant combinations include the following:

1. 90W gear oil in the differential and the transmission.
2. 90W gear oil in the differential and ATF in the transmission.
3. ATF in both the transmission and the differential.

Hypoid gears require hypoid gear lubricant of the extreme pressure type with high viscosity. Many new models recommend using ATF in manual transaxles because it is a lower viscosity and improves shifting effort and fuel economy. Most RWD vehicles use 90W gear oil in the manual transmission and differential. However, because of the reasons already noted, some manual transmissions should be filled with ATF and the differentials with 90W gear lube.

Multiviscosity lubricants are most commonly required for differentials. These lubricants provide temperature protection to –40° F. 75W–90W and 80W–90W are two of the most common multiviscosity lubricants. These lubricants are also compatible with the single-viscosity lubricants used before. Most are also suited for use on axle assemblies equipped with limited-slip or other locking-type differentials. However, most limited-slip differentials require a special oil additive that makes the fluid slicker to allow the unit to slip for cornering.

Before adding friction modifiers to the lubricant of a limited-slip differential, always check its compatibility. Using the wrong additive may cause the axle's lubricant to break down, which would destroy the axle assembly very quickly. Also pay attention to the manufacturer's recommendations. Most modern units require the use of synthetic fluid.

Gear lubricant is circulated throughout the final drive housing by the ring gear. Special troughs or gullies are used to return the lubricant to the ring and the pinion area. The housing is sealed to keep fluid in and dirt out.

Possible Sources of Leaks

To find the source of a leak, carefully inspect the assembly for wet spots. Thoroughly clean the area around the leak so the exact source can be found.

The tail of the transmission should be examined. If the extension housing seal is leaking, it can be easily replaced. However, before replacing the seal, check the extension housing bushing and replace it with the seal if it is worn. Inspect the slip yoke's surface if the seal was leaking. Surface damage on the slip yoke will destroy the seal.

An improperly installed or damaged drive pinion seal...
outer edge of the seal (Figure 7-10)...
gouges, will distort the...

SPECIAL TOOLS
Pinion flange holding tool
Catch pan
Center punch
Slide hammer
Gasket sealer
Seal driver

JOB SHEETS

Located at the end of each chapter, the Job Sheets provide a format for students to perform procedures covered in the chapter. A reference to the NATEF Task addressed by the procedure is referenced on the Job Sheet.

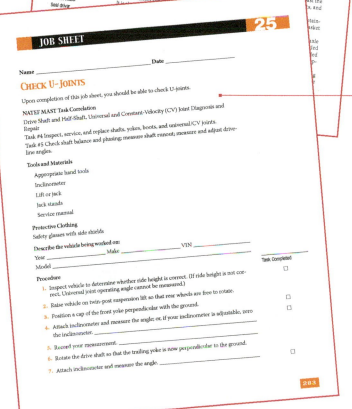

JOB SHEET 25

Name _____ Date _____

CHECK U-JOINTS

Upon completion of this job sheet, you should be able to check U-joints.

NATEF MAST Task Correlation
Drive Shaft and Half-Shaft, Universal and Constant-Velocity (CV) Joint Diagnosis and Repair
Task #4 Inspect, service, and replace shafts, yokes, boots, and universal/CV joints.
Task #5 Check shaft balance and phasing; measure shaft runout; measure and adjust driveline angles.

Tools and Materials
Appropriate hand tools
Inclinometer
Lift or jack
Jack stands
Service manual

Protective Clothing
Safety glasses with side shields

Describe the vehicle being worked on:
Year _____ Make _____ VIN _____
Model _____

Procedure

Task Completed

1. Inspect vehicle to determine whether ride height is correct. (If ride height is not correct, Universal joint operating angle cannot be measured.) ☐

2. Raise vehicle on twin-post suspension lift so that rear wheels are free to rotate. ☐

3. Position a cap of the front yoke perpendicular with the ground. ☐

4. Attach inclinometer and measure the angle; or, if your inclinometer is adjustable, zero the inclinometer. _____

5. Record your measurement. _____

6. Rotate the drive shaft so that the trailing yoke is now perpendicular to the ground. ☐

7. Attach inclinometer and measure the angle. _____

283

TERMS TO KNOW LIST

Terms in this list can be found in the Glossary at the end of the manual.

ASE-STYLE REVIEW QUESTIONS

Each chapter contains ASE-style review questions that reflect the performance-based objectives listed at the beginning of the chapter. These questions can be used to review the chapter as well as to prepare for the ASE certification exam.

CASE STUDIES

Case Studies concentrate on the ability to properly diagnose the systems. Beginning with Chapter 3, each chapter ends with a case study in which a vehicle has a problem, and the logic used by a technician to solve the problem is explained.

ASE CHALLENGE QUESTIONS

Each technical chapter ends with five ASE challenge questions. These are not more review questions, rather they test the students' ability to apply general knowledge to the contents of the chapter.

ASE PRACTICE EXAMINATION

A 50-question ASE practice exam, located in the appendix, is included to test students on the contents of the Shop Manual.

TERMS TO KNOW
(continued)

Lapping
Locking
Lugging
Neutral rollover rattle
Peening
Pitting
Rolling
Scoring
Spalling
Tail shaft

CASE STUDY

A customer brought his 2008 Toyota into the shop with a complaint of the transaxle jumping out of fifth gear. The technician verified the complaint, then proceeded to check the adjustment of the shift linkage. No problem was found after she checked the adjustment and inspected the linkage. She proceeded to check the alignment of the transaxle to the engine, and again the cause of the problem was not found.

The transaxle was removed and disassembled; again no cause for the problem was identified. The technician carefully reassembled the transaxle and installed it back in the car. During the road test, she again experienced the problem. Because she could not find a cause for the problem, she returned it to the customer and told him the problem could not be fixed. Taking this approach is a good way to lose customers and your job. True, she did check all of the right things. However, she probably did not check the internal parts of the transaxle carefully enough. Wear on the back taper of the dog clutch teeth of the speed gear and/or slider will cause this problem. Only a careful inspection will reveal this.

ASE-STYLE REVIEW QUESTIONS

1. When discussing shift problems,
   Technician A says that broken or worn engine and transaxle mounts can cause a transaxle to have shifting problems.
   Technician B says that poor shift boot alignment can cause a transaxle to jump out of gear.
   Who is correct?
   A. A only
   B. B only
   C. Both A and B
   D. Neither A nor B

2. When inspecting a transaxle's gears,
   ... that a wear pattern on the center of

4. When discussing transmission disassembly,
   Technician A says that the alignment of the parts of each synchronizer must be marked prior to disassembling the synchronizer assembly.
   Technician B says that most synchronizer hubs are splined to the shaft and are pressed on and off the shaft.
   Who is correct?
   A. A only
   B. B only
   C. Both A and B
   D. Neither A nor B

5. When disassembling a transaxle, a severely worn second gear synchronizer blocker ring is found.
   Technician A says that this should have caused the ... out of gear.

ASE CHALLENGE QUESTIONS

1. A vehicle makes a loud clicking sound during a left turn; during a right turn the clicking noise is still apparent but is less pronounced.
   Technician A says that the left inner CV joint may be the source of the noise.
   Technician B says that the left outer CV joint may be causing the noise.
   Who is correct?
   A. A only
   B. B only
   C. Both A and B
   D. Neither A nor B

2. A vehicle vibrates when under acceleration at all speeds above 10 mph; it does not vibrate during coast down.
   Technician A says that a worn inner CV joint could cause this problem.
   Technician B says that an out-of-balance tire is the likely problem.
   Who is correct?
   A. A only
   B. B only
   C. Both A and B
   D. Neither A nor B

3. A discussion concerning CV joint testing is being held.
   Technician A says that a recommended method to check CV joints is to raise the vehicle so that the lower control arms are hanging freely; at this point the engine can be accelerated while noise checks can be performed.
   Technician B says that the vehicle should be road tested so that the CV joint...

4. A half shaft is being removed from a vehicle. The outer CV joint appears to be pressed into the hub.
   Technician A says that a brass punch or hammer can be used to remove the CV joint from the hub.
   Technician B says that a puller should be used to remove the CV joint from the hub.
   Who is correct?
   A. A only
   B. B only
   C. Both A and B
   D. Neither A nor B

5. The CV joint boots on the left side of a vehicle appear to be squeezed together; the boots on the right side of the vehicle appear to be stretched apart.
   Technician A says that one or more of the engine/transaxle's mounts may be broken.
   Technician B says that one or more of the engine/transaxle's subframe mounts may be broken.
   Who is correct?
   A. A only
   B. B only
   C. Both A and B
   D. Neither A nor B

APPENDIX A
ASE PRACTICE EXAMINATION

41. When discussing viscous clutch service and inspection,
   Technician A says that viscous clutches can be tested on a bench.
   Technician B says that a viscous clutch can be serviced in the shop by adding oil to its reservoir.
   Who is correct?
   A. Technician A
   B. Technician B
   C. Both A and B
   D. Neither A nor B

42. When discussing the unsprung weight of a vehicle,
   Technician A says that installing lightweight aluminum wheels on a vehicle will reduce its sprung weight.
   Technician B says that replacing a steel hood with an aluminum hood will reduce a vehicle's unsprung weight.
   Who is correct?
   A. Technician A
   B. Technician B
   C. Both A and B
   D. Neither A nor B

43. Transfer case problems are being discussed,
   Technician A says that incorrect drive shaft angles can result in harsh engagement when shifting into 4WD.
   Technician B says that an overheated viscous coupling can cause binding when making a sharp turn on dry pavement.
   Who is correct?
   A. Technician A
   B. Technician B
   C. Both A and B
   D. Neither A nor B

44. The resistance of an electromagnetic clutch coil is 200 ohms; it should be 4 ohms.
   Technician A says that this will result in a blown fuse.
   Technician B says that the magnetic field developed by the clutch coil will be excessive.
   Who is correct?
   A. Technician A
   B. Technician B
   C. Both A and B
   D. Neither A nor B

45. A poorly operating fan motor circuit is being tested. A voltmeter that is placed across the power and ground terminals of connector to the motor indicates 0.0 volts when the fan switch is turned on.
   Technician A says that this test reading indicates that the connector is probably okay.
   Technician B says that this test is referred to as an available voltage test.
   Who is correct?
   A. Technician A
   B. Technician B
   C. Both A and B
   D. Neither A nor B

46. The amount of current being drawn by a load is about to be measured.
   Technician A says that the ammeter can be connected in the positive side of the circuit.
   Technician B says that the amperage can be measured by connecting the ammeter to the negative side of the circuit.
   Who is correct?
   A. Technician A
   B. Technician B
   C. Both A and B
   D. Neither A nor B

47. A digital ammeter being used to measure current drain is indicating 101 mA.
   Which of the following represents this reading?
   A. .101 amps
   B. .00101 amps
   C. 1.01 amps
   D. 10.1 amps

48. A vehicle towed into the shop because of a no-start condition is found to have an inoperative starter. A voltmeter connected to both of the terminals of the clutch safety switch indicates 8 volts when the clutch pedal is depressed and the ignition switch is placed in the Start position.
   Technician A says that the clutch safety switch is faulty.
   Technician B says that the resistance of the circuit is higher than normal.
   Who is correct?
   A. Technician A
   B. Technician B
   C. Both A and B
   D. Neither A nor B

SUPPLEMENTS

INSTRUCTOR RESOURCES

A robust set of Instructor Resources is available on line through Cengage's Instructors Companion Website, and on DVD. This powerful suite of classroom preparation tools includes PowerPoint slides with images and video clips that coincide with each chapter's content coverage, an image gallery with pictures from the text, theory-based worksheets in Word that provide homework or in-class assignments, the Job Sheets from the Shop Manual in Word, a NATEF correlation chart, and an Instructor's Guide in electronic format. Cengage testing powered by Cognero is available through the Instructor Companion Website; in addition, the complete set of test bank questions is available on the Instructor Resources DVD in Word format.

MindTap Automotive for Today's Technician™: Manual Transmissions and Transaxles, 6th edition

MindTap Automotive for Manual Transmissions and Transaxles provides a customized learning experience with relevant assignments that will help students learn and apply concepts while it allows instructors to measure skills and outcomes with ease.

MindTap Automotive meets the needs of today's automotive classroom, shop and student. Within the MindTap, faculty and students will find editable and submittable job sheets based on NATEF tasks. MindTap also offers students the opportunity to practice diagnostic techniques in a safe environment while strengthening their critical thinking and troubleshooting skills with the inclusion of diagnostic scenarios from Delmar Automotive Training Online (DATO). Additional engaging activities include videos, animations, matching exercises and gradable assessments.

REVIEWERS

The authors and publisher would like to extend a special thanks to the following instructors for their contributions to this text:

Ronald Alexander
Professor, Automotive Technology
Morrisville State College, Morrisville,
New York

Jeffrey P. Nidiffer
University of Northwestern Ohio
Lima, OH

Christopher Parrott
Automotive Program Director
Vatterott College Inc., Wichita Campus

William Schaefer
Pennsylvania College of Technology
Williamsport, PA

Lonnie Schulz
University of Northwestern Ohio
Lima, OH

William White
University of Northwestern Ohio
Lima, OH

MANUAL DRIVETRAINS AND AXLES

UPON COMPLETION AND REVIEW OF THIS CHAPTER, YOU SHOULD BE ABLE TO:

- Identify the major components of a vehicle's drivetrain.
- State and understand the purpose of a transmission.
- Describe the difference between a transmission and a transaxle.
- Describe the construction and operation of CVTs.
- State and understand the purpose of a clutch assembly.
- Describe the differences between a typical FWD and RWD car.

- Describe the construction of a drive shaft.
- State and understand the purpose of a U-joint and a CV joint.
- State and understand the purpose of a differential.
- Identify and describe the various gears used in modern drivetrains.
- Identify and describe the various bearings used in modern drivetrains.

INTRODUCTION

An automobile can be divided into four major systems or basic components: (1) the engine, which serves as a source of power; (2) the powertrain, or drivetrain, which transmits the engine's power to the car's wheels; (3) the chassis, which supports the engine and body and includes the brake, steering, and suspension systems; and (4) the car's body, interior, and accessories, which include the seats, heater and air conditioner, lights, windshield wipers, and other comfort and safety features.

Basically, the **drivetrain** has four primary purposes: (1) to connect and disconnect the engine's power to the wheels, (2) to select different speed ratios, (3) to provide a way to move the car in reverse, and (4) to control the power to the drive wheels for safe turning of the automobile. The main components of the manual drivetrain are the clutch, transmission, differential, and drive axles (Figure 1-1). The exact components used in a vehicle's drivetrain depend on whether the vehicle is equipped with front-wheel drive, rear-wheel drive, or four-wheel drive.

Today, most cars are **front-wheel drive (FWD)**. Power-flow through the drivetrain of FWD vehicles passes through the clutch or torque converter, through the transmission, and then moves to a front differential, the driving axles, and onto the front wheels. The transmission and differential are housed in a single unit (Figure 1-2) called a **transaxle**.

Some larger and many performance cars are **rear-wheel drive (RWD)**. Most pickup trucks and many SUVs are also RWD vehicles. Power-flow in a RWD vehicle passes through the clutch or torque converter, transmission, and the drive line (drive shaft assembly). Then it goes through the rear differential, the rear-driving axles, and onto the rear wheels.

About 35 percent of the questions on the ASE Manual Drive Trains and Axles Certification Test are based on transmissions. The remaining questions are related to the other drivetrain components.

Torque converters use fluid flow to transfer the engine's power to the transmission.

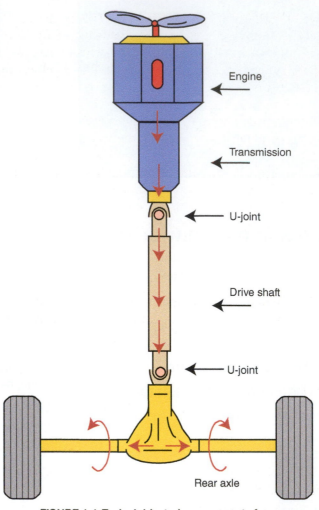

Engine

Transmission

U-joint

Drive shaft

U-joint

Rear axle

FIGURE 1-1 Typical drivetrain components for a RWD car.

FIGURE 1-2 A transaxle.

Four-wheel drive (4WD) or all-wheel drive (AWD) vehicles combine features of both rear- and front-wheel drive systems so power can be delivered to all wheels all the time or when the conditions or driver selects two-wheel or four-wheel drive. Normally, a 4WD pickup or full-size SUV has 4WD; the drivetrain is based on a RWD vehicle modified to include a front drive axle and transfer case. When a smaller SUV or car has AWD or 4WD, the drivetrain is a modified FWD system. Modifications include a rear drive axle and an assembly that transfers some of the engine's power to the rear drive axle.

There are two sets of gears in the drivetrain: the transmission and the differential. The transmission allows the gear ratios to change and provides a reverse gear. The final drive unit provides a final torque multiplication ratio. The final drive contains a differential unit that changes the power output from the transmission and allows the drive wheels to rotate at different speeds during turns, which prevents tire scuffing. Through the use of different gear ratios, torque is multiplied. Many 4WD vehicles have a third set of gears in the transfer case to provide a low range.

> **AUTHOR'S NOTE:** For many years, buyers preferred automatic transmissions to manual transmissions, but lately the number of manuals has increased. This is primarily because of the economy, performance, and "fun" factors associated with manuals. The increased popularity of automated manual transmissions means that even more manuals are on our roads, even if some manufacturers refer to these units as "automatics."

ENGINE

Although the engine is a major system by itself (Figure 1-3), its output should be considered a component of the drivetrain. The engine provides the power to drive the wheels of the vehicle. An engine develops a rotary motion or torque that, when multiplied by the transmission gears, will move the car under a variety of conditions. The engine produces power by burning a mixture of fuel and air in its combustion chambers. Combustion causes a high pressure in the cylinders, which forces the pistons downward. Connecting rods transfer the downward movement of the pistons to the crankshaft, which rotates by the force on the pistons.

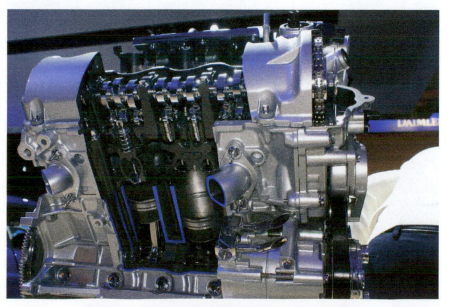

FIGURE 1-3 Major components of a four-stroke engine.

A BIT OF HISTORY

The French-built Panhard, in 1892, was the first vehicle to have its power generated by a front-mounted, liquid fueled, internal combustion engine and transmitted to the rear driving wheels by a clutch, transmission, differential, and drive shaft.

Power is the rate or speed at which work is performed.

Torque is turning or twisting force.

All automobile engines, both gasoline and diesel, are classified as internal combustion engines because the combustion or burning that creates energy takes place inside the engine. **Combustion** is the burning of an air and fuel mixture. As a result of combustion, large amounts of pressure are generated in the engine. This pressure or energy is used to power the car. The engine must be built strong enough to hold the pressure and temperatures formed by combustion.

Diesel engines have been around a long time and at one time were mostly found in big heavy-duty trucks. Today they are also commonly used in some pickup trucks and will become more common in automobiles in the future. Fuel economy concerns mean that automotive diesel applications will increase exponentially in the coming years (Figure 1-4). Although the construction of a gasoline and diesel engine are similar, their operation is quite different.

A gasoline engine relies on a mixture of fuel and air that is ignited by a spark to produce power. A diesel engine also uses fuel and air, but does not need a spark to cause ignition. Diesel engines are often called compression ignition engines. This is because its incoming air is tightly compressed, which greatly raises its temperature. The fuel is then injected into the compressed air. The heat of the compressed air ignites the fuel and combustion takes place. The following sections cover the basic parts and the major systems of a gasoline engine.

Most automotive engines are four-stroke cycle engines. The opening and closing of the intake and exhaust valves are timed to the movement of the piston. As a result, the engine passes through four different events or strokes during one combustion cycle. These four are the intake, compression, power, and exhaust strokes.

On the intake stroke, the piston moves downward, and a charge of air/fuel mixture is introduced into the cylinder. As the piston travels upward, the air/fuel mixture is compressed in preparation for burning. Just before the piston reaches the top of the cylinder, ignition occurs and combustion starts. The pressure of expanding gases forces the piston downward on its power stroke. When it reciprocates, or moves upward again, the piston is on the exhaust stroke. During the exhaust stroke, the piston pushes the burned gases out of the cylinder. As long as the engine is running, this cycle of events repeats itself, resulting in the production of engine torque.

FIGURE 1-4 An automotive diesel engine.

Hybrid Electric Vehicles

Hybrid vehicles use a combination of a gas or diesel engine together with an electric motor. These vehicles are commonly placed into different classifications. A **mild hybrid** uses electric motor assistance under load. The electric motor is typically used as a starter and alternator. A **parallel hybrid** uses the electric motor and engine to power the drive wheels. The vehicle can be driven on the electric motor only under certain circumstances. The Toyota Prius is one example of a parallel hybrid. A **series hybrid**, such as the Chevrolet Volt, uses a small internal combustion engine to charge the batteries that power the electric motor. The engine is not used to power the drive wheels. Most hybrid vehicles use planetary systems in their transmission designs.

Engine Torque

The rotating or turning effort of the engine's crankshaft is called **engine torque**. Engine torque is measured in pound-feet (lb.-ft.). However, it is common terminology to discuss torque in foot-pounds (ft.-lbs.). The metric or SI measurement of torque is expressed in Newton-meters (Nm). Engines produce a maximum amount of torque when operating within a narrow range of engine speeds. When an engine reaches the maximum speed within that range, torque is no longer increased.

The amount of **torque** produced in relation to engine speed is called the engine's torque curve. Ideally, the engine should always be run when it is providing a maximum amount of torque (Figure 1-5). This allows the engine to provide the required amount of power while using a minimum amount of fuel.

As a car is climbing up a steep hill, its driving wheels slow down because of the increased amount of work it must do; this causes engine speed to decrease as well as reduce the engine's torque. The driver must downshift the transmission or press down harder on the gas pedal, which increases engine speed and allows the engine to produce more torque (Figure 1-6). When the car reaches the top of the hill and begins to go down, its speed and the speed of the engine rapidly increase. The driver can now upshift or let up on the gas pedal, which allows the engine's speed to decrease and places it back into its peak within the torque curve.

Torque Multiplication

Measurements of **horsepower** indicate the amount of work being performed and the rate at which it is being done. The drive line can transmit power and multiply torque, but it cannot multiply power. When power flows through one gear to another, the torque is multiplied in

> To convert foot-pounds to Newton-meters, multiply the number of foot-pounds by 1.355.

> Downshifting is the shifting to a higher numerical ratio gear.

> One horsepower is the equivalent of moving 33,000 pounds 1 foot in 1 minute.

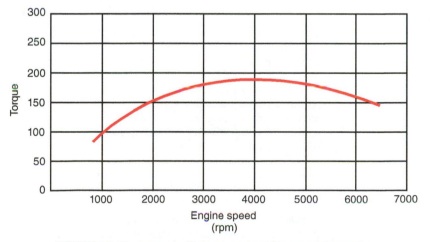

FIGURE 1-5 The amount of torque produced by an engine varies with the speed of the engine.

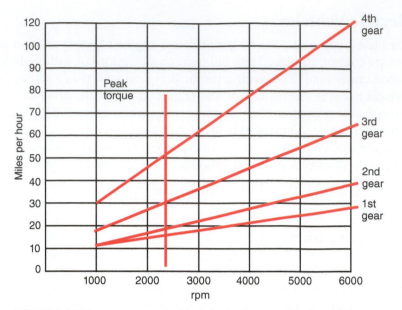

FIGURE 1-6 If the car represented by this chart were going up a hill in fourth gear and the hill caused the car's speed to drop to 30 mph, the engine would labor heavily because it would be operating below its peak torque. To overcome the hill and allow the car to increase speed, the driver should place the car into third gear.

proportion to the different gear sizes. Torque is multiplied, but the power remains the same, as the torque is multiplied at the expense of rotational speed. The engine's horsepower is determined by torque and engine speed. Maximum horsepower occurs when the engine is operating at a high speed and producing close to its maximum torque. Engine horsepower can be calculated by using a mathematical formula:

$$hp = T\,rpm\,/\,5252$$

where hp = horsepower; T = the amount of torque produced by an engine; rpm = the speed of the engine when it is producing the torque; and 5252 being a mathematical constant. Figure 1-7 shows this relationship between horsepower and torque.

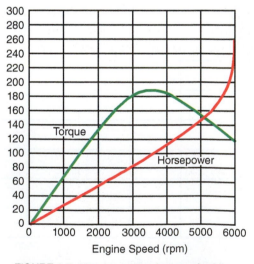

FIGURE 1-7 After an engine has reached its peak torque, its horsepower output increases with an increase in engine speed but is independent of engine torque.

TRANSMISSIONS

The transmission is mounted to the rear of the engine and is designed to allow the car to move forward and in reverse. It also has a neutral position. In this position, the engine can run without applying power to the drive wheels. Therefore, although there is input to the transmission when the vehicle is in neutral, there is no output from the transmission because the driving gears are not engaged to the output shaft. Simply put, a transmission is "in gear" whenever the input shaft is connected to the output shaft.

There are two basic types of transmissions and transaxles: automatic and manual. Automatic transmissions use a combination of a torque converter and a planetary gear system to change gear ratios automatically. A manual transmission is an assembly of gears and shafts (Figure 1-8) that transmits power from the engine to the drive axle and changes in gear ratios are controlled by the driver. A clutch assembly is used to connect and disconnect the engine from the transmission.

By moving the shift lever, various gear and speed ratios can be selected. The gears in a transmission are selected to give the driver a choice of both speed and torque. Lower gears allow for lower vehicle speeds but more torque. Higher gears provide less torque but higher vehicle speeds. Gear ratios state the ratio of the number of teeth on the driven gear to the number of teeth on the drive gear. There often is much confusion about the terms *high* and *low gear ratios*. A gear ratio of 4:1 is lower than a ratio of 2:1. Although numerically the 4 is higher than the 2, the 4:1 gear ratio allows for lower speeds and, hence, is a low gear ratio.

Manual transmissions are commonly called standard shift or "stick-shift" transmissions.

Shift levers are typically called "shifters."

First and second gears are the low gears in a typical transmission, whereas fifth and sixth are the high gears.

> **AUTHOR'S NOTE:** Sometimes we are our own worst enemies, especially when talking about cars. Gear ratios are a great example. *The conversations:*
>
> "So and so put a low gear in her car and now it really flies!" "So and so put in a high gear and now he claims his gas mileage is much better."
>
> "That's great, you can put in a higher gear and go faster and get better mileage!" "Yeah, but it will take off like a slug."
>
> "A 4.10:1 provides more torque multiplication than a 2.73:1 gear, but a 4 is higher than a 2 so the 4.10 gear must be the higher gear of the two!" "*No!* The 4.10 gear is lower."
>
> Gears are used in transmissions, transfer cases, and differential units to multiply torque. We most commonly think of ratios in the final drive/differential unit because we can change ring and pinion gears without too much trouble. Just remember that higher numerical gears (e.g., 4:10s) provide more torque but lower maximum speed. Common terminology leads us to call these "low" gears.

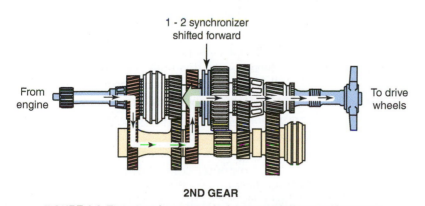

2ND GEAR

FIGURE 1-8 The gears in a transmission transmit the rotating power from the engine.

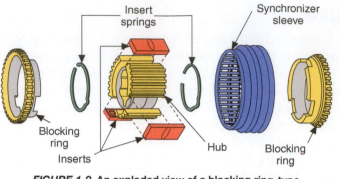

FIGURE 1-9 An exploded view of a blocking ring-type synchronizer assembly.

Today, most manual transmissions have five or six forward speeds. These speeds or gears are identified as first, second, third, fourth, fifth, and sixth. Different gear ratios are necessary because an engine develops relatively little power at low engine speeds. The engine must be turning at a fairly high speed before it can deliver enough power to get the car moving. Through selection of the proper gear ratio, torque applied to the drive wheels can be multiplied. The top gear will be an overdrive ratio designed to reduce engine rpm at cruising speed and increase fuel economy. A transmission design that has been used only in pure racing is now available on a few cars. The driver changes gears with the gearshift or paddles located on the steering wheel. The **clutch** is automatically engaged and disengaged by the system, based on the numerous inputs to the control module.

Automotive manual transmissions are constant mesh, fully synchronized transmissions. "Constant mesh" means that the transmission gears are constantly in mesh, regardless of whether the car is stationary or moving. "Fully synchronized" refers to a mechanism of blocker rings and synchronizers (Figure 1-9) used to bring the rotating shafts and gears of the transmission to one speed for smooth up- and downshifting.

> Synchronizer rings serve as cone clutches that use friction to match speed gear rpm to output shaft rpm.

Continuously Variable Transmission

Another unconventional transmission design, the **continuously variable transmission (CVT)**, is found on some late-model cars and small SUVs. Basically, a CVT (Figure 1-10) is a transmission without fixed forward speeds. The gear ratio varies with engine speed and load. These transmissions are, however, fitted with a one-speed reverse gear. Some of these automatic-like transaxles do not have a torque converter; rather, they use a manual transmission-type flywheel with a start clutch. Instead of relying on gear sets to provide drive ratios, a CVT uses a belt and pulleys.

One pulley is the driven member and the other is the driver. Each pulley has a moveable face and a fixed face. When the moveable face moves, the effective diameter of the pulley changes. The change in effective diameter changes the effective pulley (gear) ratio. A steel belt links the driven and drive pulleys.

To achieve a low pulley ratio, high hydraulic pressure works on the moveable face of the driven pulley to make it larger. In response to this high pressure, the pressure on the drive pulley is reduced. Because the belt links the two pulleys and proper belt tension is critical, the drive pulley reduces just enough to keep the proper tension on the belt. The increase of pressure at the driven pulley is proportional to the decrease of pressure at the drive pulley. The opposite is true for high pulley ratios designed to increase speed output with an overdrive ratio. Low pressure causes the driven pulley to decrease in size, whereas high pressure increases the size of the drive pulley. Computer control of pulley size factors vehicle speed, engine load, and other variables into determining the correct ratio for any given condition.

Different speed ratios are available any time the vehicle is moving. Because the size of the drive and driven pulleys can vary greatly, vehicle loads and speeds can be changed without

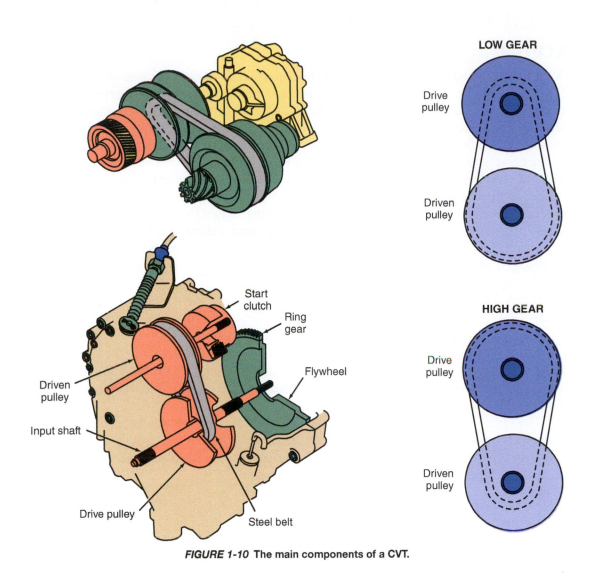

LOW GEAR

Drive pulley

Driven pulley

HIGH GEAR

Drive pulley

Driven pulley

Start clutch

Ring gear

Flywheel

Driven pulley

Input shaft

Drive pulley

Steel belt

FIGURE 1-10 The main components of a CVT.

changing the engine's speed. With this type of transmission, attempts are made to keep the engine operating at its most efficient speed. This increases fuel economy and decreases emissions.

Self-Shifting Manual Transmissions

Self-shifting manual transmissions are currently available on some passenger cars. These units were first and are used regularly in Formula One race cars. These transmissions work like typical manual transmissions, except electronic or hydraulic actuators shift the gears and work the clutch. There is no clutch pedal. The driver shifts the gears with electronic control using buttons or paddles on the steering wheel or a console-mounted shifter. Some units can also be operated in a fully automatic mode. It is important to realize that these are not automatic transmissions with manual controls!

These transmissions have computer-controlled actuators connected to the shift forks and a clutch actuator. The computer is programmed to shift the transmission automatically at the correct time, in the correct sequence, and to activate the clutch and allow for precise shifting when the driver selects the automatic mode.

The driver can also control gear changes by using the shifting mechanism. There is no gearshift linkage or cable; instead, a sensor at the shifter sends a signal to the computer. The computer, in turn, commands the actuators to engage or disengage the clutch and the gears with very fast response times. Engine torque is controlled during the shift by directly controlling the throttle or ignition/fuel injection system to provide smooth shifts.

CLUTCH

The flywheel also serves to dampen crankshaft vibrations, add inertia to the rotation of the crankshaft, and as the mount for a large gear for the starter motor.

A **flywheel**, which is a large and heavy disc, is attached to the rear end of the crankshaft (Figure 1-11). The rear face of the flywheel is machined flat and smooth to serve as the driving member of the clutch. When the clutch is engaged, the rotary motion of the engine's crankshaft is transferred from the flywheel, through the clutch, to the transmission. The rotary motion is then delivered by the transmission to the differential, then transferred by axle shafts to the tires, which push against the ground to move the car.

The clutch assembly is located immediately to the rear of the engine and is directly connected to the engine's crankshaft at the flywheel. Its purpose is to engage and disengage the engine's power smoothly to the rest of the drivetrain. When the driver depresses the clutch pedal, the clutch assembly is disengaged and the engine is disconnected from the rest of the drivetrain. This allows the engine to run when the car is standing still. The clutch is said to be engaged when the clutch pedal is up from the floor and the engine is connected to the drivetrain.

If two discs are attached to a shaft and are not touching each other, we can spin one disc as fast as we want to without affecting the other disc. As the discs move together, however, the spinning disc will cause the other to begin to turn. When the discs are held tightly together, both discs will turn as if they were one. Automotive clutches use this two-disc system. One of the discs is the flywheel. The other is called the pressure plate and is bolted to the flywheel. Sandwiched between the two discs is the clutch disc (Figure 1-12). Friction causes the clutch disc to rotate with the other two discs.

The clutch disc is made of a high frictional material that provides for a solid connection between the two discs. The discs are forced together by strong springs, which clamp the clutch disc between the pressure plate and flywheel. The spring pressure is released by pushing down on the clutch pedal, which allows the discs to separate.

⚠️ **WARNING:** **Some early clutch discs were made with an asbestos lining. Asbestos has been found to cause lung problems and cancer. Use approved respiratory gear and proper cleaning methods when handling and working around these clutch discs.**

FIGURE 1-11 **A typical flywheel mounted to the rear of an engine's crankshaft.**

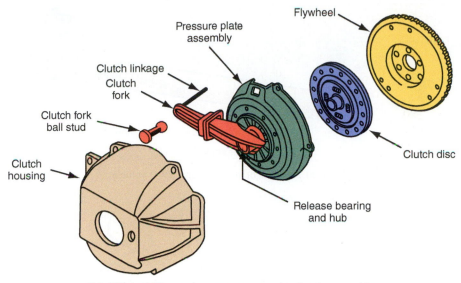

FIGURE 1-12 The major components of a clutch assembly.

DRIVE LINE

The car's drive shaft and its joints are often called the **drive line**. The drive line transmits torque from the transmission to the driving wheels. RWD cars use a long drive shaft that connects the transmission to the rear axle. The engine and complete drive line of FWD cars are located between the front driving wheels. The transmission section of a transaxle is practically identical to RWD transmissions. However, a transaxle also contains the differential gear sets and the connections for the drive axles (Figure 1-13).

Drive Line for RWD Cars

A **drive shaft** is a steel or aluminum tube normally connected to at least two universal joints and a slip joint (Figure 1-14). The drive shaft transfers power from the transmission output shaft to the rear drive axle. A differential in the axle housing transmits the power to the rear wheels, which then move the car forward or backward.

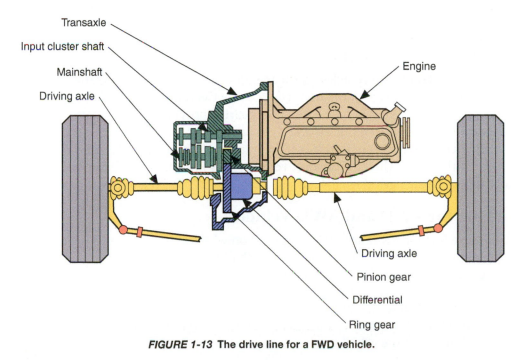

FIGURE 1-13 The drive line for a FWD vehicle.

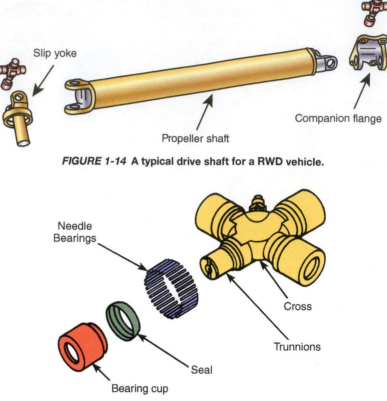

FIGURE 1-14 A typical drive shaft for a RWD vehicle.

FIGURE 1-15 An exploded view of a universal joint.

Drive shafts differ in construction, lining, length, diameter, and type of slip joint. Typically, the drive shaft is connected at one end to the transmission and at the other end to the rear axle, which moves up and down with wheel and spring movement.

Drive shafts are typically made of hollow, thin-walled steel or aluminum tubing with the universal joint yokes welded at either end. **Universal joints (U-joint)** allow the drive shaft to change angles in response to the movements of the rear axle assembly (Figure 1-15). As the angle of the drive shaft changes, its length also must change. The slip yoke normally fitted to the front universal joint allows the shaft to remain in place as its length requirements change.

> Universal joints are most often called U-joints.

Drive Line for FWD Vehicles

A transaxle is used on FWD vehicles. It is made up of a transmission and differential housed in a single unit. The gear sets in the transaxle provide the required gear ratios and direct the power-flow into the differential. The differential gearing provides the final gear reduction and splits the power-flow between the left and right drive axles.

The drive axles extend from the sides of the transaxle. The outer ends of the axles are fitted to the hubs of the drive wheels. **Constant velocity (CV) joints** mounted on each end of the drive axles allow for changes in length and angle without affecting the power-flow to the wheels.

Drive Line for 4WD and AWD Vehicles

The basic drive line for 4WD and AWD depends on whether the system uses a transfer case. If the drivetrain is based on a RWD, the output shaft from the transmission is connected to the transfer case. The transfer case is connected, by drive shafts, to the rear and front drive axle assemblies (Figure 1-16). If the vehicle is based on a FWD vehicle, a differential (or similar component) is attached to the output of the transaxle and a drive shaft connects that unit to the rear drive axles. Of course, there are many variations to this basic setup, all depending on the make and model of the vehicle.

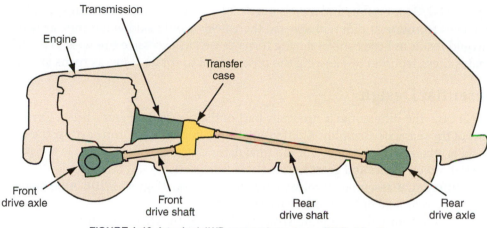

FIGURE 1-16 A typical 4WD system based on a RWD drive line.

DIFFERENTIALS

The rear axle housing encloses the complete RWD axle assembly. In addition to housing the parts, the axle housing also serves as a place to mount the vehicle's rear suspension and braking system. The rear axle assembly serves two other major functions: it changes the direction of the power-flow 90 degrees and acts as the final gear reduction unit.

The rear axle consists of two sets of gears: the ring and pinion gear set and the **differential** gears. When torque leaves the transmission, it flows through the drive shaft to the ring and pinion gears, where it is further multiplied (Figure 1-17). By considering the engine's torque curve, the car's weight, and tire size, manufacturers are able to determine the best rear axle gear ratios for proper acceleration, hill-climbing ability, fuel economy, and noise-level limits.

The primary purpose of the differential gear set is to allow a difference in driving wheel speed when the vehicle is rounding a corner or curve. The differential also transfers torque equally to both driving wheels when the vehicle is traveling in a straight line (Figure 1-18).

The torque on the ring gear is transmitted to the differential, where it is sent to the drive wheel that requires less turning torque. When the car is traveling in a straight line, both driving wheels travel the same distance at the same speed. However, when the car is making a turn, the outer wheel must travel farther and faster than the inner wheel. For example, if a car is steered into a 90-degree turn to the left and the inner wheel turns on a 20-foot radius, the inner wheel travels about 31 feet. The outer wheel, which is nearly 5 feet from the inner wheel,

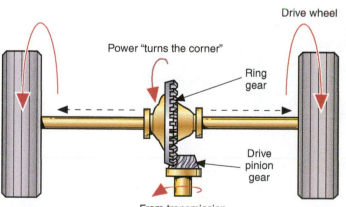

FIGURE 1-17 The gears of a final drive unit not only multiply torque but also transmit power "around the corner" to the drive wheels.

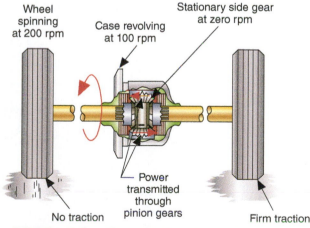

FIGURE 1-18 The differential allows for one drive wheel to turn at a different speed than the other while the car is turning and allows for more power to the wheel with less traction.

turns on a 24-2/3 foot radius and travels nearly 39 feet. Without some means for allowing the drive wheels to rotate at different speeds, the wheels would skid when the car was turning. This would result in little control during turns as well as excessive tire wear. The differential eliminates these troubles by allowing the outer wheel to rotate faster as turns are made.

Differential Design

On FWD cars, the differential is normally part of the transaxle (Figure 1-19). On RWD cars, it is part of the rear axle assembly. A differential is located in a cast iron casting, the differential case, and is attached to the center of the final drive ring gear. Located inside the case are the differential pinion shafts and gears and the axle side gears.

The differential assembly revolves with the ring gear (Figure 1-20). Axle side gears are splined to the rear axle or front axle drive shafts.

When an automobile is moving straight ahead, both wheels are free to rotate. Engine power is applied to the drive pinion gear, which rotates the ring gear. Beveled pinion gears are carried around by the ring gear and case and rotate as one unit. Each side gear and axle turns at the same speed.

When the car turns a sharp corner, only one wheel rotates freely. Torque still comes in on the drive pinion gear and rotates the ring gear and case, carrying the beveled pinions around with it. However, one axle doesn't easily rotate and the slower axle provides resistance and the beveled pinions are forced to rotate on their own axis and "walk around" their gear. The other side is forced to rotate faster because it is subjected to the turning force of the ring gear, which is transmitted through the pinions.

During one revolution of the ring gear, one differential pinion gear makes two revolutions and the other "walks around" the gear. As a result, when the drive wheels have unequal resistance applied to them, the wheel with the least resistance turns more revolutions. As one wheel turns faster, the other turns proportionally slower.

When the vehicle turns a corner or a curve, the differential pinion gears rotate around the differential pinion shaft. The differential pinion gears allow the inside axle shaft and driving wheel to slow down. On the opposite side, the pinion gears allow the outside wheels to accelerate. Both driving wheels resume equal speeds when the vehicle completes the corner or curve. This differential action improves vehicle handling and reduces driving wheel tire wear.

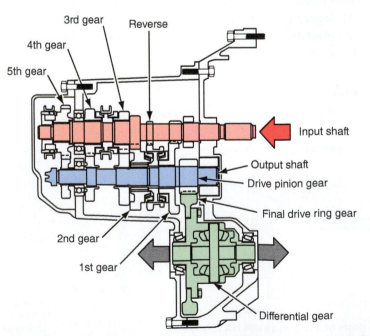

FIGURE 1-19 The final drive and differential assembly is part of the transaxle.

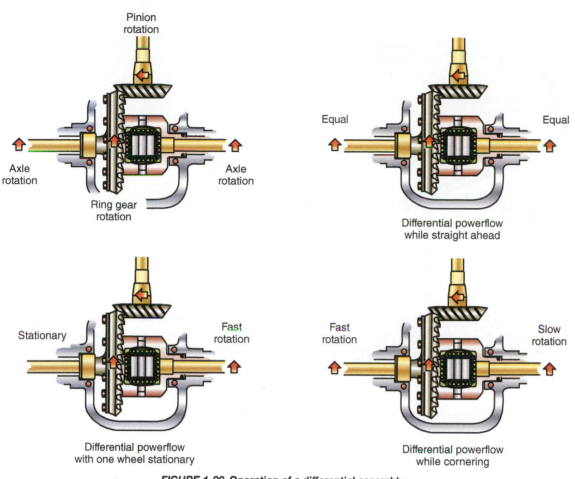

Pinion
rotation

Axle
rotation

Axle
rotation

Ring gear
rotation

Equal

Equal

Differential powerflow
while straight ahead

Stationary

Fast
rotation

Differential powerflow
with one wheel stationary

Fast
rotation

Slow
rotation

Differential powerflow
while cornering

FIGURE 1-20 Operation of a differential assembly.

To prevent a loss of power on slippery surfaces, some differentials are fitted with clutches that allow the wheel with traction to continue to receive torque. These differentials are referred to as *limited-slip, traction-lock,* or *Positraction* differentials. Rather than clutch discs, some units use cone clutches or gears to restrict the normal differential action and deliver torque to the nonslipping wheel. A true locking differential has no differential action at all. These units are used in some racing and off-road applications. Some AWD vehicles have a driver-controlled differential locking feature to aid traction in adverse conditions.

DRIVING AXLES

Driving axles are solid steel shafts that transfer the torque from the differential to the driving wheels. A separate axle shaft is used for each driving wheel.

In a RWD vehicle, the driving axles are part of the rear axle assembly and are enclosed within the hollow horizontal tubes of the axle housing. The housing protects and supports these parts. Each drive axle is connected to the side gears in the differential. The inner ends of the axles are splined to fit into the side gears. As the side gears are turned, the axles to which they are splined turn at the same speed. A bearing inside the axle housing supports the outer end of the axle shaft (Figure 1-21). This bearing, called the axle bearing, allows the axle to rotate smoothly inside the axle housing.

The drive wheels are attached to the outer ends of the axles. The outer end of each axle has a flange mounted to it. A **flange** is a rim for attaching one part to another part. The flange, fitted with studs, at the end of an axle holds the wheel in place. **Studs** are threaded shafts, resembling bolts without heads. One end of the stud is screwed or pressed into the flange. The wheel fits over the studs and a nut, called the **lug nut**, is tightened over the open

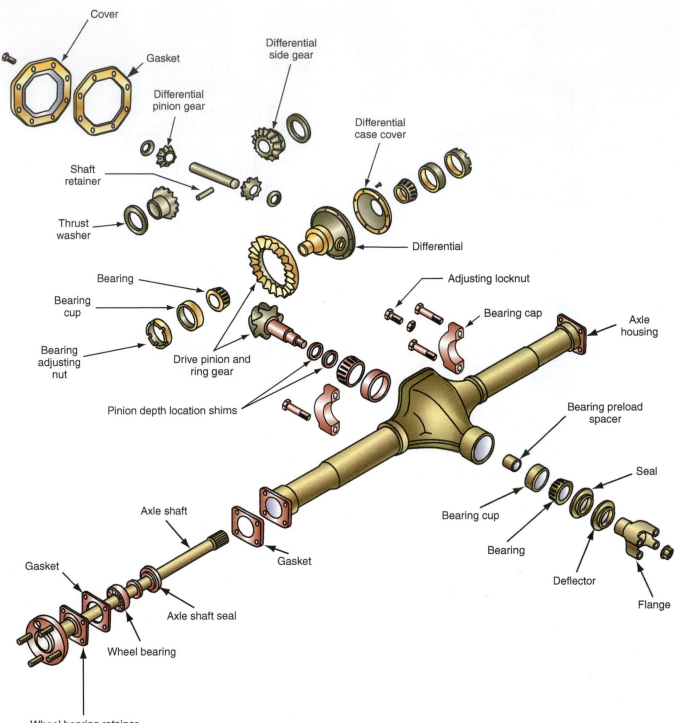

Cover

Gasket

Differential side gear

Differential pinion gear

Differential case cover

Shaft retainer

Thrust washer

Differential

Bearing

Adjusting locknut

Bearing cup

Bearing cap

Axle housing

Bearing adjusting nut

Drive pinion and ring gear

Pinion depth location shims

Bearing preload spacer

Seal

Bearing cup

Bearing

Axle shaft

Deflector

Gasket

Flange

Gasket

Axle shaft seal

Wheel bearing

Wheel bearing retainer

FIGURE 1-21 Axle shafts are supported by the axle housing and are driven by the differential side gears. The vehicle's wheels are bolted to the axle shaft's outer flange.

end of the stud. This holds the wheels in place as they rotate with the axles, allowing the vehicle to move.

The drive axles of a FWD vehicle extend from the sides of the transaxle to the drive wheels. CV joints are fitted to the axles to allow the axles to move with the car's suspension and steering systems. They are designed to allow the wheels to steer and to move up and down with the suspension.

The differential side gears are connected to inboard CV joints by splines. A short shaft extends from the differential to the inner CV joint. Connecting the inner CV joint and the outer CV joint is the axle shaft. Extending from the outer CV joint is a short shaft that fits into the hub of the wheels.

Some rear drive axle units are mounted to an independent suspension and the drive CV axle assembly is axle shafts are similar to that of those in a FWD vehicle.

FOUR-WHEEL-DRIVE SYSTEMS

Four-wheel-drive (4WD) vehicles designed for off-road use are normally RWD vehicles equipped with a **transfer case** (Figure 1-22), a front drive shaft, and a front differential and drive axles. Some 4WD vehicles use three drive shafts. One short drive shaft connects the output of the transmission to the transfer case. The output from the transfer case is then sent to the front and rear axles through separate drive shafts.

Some cars are equipped with AWD to improve the handling characteristics of the car. Many of these cars are FWD models converted to 4WD. Normally, FWD cars are modified by adding a transfer case, a rear drive shaft, and a rear axle with a differential. Although this is the typical modification, some cars are equipped with a center differential in place of the transfer case. This differential unit allows the rear and front wheels to turn at different speeds and with different amounts of torque.

The transfer case is usually mounted to the side or rear of the transmission. When a drive shaft is not used to connect the transmission to the transfer case, a chain, or gear drive, within the transfer case receives the engine's power from the transmission and transfers it to the drive shafts leading to the front and rear drive axles.

The transfer case itself is constructed similar to a transmission. It uses shift forks to select the operating mode and the splines, gears, shims, bearings, and other components found in transmissions. The housing is filled with lubricant that cuts friction on all moving parts. Seals hold the lubricant in the case and prevent leakage around shafts and yokes. Shims set up the proper clearance between the internal components and the case.

An electric switch or shift lever located in the passenger compartment controls the transfer case so that power is directed to the axles selected by the driver. Power typically can be directed to all four wheels, two wheels, or to none of the wheels in the neutral position. On many vehicles, the driver can also select a low-speed range for extra torque when traveling in very adverse conditions.

The rear drive axle of a 4WD vehicle is identical to those used in 2WD vehicles. The front drive axle is also like a conventional rear axle, except that it is modified to allow the front wheels to steer. Further modifications are also necessary to adapt the axle to the vehicle's suspension system. The differential units housed in the axle assemblies are similar to those found in a RWD vehicle.

FIGURE 1-22 **A transfer case.**

Types of Gears

Gears are normally used to transmit torque from one shaft to another. These shafts may operate in line, parallel to each other, or at an angle to each other. These different applications require a variety of gear designs, which vary primarily in the size and shape of the teeth.

Idler gears are gears that do not drive something. They simply are used to reverse the direction of rotation (Figure 1-23). If three gears were connected, the center gear would be considered an idler gear.

In order for gears to mesh, they must have teeth of the same size and design. Meshed gears have at least one pair of teeth engaged at all times. Some gear designs allow for contact between more than one pair of teeth. Gears normally are classified by the type of teeth they have and by the surface on which the teeth are cut.

Automobiles use a variety of gear types to meet the demands of speed and torque. The most basic type of gear is the **spur gear**, which has its teeth parallel to and in alignment with the center of the gear (Figure 1-24). Early transmissions used straight-cut spur gears, which were easier to machine but were noisy and difficult to shift. Today these gears are used mainly for slow speeds to avoid excessive noise and vibration. They commonly are used as reverse gears and in simple devices such as hand or powered winches.

Helical gears are like spur gears except that their teeth have been twisted at an angle from the gear centerline (Figure 1-25). These gears are cut in a helix, which is a form of curve. This curve is more difficult to machine but is used because it reduces gear noise. Engagement of these gears begins at the tooth tip of one gear and rolls down the trailing edge of the teeth. This angular contact tends to cause side thrusts that are absorbed by a thrust bearing or bearings. Helical gears are widely used in transmissions today because they are quieter at all speeds and are durable.

> **AUTHOR'S NOTE:** Despite the advantages that helical gears have over spur gears, most racing transmissions are constructed with spur gears. The axial thrust and sliding friction generated by helical gears can create excess heat and frictional losses. The noise generated by spur gears is not an important factor on the track.

Bevel spur gears are shaped like a cone with its top cut off. The teeth point inward toward the peak of the cone. These gears permit the power flow to "turn a corner." Spiral bevel gears were developed for use when higher speed and strength as well as a change in the angle of the power-flow were required (Figure 1-26A). Their teeth are cut obliquely on the angular faces of the gears.

FIGURE 1-23 An idler gear between the input and output gears. The teeth of the idler gear are not used to calculate the gear ratio.

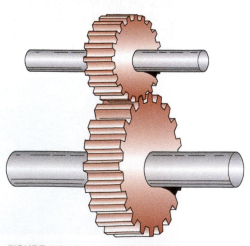

FIGURE 1-24 Spur gears have teeth cut straight across the gear's edge and parallel to the shaft.

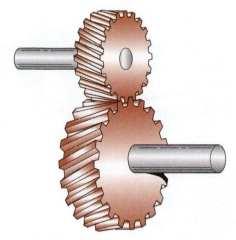

FIGURE 1-25 Helical gears have teeth cut at an angle to the gear's axis of rotation.

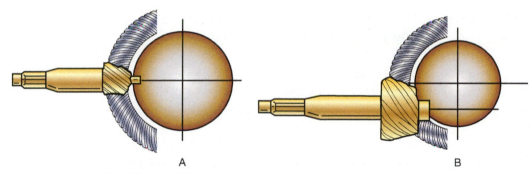

A   B

FIGURE 1-26 (A) Spiral bevel differential gears and (B) hypoid gears.

Spiral bevel gear sets are also used in many racing differentials. The most commonly used spiral bevel gear set is the ring and pinion gears used in heavy truck differentials. Straight cut B bevel type gears are also used for slow-speed applications, which are not subject to high-impact forces. Handwheel controls that must operate some remote device at an angle use straight bevel gears.

The **hypoid gear** resembles the spiral bevel gear, but the pinion drive gear is located below the center of the ring gear. Its teeth and general construction are similar to that of the spiral bevel gear. The most common use for hypoid gears is in modern differentials (Figure 1-26B). Here, they allow for lower body styles by lowering the transmission drive shaft.

A variation of hypoid gears are amboid gears. Rather than having the pinion ride below the centerline of the ring gear, the pinion is set above the centerline. These units are often referred to as "high pinion" axles. They minimize the required drive line angle on short 4WD vehicles and help to reduce drive line vibration during operation (Figure 1-27).

The **worm gear** is actually a screw capable of high-speed reductions in a compact space. Its mating gear has teeth that are curved at the tips to permit a greater contact area. Power is supplied to the worm gear, which drives the mating gear. Worm gears usually provide right-angle power flows. The most common use for the worm gear is in applications in which the power source operates at high speed and the output is at slow speed with high torque. Many steering mechanisms use a worm gear connected to the steering shaft and wheel and a partial (sector) gear connected to the steering linkage. Small power hand tools frequently use a high-speed motor with a worm gear drive. A notable drivetrain application of the use of worm gears can be found in the Torsen limited-slip differential (Figure 1-28).

Rack and pinion gears convert straight-line motion into rotary motion, and vice versa. Rack and pinion gears also change the angle of power-flow with some degree of speed change.

Some tractors and off-road vehicles use worm gears as final drive gears.

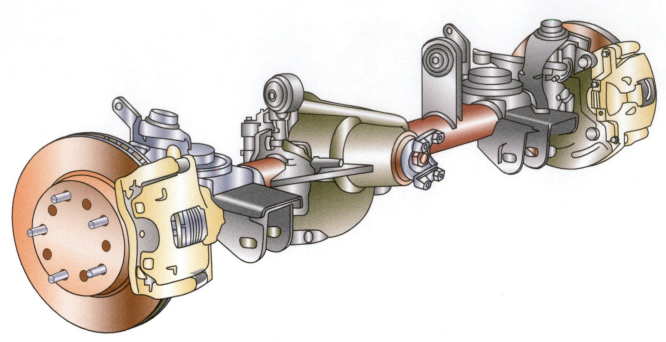

FIGURE 1-27 A high-pinion axle assembly.

The teeth on the rack are cut straight across the shaft, whereas those on the pinion are cut like a spur gear. These gear sets can provide control of arbor presses and other devices when slow speed is involved. Rack and pinion gears are also commonly used in automotive steering systems.

Internal gears have their teeth pointing inward and mesh with an external gear. This type of gear set is used in many different applications. The gear set can provide for torque multiplication without a change in direction.

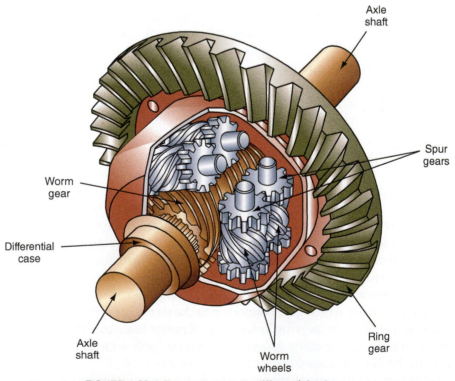

Axle shaft

Spur gears

Worm gear

Differential case

Axle shaft

Worm wheels

Ring gear

FIGURE 1-28 A Torsen limited slip differential using worm gear components.

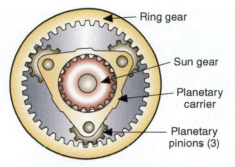

FIGURE 1-29 **A simple planetary gear set.**

Planetary gear sets are commonly used in automatic transmissions, transfer cases to provide low range, and in many hybrid-vehicle CVTs. These are gear sets in which an outer ring gear has internal teeth, which mate with teeth on smaller external tooth pinion gears.

A simple planetary gear set consists of three parts: a sun gear, a carrier with planetary pinions mounted to it, and an internally toothed ring gear or **annulus**. The **sun gear** is located in the center of the assembly (Figure 1-29). It can be either a spur or helical gear design. It meshes with the teeth of the planetary pinion gears. Planetary pinion gears are small gears fitted into a framework called the **planetary carrier**. The planetary carrier can be made of cast iron, aluminum, or steel plate and is designed with a shaft for each of the planetary pinion gears. (For simplicity, planetary pinion gears are called **planetary pinions**.)

The planetary pinions rotate on needle bearings positioned between the planetary carrier shaft and the planetary pinions. The carrier and pinions are considered one unit—the midsize gear member. The planetary pinions surround the center axis of the sun gear and are surrounded by the annulus or ring gear, which is the largest part of the gear set. The ring gear acts like a band to hold the entire gear set together and provides great strength to the unit.

Many changes in speed and torque are possible with a planetary gear set. These depend on which parts are held stationary and which are driven. Any one of the three members can be used as the driving or input member. At the same time, another member might be kept from rotating and thus becomes the reaction, held, or stationary member. The third member then becomes the driven or output member. Depending on which member is the driver, which is held, and which is driven, either a torque increase (**underdrive**) or a speed increase (**overdrive**) is produced by the planetary gear set. Power transfer through a planetary gear set is only possible when one of the members is held at rest, or if two of the members are locked together.

Planetary gears are widely used in different applications because of their many advantages over other gear arrangements. The gear set is strong because the gear load is spread over several gears, reducing stress and wear on any one gear. The arrangement is compact and works on a single axis. Multiple ratios, reverse and neutral, are available from a planetary gear set.

TYPES OF BEARINGS

Gears are either securely attached to a shaft or designed to move freely on the shaft. The ease with which the gears rotate on the shaft or the shaft rotates with the gears partially determines the amount of power needed to rotate them. If they rotate with great difficulty because of high friction, much power is lost. High friction also will cause excessive wear to the gears and shaft. To reduce the friction, bearings are fitted to the shaft or gears.

The simplest type of bearing is a cylindrical hole formed in a piece of material, into which the shaft fits freely. The hole is usually lined with a brass or bronze lining, or **bushing**, which not only reduces the friction but also allows for easy replacement when wear occurs. Bushings are usually tight in the hole in which they fit.

Ball- or roller-type bearings are used wherever friction must be minimized. With these types of bearings, rolling friction replaces the sliding friction that occurs in plain bearings.

Bushings often are referred to as plain bearings.

FIGURE 1-30 A Torrington thrust bearing.

End play is the end-to-end move-ment of a shaft or component within its housing.

A **journal** is the area on a shaft that rides on the bearing.

A BIT OF HISTORY

There is a reason that thrust bearings are often called Torrington bearings. The Excelsior Needle Company of Torrington, Connecticut, began manufacturing sewing machine needles in the mid-1800s. That company eventu-ally became the Torrington Bearing Company and was a major manufacturer of needle thrust bearings beginning in the 1930s.

Typically, two bearings are used to support a shaft instead of a single long bushing. Bearings have four purposes: support a load, maintain alignment of a shaft, reduce rotating friction, and control **end play**.

Most bearings are capable of withstanding only those loads that are perpendicular to the axis of the shaft. Such loads are called **radial loads** and bearings that carry them are called radial or **journal** bearings.

To prevent the shaft from moving in the axial direction, shoulders or collars may be formed on it or secured to it. If both collars are made integral with the shaft, the bearings must be split or made into halves, and the top half or cap bolted in place after the shaft has been put in place. The collars or shoulders withstand any end thrusts, and bearings designed this way are called **thrust bearings** (or Torrington bearings). Most thrust bearings are similar to roller and ball bearings except the plates that the rollers or balls ride between are designed as flat washers. These rollers or balls reduce side-to-side load (Figure 1-30).

A single row journal or radial **ball bearing** has an inner race made of a ring of case-hardened steel with a groove or track formed on its outer circumference on which a number of hardened steel balls can run (Figure 1-31). The outer race is another ring, which has a track on its inner circumference. The balls fit between the two tracks and roll around in the tracks as either race turns. The balls are kept from rubbing against each other by some form of cage. These bearings can withstand radial loads and also can withstand a considerable amount of axial thrust. Therefore, they often are used as combined journal and thrust bearings.

A bearing designed to take only radial loads has only one of its races machined with a track for the balls. Other bearings are designed to take thrust loads in only one direction. If this type of bearing is installed incorrectly, the slightest amount of thrust will cause the bearing to come apart.

Another type of ball bearing uses two rows of balls. These are designed to withstand considerable amounts of radial and axial loads. Constructed like two single-row ball bearings joined together, these bearings often are used in rear axle assemblies.

Roller bearings are used wherever it is desirable to have a large bearing surface and low amounts of friction. Large bearing surfaces are needed in areas of extremely heavy loads. The rollers are usually fitted between a journal of a shaft and an outer race. As the shaft rotates, the rollers turn and rotate in the race.

Roller bearings are constructed almost the same way as ball bearings, but they have cylindrical-shaped bearings instead of spherical-shaped bearings. This gives them a greater load-bearing area. They are used in many wheel bearing applications.

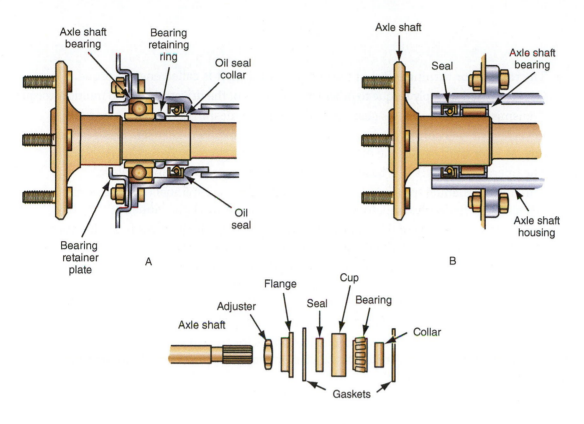

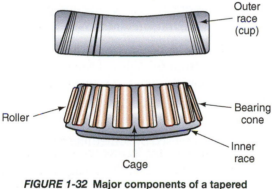

FIGURE 1-31 (A) Ball-type bearings supporting an axle shaft. (B) Straight roller bearing installed in an axle housing. (C) Tapered roller drive axle bearing.

FIGURE 1-32 Major components of a tapered roller bearing assembly.

Tapered roller bearings are the most commonly used type of bearing in drivetrains. These bearings use long tapered rollers, which are fitted into a tapered cone. This assembly rides in a tapered cup (Figure 1-32). The basic principle behind the design of a tapered roller bearing is that the apexes of the tapered surfaces converge to a common axis. The angular shape of the bearing increases the frictional surface area and limits shaft end play.

SUMMARY

- The drivetrain has four primary purposes: to connect the engine's power to the drive wheels, to select different speed ratios, to provide a way to move the vehicle in reverse, and to control the power to the drive wheels for safe turning of the vehicle.

- The main components of the drivetrain are the clutch, transmission, differential, and drive axles.

SUMMARY

- The rotating or turning effort of the engine's crankshaft is called engine torque.
- Gears are used to apply torque to other rotating parts of the drivetrain and to multiply torque.
- Transmissions offer various gear ratios through the meshing of various-sized gears.
- Reverse gear is accomplished by adding a third gear to a two-gear set. This gear, the reverse idler gear, causes the driven gear to rotate in reverse.
- The operation of a CVT is based on a steel belt linking two variable pulleys.
- Self-shifting manual transmissions are available. They are manual transmissions that use electronic or hydraulic actuators to shift the gears and work the clutch.
- Connected to the rear of the crankshaft is the flywheel, which serves many functions, including acting as the driving member of the clutch assembly.
- The clutch assembly is comprised of another driving disc, the pressure plate, and a driven disc, the clutch disc.
- The clutch disc is mounted to the input shaft of the transmission and carries the engine's torque to the transmission when the clutch assembly is engaged.
- In FWD cars, the transmission and drive axle is located in a single assembly called a transaxle. In RWD cars, the drive axle is connected to the transmission through a drive shaft.
- The drive shaft and its joints are called the drive line of the car.
- Universal joints allow the drive shaft to change angles in response to movements of the car's suspension and rear axle assembly.
- The rear axle housing encloses the entire rear-wheel driving axle assembly.
- The primary purpose of the differential is to allow a difference in driving wheel speed when the vehicle is rounding a corner or curve. The ring and pinion in the drive axle also multiples the torque it receives from the transmission.
- On FWD cars, the differential is part of the transaxle assembly.
- The drive axles on FWD cars extend from the sides of the transaxle to the drive wheels. CV joints are fitted to the axles to allow the axles to move with the car's suspension.
- 4WD vehicles typically use a transfer case, which relays engine torque to both a front and rear driving axle.
- Bearings are used to reduce the friction caused by something rotating within something else.

REVIEW QUESTIONS

Short Answer Essays

1. What are the primary purposes of a vehicle's drivetrain?

2. What is the basic advantage of a limited-slip differential?

3. What is the purpose of an idler gear?

4. Why are transmissions equipped with many different forward gear ratios?

5. What is the primary difference between a transaxle and a transmission?

6. Why are U-joints and CV joints used in the drive line?

7. What does a differential (final drive) unit do to the torque it receives?

8. What is the purpose of the clutch assembly? How does it work?

9. What kind of gears are commonly used in today's automotive drivetrains?

10. When are ball- or roller-type bearings used?

Fill-in-the-Blanks

1. The main components of the drivetrain are the
   _____, _____ and
   _____ _____.

2. The rotating or turning effort of the engine's
   crankshaft is called _____

3. Gears are used to apply torque to other rotating parts
   of the drivetrain and to _____ torque.

4. _____ _____, especially
   those that are used off-road, can deliver power
   to all four wheels. The driver can select two-wheel
   or four-wheel drive. Some four-wheel drive
   vehicles are always engaged in four-wheel drive;
   these vehicles are commonly said to have
   _____-_____ four-wheel
   drive or _____-_____-
   _____.

5. The car's drive shaft and its joints are often called the
   _____ _____.

6. The simplest type of bearing is a _____.

7. Reverse gear is accomplished by adding an
   _____ _____ to a
   two-gear set.

8. The clutch assembly is comprised of two driving discs,
   called the _____ and _____
   _____, and a driven disc, called
   the _____ _____.

9. In FWD cars, the transmission and drive axle
   is located in a single assembly called a
   _____.

10. In RWD cars, the drive axle is connected to the
    transmission through a _____.

ASE-Style Review Questions

1. When discussing the purposes of a drivetrain,
   Technician A says that it connects the engine's power
   to the drive wheels.
   Technician B says that it controls the power to the
   drive wheels for safe turning of the vehicle.
   Who is correct?
   A. A only
   B. B only
   C. Both A and B
   D. Neither A nor B

2. When discussing gears and pulleys,
   Technician A says that gears are used to apply torque
   to other rotating parts of the drivetrain.

Technician B says that variable pulleys are used to
multiply torque or increase speed in many CVTs.
Who is correct?
A. A only
B. B only
C. Both A and B
D. Neither A nor B

3. The type of bearing that can only control loads
   perpendicular to the shaft is a:
   A. roller bearing
   B. tapered roller bearing
   C. thrust bearing
   D. radial bearing

4. When discussing reverse gear,
   Technician A says reverse is accomplished by adding a
   third gear.
   Technician B says the reverse idler gear causes the
   driven gear to rotate in reverse.
   Who is correct?
   A. A only
   B. B only
   C. Both A and B
   D. Neither A nor B

5. When discussing the purpose of a flywheel,
   Technician A says that it increases engine torque.
   Technician B says that it acts as the driving member
   of the clutch assembly.
   Who is correct?
   A. A only
   B. B only
   C. Both A and B
   D. Neither A nor B

6. When discussing clutch components,
   Technician A says that the clutch disc is splined to the
   input shaft of the transmission.
   Technician B says that the clutch disc carries the
   engine's torque to the transmission when the clutch
   assembly is engaged.
   Who is correct?
   A. A only
   B. B only
   C. Both A and B
   D. Neither A nor B

7. When discussing universal joints,
   Technician A says that they eliminate vibrations caused
   by the power pulses of the engine.
   Technician B says that they allow the drive shaft to
   change angles in response to movements of the car's
   suspension and rear axle assembly.
   Who is correct?
   A. A only
   B. B only
   C. Both A and B
   D. Neither A nor B

8. When discussing the purpose of a differential,

Technician A says that it allows for equal wheel speed when the vehicle is rounding a corner or curve.

Technician B says that the ring and pinion in the drive axle multiplies the torque it receives from the transmission.

Who is correct?

A. A only
B. B only
C. Both A and B
D. Neither A nor B

9. When discussing FWD vehicles,

Technician A says that the differential is normally part of the transaxle assembly.

Technician B says that the drive axles extend from the sides of the transaxle to the drive wheels.

Who is correct?

A. A only
B. B only
C. Both A and B
D. Neither A nor B

10. When discussing 4WD components,

Technician A says that 4WD vehicles typically use a transfer case to transfer engine torque to both a front and a rear driving axle.

Technician B says that 4WD vehicles normally have two clutch assembles, two drive shafts, and three differentials.

Who is correct?

A. A only
B. B only
C. Both A and B
D. Neither A nor B

Chapter 2

DRIVETRAIN THEORY

UPON COMPLETION AND REVIEW OF THIS CHAPTER, YOU SHOULD BE ABLE TO:

- Describe how all matter exists.
- Explain what energy is and how energy is converted.
- Explain the forces that influence the design and operation of an automobile.
- Describe and apply Newton's laws of motion to an automobile.
- Define *friction* and describe how it can be minimized.
- Describe the various types of simple machines.

- Explain how a set of gears can increase torque.
- Define and determine the ratio between two meshed gears.
- Explain the difference between torque and horsepower.
- Explain Pascal's Law and give examples of where it is applied to an automobile.
- Explain how heat affects matter.
- Describe the origin and practical applications of electromagnetism.

INTRODUCTION

This chapter contains many of the things you have learned or will learn in other courses. It is not my intent to present this material in lieu of those other courses, but instead to emphasize those things that you will need to gain employment and be successful in an automotive career. Many of the facts presented in this chapter are addressed again, in greater detail, according to the topic covered. This chapter contains the theories that are the basis for the rest of the contents in this book. I highly recommend that you make sure you understand the contents of this chapter. Move to the end of the chapter review questions and if you have difficulty answering the questions, study the appropriate content in the chapter until you have a clear understanding and are able to answer the questions correctly.

MATTER

Matter is anything that occupies space. All matter exists as a gas, liquid, or solid. Gases and liquids are considered fluids because they move or flow easily and easily respond to pressure. A gas has neither a shape nor volume of its own and tends to expand without limits. A liquid takes a shape and has volume. A solid is matter that does not flow.

Atoms and Molecules

All matter consists of countless tiny particles called **atoms**. An atom may be defined as the smallest particle of an element in which all the chemical characteristics of the element are present. Atoms are so small they cannot be seen with an electron microscope, which magnifies millions of times. A substance with only one type of atom is referred to as an **element**. Over 100 elements are known to exist at present and, of the known elements, 92 occur naturally. The remaining elements have been manufactured in laboratories.

Small, positively charged particles called protons are located in the center, or nucleus, of each atom. In most atoms, neutrons are also located in the nucleus. Neutrons have no electrical charge, but they add weight to the atom. The positively charged protons tend to repel each other, and this repelling force could destroy the nucleus. The presence of the neutrons with the protons in the nucleus cancels the repelling action of the protons and keeps the nucleus together. Electrons are small, very light particles with a negative electrical charge. Electrons move in orbits around the nucleus of an atom.

A proton is approximately 1,840 times heavier than an electron, and this makes the electron much easier to move. Because the electrons are orbiting around the nucleus, centrifugal force tends to move the electrons away from the nucleus. However, the attraction between the positively charged protons and the negatively charged electrons holds the electrons in their orbits. The atoms of the different elements have different numbers of protons, electrons, and neutrons. Some of the lighter elements have the same number of protons and neutrons in the nucleus, but many of the heavier elements have more neutrons than protons.

The simplest atom is the hydrogen (H) atom, which has one proton in the nucleus and one electron orbiting around the nucleus (Figure 2-1). The nucleus of a copper (CU) atom contains 29 protons and 34 neutrons, whereas 29 electrons orbit in different rings around the nucleus. Because 2, 8, and 18 electrons are the maximum number of electrons on the first three electron rings next to the nucleus, the fourth ring must have one electron (Figure 2-2). The outer ring of an atom is called the valence ring, and the number of electrons on this ring determines the electrical characteristics of the element. Elements are listed on the atomic scale, or periodic chart, according to their number of protons and electrons. For example, hydrogen is number 1 on this scale, and copper is number 29.

Water is a compound that contains oxygen and hydrogen atoms. The chemical symbol for water is H_2O. This chemical symbol indicates that each molecule of water contains two atoms of hydrogen and one oxygen atom (Figure 2-3).

Ions

An **ion** is an atom or molecule that has lost or gained one or more electrons. As a result, it has a negative or positive electrical charge. A negatively charged ion has more electrons than it has protons. The opposite is true of positively charged ions, which have fewer electrons than protons. Ions are denoted in the same way as other atoms and molecules except a

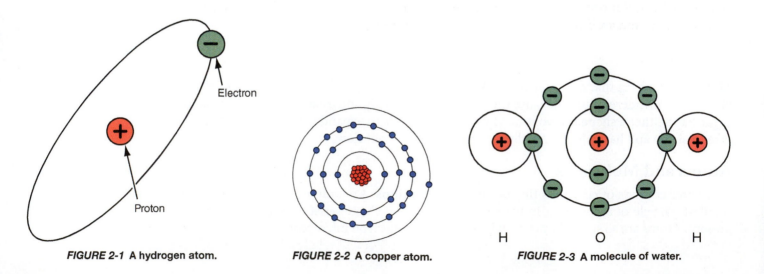

FIGURE 2-1 A hydrogen atom.   FIGURE 2-2 A copper atom.   FIGURE 2-3 A molecule of water.

superscript that shows the electrical charge and the number of electrons gained or lost. For example, hydrogen with a positive charge (H^+) and oxygen with a negative charge (O^{2-}) is called an **oxide**.

Considered by scientists as the fourth state of matter, **plasma** refers to an ionized gas that has about an equal amount of positive ions and electrons. The electrons travel with the nucleus of the atoms but can move freely and are not bound to it. The gas at this point no longer behaves as a gas. It now has electrical properties and creates a magnetic field, which radiate light and other forms of electromagnetic energy. It typically takes the form of gas-like clouds and is the basis of most stars. In fact, our sun is really just a large piece of plasma. Plasmas are the most common form of matter in the universe. Plasma in the stars and in the space between them occupies nearly 99 percent of the visible universe.

Plasma does not exist as solid, liquid, or gas; it is different and has a much different temperature range and is denser than other states of matter. Technicians may be familiar with plasma cutters that use a pressurized inert gas (or air) and an electrical arc to form plasma that is hot enough to melt and cut steel and other metals.

States of Matter

The particles of a solid are held together in a rigid structure. When a solid dissolves into a liquid, its particles break away from this structure and mix evenly in the liquid, forming a **solution**. When they are heated, most liquids **evaporate**. This means that the atoms or molecules of which they are made break free from the body of the liquid to become gas particles. When all of the liquid in a solution has evaporated, the solid is left behind. The particles of the solid normally arrange in a structure called a crystal.

Absorption and Adsorption. Not all solids will dissolve in a liquid; rather, the liquid will be either absorbed or adsorbed. The action of a sponge serves as the best example of absorption. When a dry sponge is put into water, the water is absorbed by the sponge. The sponge does not dissolve, the water merely penetrates into the sponge and the sponge becomes filled with water. There is no change to the atomic structure of the sponge, nor does the structure of the water change. If we take a glass and put it into water, the glass does not absorb the water. The glass, however, still gets wet as a thin layer of water adheres to the glass. This is adsorption. Materials that *absorb* fluids are **permeable** substances. **Impermeable** substances *adsorb* fluids. Some materials are impermeable to most fluids whereas others are impermeable to just a few.

ENERGY

Energy may be defined as the ability to do work. Because all matter consists of atoms and molecules that are in constant motion, all matter has energy. Energy is not matter, but it affects the behavior of matter. Everything that happens requires energy, and energy comes in many forms.

Each form of energy can change into other forms. However, the total amount of energy never changes; it can only be transferred from one form to another, not created or destroyed. This is known as the "Principle of the Conservation of Energy."

Kinetic and Potential Energy

When energy is released to do work, it is called **kinetic energy**. This type of energy also may be referred to as energy in motion. Stored energy may be called **potential energy**.

There are many components and systems that have potential energy and, at times, kinetic energy. The ignition system is a source for high electrical energy. The heart of the ignition system is the ignition coil, which has much potential energy when it has current flow through it. When it is time to fire a spark plug, current flow is stopped and energy is released and becomes kinetic energy as it creates a spark across the gap of a spark plug.

Energy Conversion

Energy conversion occurs when one form of energy is changed to another form. Because energy is not always in the desired form, it must be converted to a form we can use. Some of the most common automotive energy conversions are discussed here.

Chemical to Thermal Energy. Chemical energy in gasoline or diesel fuel is converted to thermal energy when the fuel burns in the engine cylinders.

Chemical to Electrical Energy. The chemical energy in a battery (Figure 2-4) is converted to electrical energy to power many of the accessories on an automobile. In some hybrid and all electric vehicles, the battery is used to power the drive wheels.

Electrical to Mechanical Energy. In an automobile, the battery supplies electrical energy to the starting motor, and this motor converts the electrical energy to mechanical energy to crank the engine.

Thermal to Mechanical Energy. The thermal energy that results from the burning of the fuel in the engine is converted to mechanical energy. The firing impulses on the pistons create rotational motion of the crankshaft. That motion is transferred to the drivetrain and is used to move the vehicle.

Mechanical to Electrical Energy. The generator is driven by mechanical energy from the engine. The generator converts this energy to electrical energy, which powers the electrical accessories on the vehicle and recharges the battery.

Electrical to Radiant Energy. Radiant energy is light energy. In the automobile, electrical energy is converted to thermal energy, which heats up the inside of light bulbs so that they illuminate and release radiant energy.

Solar to Electrical Energy. Solar energy can be converted to electrical energy through the use of photovoltaic panels. Automotive applications include roof panels that power cooling fans while the vehicle is parked.

Kinetic to Thermal Energy. To stop a vehicle, the brake system must change the kinetic energy of the moving vehicle to kinetic and static thermal or heat energy. This is the result of friction, which is discussed later in this chapter.

Kinetic to Mechanical to Electrical Energy. Hybrid vehicles have a system, called regenerative braking, that uses the energy of the moving vehicle (kinetic) to rotate a generator. The mechanical energy used to operate the generator is used to provide electrical energy to charge the batteries (Figure 2-4) or power the electric drive motor.

A BIT OF HISTORY

A Kinetic Energy Recovery System (KERS) has been used by some Formula 1 racing teams beginning in 2008. The system captures energy from braking forces and stores it in a high-rpm flywheel assembly (mechanical system) or a supercapacitor (electrical system). The energy is reused on acceleration. Various automakers are researching KERS systems for production use.

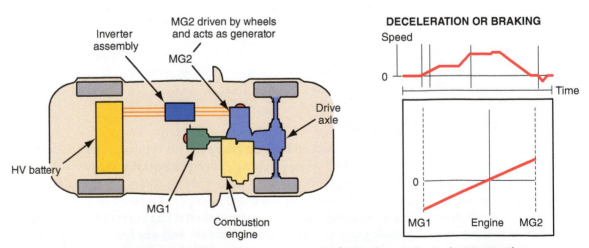

FIGURE 2-4 The power-flow for a hybrid with two motor/generators during brake regeneration.

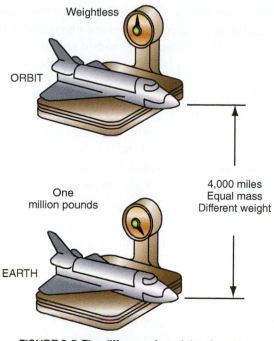

FIGURE 2-5 The difference in weight of a space shuttle on Earth and in space.

Mass and Weight

Mass is the amount of matter in an object. **Weight** is a force and is measured in pounds or kilograms. Gravitational force gives the mass its weight. As an example, a spacecraft can weigh 500 tons (1 million pounds) here on Earth where it is affected by Earth's gravitational pull. In outer space, beyond Earth's gravity and atmosphere, the spacecraft is nearly weightless (Figure 2-5).

Automobile specifications list the weight of a vehicle primarily in two ways. **Gross weight** is the total weight of the vehicle when it is fully loaded with passengers and cargo. **Curb weight** is the weight of the vehicle when it is not loaded with passengers or cargo.

To convert kilograms into pounds, simply multiply the weight in kilograms by 2.2046. For example, if something weighs 5 kilograms, the equation would be ($5 \times 2.2046 = 11.023$ pounds). If you want to express the answer in pounds and ounces, you will convert the .023 pounds into ounces. Because there are 16 ounces in a pound, multiply 16 by 0.023 ($16 \times 0.023 = 0.368$ ounces). Therefore, 5 kilograms is equal to 11 pounds and 0.368 ounces.

Size

The size of something is related to its mass. The size of an object defines how much space it occupies. Size dimensions are typically stated in terms of its length, width, and height. *Length* is a measurement of how long something is from one end to another. *Width* is a measurement of how wide something is from one side to another. *Height* is a measurement of the distance from something's bottom to its top. All three of these dimensions are measured in inches, feet, yards, and miles in the English system and meters in the metric system.

To convert a meter into feet, you multiply the number of meters by 3.281. If you wanted to convert the feet into inches, simply multiply your answer in feet by 12. For example, to convert 0.01 mm into inches, begin by converting 0.01 mm into meters. Because 1 mm is equal to 0.001 meters, you need to multiply 0.01 by 0.001 ($0.001 \times 0.01 = 0.00001$). Then multiply 0.00001 meters by 3.281 ($0.00001 \times 3.281 = 0.00003281$ feet). Now convert feet into inches by multiplying by 12 ($0.00003281 \times 12 = 0.00039372$ inches).

An easier way to do this would be using the conversion factor that states that 1 mm is equal to 0.03937 inches. To use this conversion factor, multiply 0.01 mm by 0.03937 ($0.01 \times 0.03937 = 0.0003937$ inches).

Sometimes distance measurements are made with a rule that has fractional rather than decimal increments. Most automotive specifications are given decimally; therefore, fractions need to be converted into decimals. It is also easier to add and subtract dimensions if they are expressed in decimal form rather than in fractions. For example, you want to find the rolling circumference of a tire. You have found the diameter of the tire to be 20-⅜ inches. The distance around the tire is the circumference and it is equal to the diameter multiplied by a constant called Pi (π). Pi is equal to approximately 3.14; therefore the circumference of the tire is equal to 20-⅜ inches multiplied by 3.14. This calculation is much easier if you convert the 20-⅜ inches into a whole number and a decimal. To convert the ⅜ to a decimal, divide the 3 by 8 (3 ÷ 8 = 0.375). Therefore, the diameter of the tire is 20.375 inches. Now multiply the diameter by π (20.375 × 3.14 = 63.98). The circumference of the tire is nearly 64 inches.

Ratios. Often automotive features are expressed as ratios. A ratio expresses the relationship between two things. If something is twice as large as another, there is a ratio of 2:1. Sometimes ratios are used to compare the movement of an object. For example, if a 1-inch movement by something causes something else to move 2 inches, there is a travel ratio of 1:2.

FORCE

A **force** is a push or pull, and can be large or small. Force can be applied to objects by direct contact or from a distance. Gravity and electromagnetism are examples of forces that are applied from a distance. Forces can be applied from any direction and with any intensity. For example, if a pulling force on an object is twice that of the pushing force, the object will be pulled at one-half of the pulling force. When two or more forces are applied to an object, the combined force is called the resultant. The resultant is the sum of the size and direction of the forces. For example, when a mass is suspended by two lengths of wire, each wire should carry half the weight of the mass. If we move the attachment of the wires so they are at an angle to the mass, the wires now carry more force. The wires carry the force of the mass plus a force that pulls against the other wire.

Automotive Forces

When a vehicle is sitting still, gravity exerts a downward force on the vehicle. The ground exerts an equal and opposite upward force and supports the vehicle. When the engine is running and its power output transferred to the vehicle's drive wheels, the wheels exert a force against the ground in a horizontal direction. This force causes the vehicle to move, but it is opposed by the mass of the vehicle (Figure 2-6). To move the vehicle faster, the force supplied by the wheels must increase beyond the opposing forces. As the vehicle does move faster, it pushes against the air as it travels. This becomes a growing opposing force and the force at the drive wheels must overcome the force in order for the vehicle to increase speed. After the vehicle has achieved the desired speed, no additional force is required at the drive wheels.

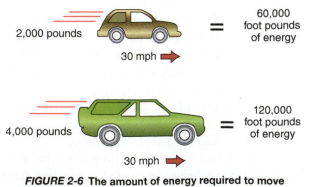

FIGURE 2-6 The amount of energy required to move a vehicle depends on its mass.

Force–Balanced and Unbalanced. When the applied forces are balanced and there is no overall resultant force, the object is said to be in **equilibrium**. An object sitting on a solid flat surface is in equilibrium because its weight is supported by the surface and there is no resultant force. If the surface is put on an angle, the object will tend to slide down the surface. If the surface is at a slight angle, the force will cause the object to slowly slide down the surface. If the surface is at a severe angle, the downward force will cause the object to quickly slide down the slope. In both cases, the surface is still supplying the force needed to support the object but the pull of gravity is greater and the resultant force causes the object to slide down the slope.

Turning Forces. Forces can cause rotation as well as straight-line motion. A force acting on an object that is free to rotate will have a turning effect, or turning force. This force is equal to the size of the force multiplied by the distance of the force from the turning point around which it acts.

Centrifugal/Centripetal Forces

When an object moves in a circle, its direction is continuously changing and all changes in direction require a force. The forces required to maintain circular motion are called **centripetal force** and **centrifugal force**. The size of these forces depends on the size of the circle and the mass and speed of the object.

Centripetal force tends to pull the object toward the center of the circle, whereas centrifugal force tends to push the object away from the center. The centripetal force that keeps an object whirling around on the end of a string is caused by **tension** in the string. If the string breaks, there is no longer string tension and centripetal force and the object will fly off in a straight line because of the centrifugal force on it. Gravity is the centripetal force that keeps the planets orbiting around the sun. Without this centripetal force, Earth would move in a straight line through space.

Tension is the result of forces applied to a material that strain the composition or construction of that material.

Pressure

Pressure is a force applied against an object and is measured in units of force per unit of surface area (pounds per square inch or kilograms per square centimeter). Mathematically, pressure is equal to the applied force divided by the area over which the force acts. Consider two 10-pound weights sitting on a table; one occupies an area of 1 square inch and the other an area of 4 square inches. The pressure exerted by the first weight would be 10 pounds per 1 square inch, or 10 psi. The other weight, although it weighs the same, will exert only 2.5 psi (10 pounds per 4 square inches = $10 \div 4 = 2.5$). This illustrates an important concept: A force acting over a large area will exert less **pressure** than the same force acting over a small area.

Because pressure is a force, all principles of force apply to pressure. If more than one pressure is applied to an object, the object will respond to the resultant force. Also, all matter (liquids, gases, and solids) tends to move from an area of high pressure to an area of low pressure.

MOTION

When the forces on an object do not cancel each other out, they will change the motion of the object. The object's speed, direction of motion, or both will change. The greater the mass of an object, the greater the force needed to change its motion. This resistance to change in motion is called **inertia**. Inertia is the tendency of an object at rest to remain at rest, or the tendency of an object in motion to stay in motion. The inertia of an object at rest is called static inertia, whereas dynamic inertia refers to the inertia of an object in motion. Inertia exists in liquids, solids, and gases. When you push and move a parked vehicle, you overcome the static inertia of the vehicle. If you catch a ball in motion, you overcome the dynamic inertia of the ball.

When a force overcomes static inertia and moves an object, the object gains momentum. **Momentum** is the product of an object's weight times its speed. Momentum is a type of mechanical energy. An object loses momentum if another force overcomes the dynamic inertia of the moving object.

Rates

Speed is the distance an object travels in a set amount of time. It is calculated by dividing distance covered by time taken. We refer to the speed of a vehicle in terms of miles per hour (mph) or kilometers per hour (km/h). **Velocity** is the speed of an object in a particular direction. **Acceleration**, which only occurs when a force is applied, is the rate of increase in speed. Acceleration is calculated by dividing the change in speed by the time it took for that change. **Deceleration** is the reverse of acceleration, as it is the rate of a decrease in speed.

Newton's Laws of Motion

How forces change the motion of objects was first explained by Sir Isaac Newton. These explanations are known as Newton's Laws. Newton's first law of motion is called the Law of Inertia. It states that an object at rest tends to remain at rest and an object in motion tends to remain in motion, unless some force acts on it. When a car is parked on a level street, it remains stationary unless it is driven or pushed.

Newton's second law states that when a force acts on an object the motion of the object will change. This change in motion is equal to the size of the force divided by the mass of the object on which it acts. Trucks have a greater mass than cars. Because a large mass requires a larger force to produce a given acceleration, a truck needs a larger engine than a car.

Newton's third law says that for every action there is an equal and opposite reaction. A practical application of this law occurs when the wheel on a vehicle strikes a bump in the road surface. This action drives the wheel and suspension upward with a certain force, and a specific amount of energy is stored in the spring. After this action occurs, the spring forces the wheel and suspension downward with a force equal to the initial upward force caused by the bump.

Friction

Friction is a force that slows or prevents motion of two moving objects or surfaces that touch. Friction may occur in solids, liquids, and gases. It is the joining or bonding of the atoms at each of the surfaces that causes the friction. When you attempt to pull an object across a surface, the object will not move until these bonds have been overcome. Smooth surfaces produce little friction; therefore, only a small amount of force is needed to break the bonds between the atoms. Rougher surfaces produce a larger friction force because stronger bonds are made between the two surfaces (Figure 2-7). To move an object over a rough surface, such as sandpaper, a great amount of force is required. Friction causes heat to build on the surfaces and can cause wear.

Lubrication. Friction can be reduced in two main ways: by lubrication or by the use of rollers. The presence of oil or another fluid between two surfaces keeps the surfaces apart. Because fluids (liquids and gases) flow, they allow movement between surfaces. The fluid keeps the surfaces apart, allowing them to move smoothly past one another (Figure 2-8).

Rollers. Rollers placed between two surfaces keep the surfaces apart. An object placed on rollers will move smoothly if pushed or pulled. Rollers actually use friction to grip the surfaces and produce rotation. Instead of sliding against one another, the surfaces produce turning forces, which cause each roller to spin. This leaves very little friction to oppose motion. Bearings are a type of roller used to reduce the friction between moving parts such as a wheel and its axle. As the wheel turns on the axle, the balls (Figure 2-9) in the bearing roll around inside the bearing, drastically reducing the friction between the wheel and axle.

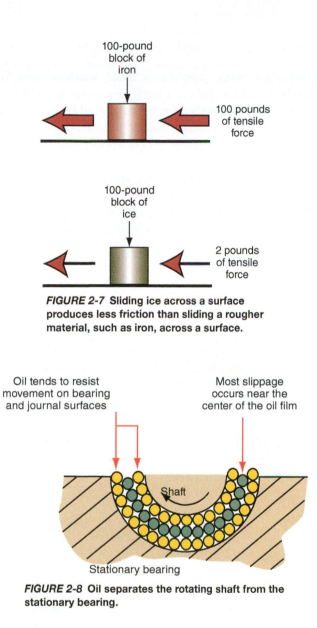

FIGURE 2-7 **Sliding ice across a surface produces less friction than sliding a rougher material, such as iron, across a surface.**

FIGURE 2-8 **Oil separates the rotating shaft from the stationary bearing.**

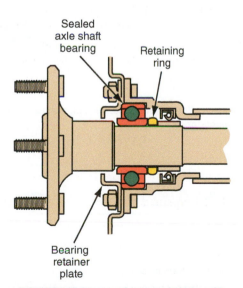

FIGURE 2-9 **The bearings support the axle while allowing it to rotate within them.**

WORK

When a force moves a certain mass a specific distance, **work** is done. When work is accomplished, the mass may be lifted or slid on a surface against a resistance or opposing force (Figure 2-10). Work is equal to the applied force multiplied by the distance the object moved (force × distance = work) and is measured in foot-pounds (Figure 2-11), watts, or Newton-meters (N·m). For example, if a force moves a 3,000-pound car 50 feet, 150,000 foot-pounds of work was done.

During work, a force acts on an object to start, stop, or change the direction of the object. It is possible to apply a force to an object and not move the object. For example, you may push with all your strength on a car stuck in a ditch, and not move the car. Under this condition, no work is done. Work is only accomplished when an object is started, stopped, or redirected by a force.

Simple Machines

A machine is any device that can be used to transmit a force and, in doing so, change the amount of force or its direction. The force applied to a machine is called the effort, whereas the force it overcomes is called the **load**. The effort is often smaller than the load because a small effort can overcome a heavy load if the effort is moved a larger distance. The machine is then said to give a **mechanical advantage**. Although the effort will be smaller when using a machine, the amount of work done, or energy used, will be equal to or greater than that without the machine.

Inclined Plane. The force required to drag an object up a slope (Figure 2-12) is less than that required to lift it vertically. However, the overall distance moved by the object is greater when pulled up the slope than if it were lifted vertically. A screw is like an inclined plane wrapped

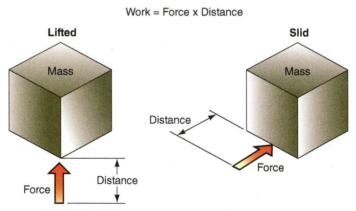

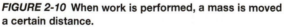

FIGURE 2-10 When work is performed, a mass is moved a certain distance.

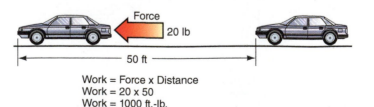

Work = Force x Distance
Work = 20 x 50
Work = 1000 ft.-lb.

FIGURE 2-11 1,000 foot-pounds of work.

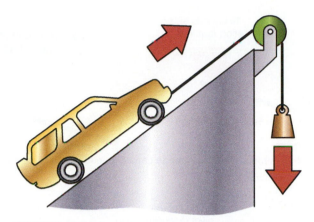

FIGURE 2-12 It takes less energy to pull a mass up an
inclined plane than lifting the mass vertically would require.

around a shaft. The force that turns the screw is converted to a larger one, which moves a
shorter distance and drives the screw in.

Pulleys. A **pulley** is a wheel with a grooved rim in which a rope, belt, or chain runs to raise
something by pulling on the other end of the rope, belt, or chain. A simple pulley changes the
direction of a force but not its size. Also, the distance the force moves does not change. By
using several pulleys connected together as a block and tackle, the size of the force can be
changed, too, so that a heavy load can be lifted using a small force. With a double pulley,
the applied force required to move an object can be reduced by one-half, but the distance the
force must be moved is doubled. A quadruple pulley can reduce the force by four times, but
the distance will be increased by four times. Pulleys of different sizes can change the amount
of required applied force, as well as the speed or distance the pulley needs to travel to accomplish work (Figure 2-13).

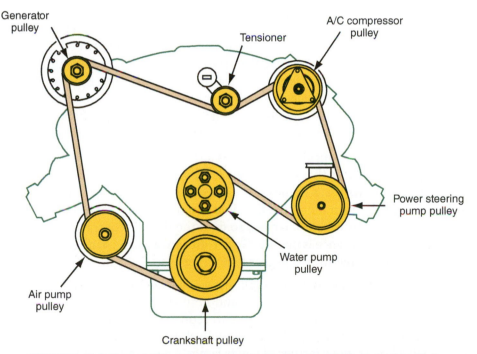

FIGURE 2-13 Pulleys of different sizes can change the amount of required applied
force and the speed or distance the pulley needs to travel to accomplish work.

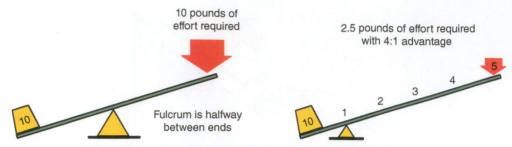

10 pounds of effort required

2.5 pounds of effort required with 4:1 advantage

Fulcrum is halfway between ends

FIGURE 2-14 A mechanical advantage can be gained with a class one lever.

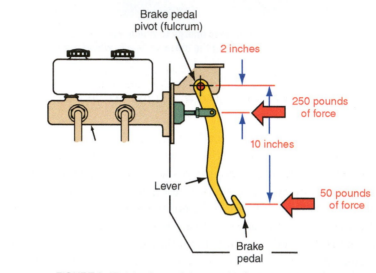

Brake pedal pivot (fulcrum)

2 inches

250 pounds of force

10 inches

Lever

50 pounds of force

Brake pedal

FIGURE 2-15 A brake pedal assembly is an example of a class two lever.

Levers. A **lever** is a device made up of a bar turning about a fixed pivot point, called the fulcrum, that uses a force applied at one point to move a mass on the other end of the bar. Types of levers are divided into classes. In a class one lever, the fulcrum is between the effort and the load (Figure 2-14). The load is larger than the effort, but it moves through a smaller distance. A pair of pliers is an example of a class one lever. In a class two lever, the load is between the fulcrum and effort. Here again, the load is greater than the effort and moves through a smaller distance (Figure 2-15). In a class three lever, the effort is between the fulcrum and the load. In this case, the load is less than the effort, but it moves through a greater distance.

Gears. A **gear** is a toothed wheel that becomes a machine when it is meshed with another gear. The action of one gear is that of a rotating lever and moves the other gear meshed with it. Based on the size of the gears in mesh, the amount of force applied from one gear to the other can be changed. Keep in mind that this does not change the amount of work performed by the gears, as although the force changes, so does the distance of travel (Figure 2-16). The relationship of force and distance is inverse. **Gear ratios** express the mathematical relationship (diameter and number of teeth) of one gear to another.

Wheels and Axles. The most obvious application of a wheel and axle is a vehicle's tires and wheels. These units revolve around an axle and limit the amount of area of a vehicle that contacts the road. Wheels function as rollers to reduce the amount of friction between a vehicle and the road. Basically, the larger the wheel, the less force is required to turn it. However, the wheel must move farther as it gets larger. An example of this is a steering wheel. A steering wheel that is twice the size of another will require one-half the force to turn it but also will require twice the distance to accomplish the same work.

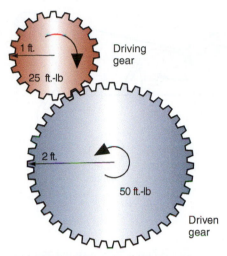

FIGURE 2-16 When a small gear drives a larger gear, the larger gear turns with more force but travels less, therefore the amount of work stays the same.

Torque

Torque is a force that tends to rotate or turn things and is measured by the force applied and the distance traveled. The technically correct unit of measurement for torque is pounds per foot (lb.-ft.). However, it is rather common to see torque stated in terms of foot-pounds (ft.-lb.). In the metric or SI system, torque is stated in Newton-meters (N·m) or kilogram-meters (kg-m).

An engine creates torque and uses it to rotate the crankshaft. The combustion of gasoline and air in the cylinder creates pressure against the top of a piston. That pressure creates a force on the piston and pushes it down. The force is transmitted from the piston to the connecting rod, and from the connecting rod to the crankshaft. The engine's crankshaft rotates with a torque that is transmitted through the drive train to turn the drive wheels of the vehicle.

Torque is force times leverage, the distance from a pivot point to an applied force. Torque is generated any time a wrench is turned with force. If the wrench is a foot long, and you put 20 lb. of force on it, 20 lb. per foot are being generated. To generate the same amount of torque when exerting only 10 lb. of force, the wrench needs to be 2 ft. long (Figure 2-17). To have torque, it is not necessary to have movement. When you pull a wrench to tighten a bolt, you supply torque to the bolt. If you pull on a wrench to check the torque on a bolt, and the bolt torque is sufficient, torque is applied to the bolt but no movement occurs. If the bolt turns

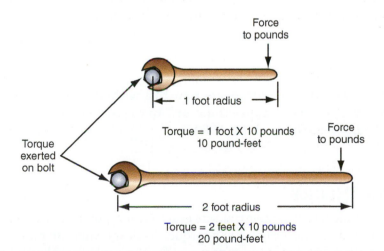

FIGURE 2-17 The amount of torque applied to a wrench is changed by the length of the wrench.

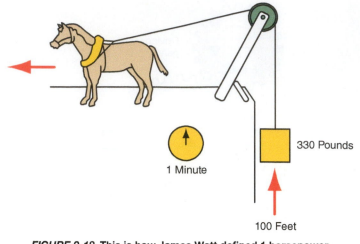

FIGURE 2-18 This is how James Watt defined 1 horsepower.

during torque application, work is done. When a bolt does not rotate during torque application, no work is accomplished.

Power

Power is a measurement for the rate, or speed, at which work is done. The metric unit for power is the watt. A watt is equal to 1 N·m per second. You can multiply the amount of torque in Newton-meters by the rotational speed to determine the power in watts. Power is a unit of speed combined with a unit of force. For example, if you were pushing something with a force of 1 N·m and it moved at a speed of 1 meter (m) per second, the power output would be 1 watt.

Horsepower

Horsepower (hp) is the rate at which torque is produced. James Watt is credited with being the first person to calculate horsepower and power. He measured the amount of work that a horse could do in a specific time. He found that a horse could move 330 lb. 100 ft. in 1 minute (Figure 2-18). Therefore, he determined that one horse could do 33,000 ft.-lb. of work in 1 minute. Thus, 1 hp is equal to 33,000 ft.-lb. per minute, or 550 ft.-lb. per second. Two hp could do this same amount of work in one-half minute. If you push a 3,000-lb. (1,360-kg) car for 11 ft. (3.3 m) in one-quarter minute, you produce 4 hp.

An engine that produces 300 lb.-ft. torque at 4,000 rpm produces 228 horsepower at 4,000 rpm. This is based on the formula that horsepower is equal to torque multiplied by engine speed and that quantity divided by 5252 ([torque × engine speed] ÷ 5252 = horsepower). The constant, 5252, is used to convert the rpm for torque and horsepower into revolutions per second.

> **AUTHOR'S NOTE:** Manufacturers are now rating their engines' output in watts. One horsepower is equal to approximately 746 watts. Therefore, a 228-hp engine is rated at 170,088 watts, or about 170 kW.

BASIC GEAR THEORY

The primary components of the drive train are gears. Gears apply torque to other rotating parts of the drivetrain and are used to multiply torque. As gears with different numbers of teeth mesh, each rotates at a different speed and torque. Torque is calculated by multiplying the force by the distance from the center of the shaft to the point where the force is exerted (Figure 2-19).

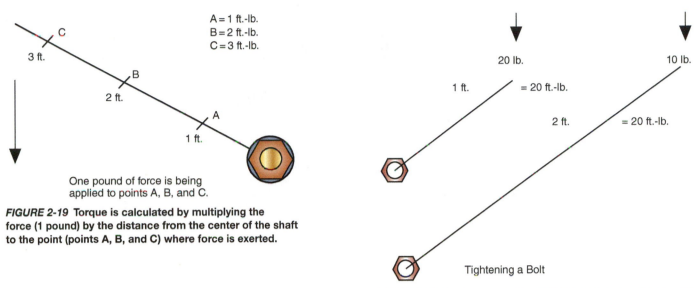

A = 1 ft.-lb.
B = 2 ft.-lb.
C = 3 ft.-lb.

3 ft.

2 ft.

1 ft.

One pound of force is being
applied to points A, B, and C.

FIGURE 2-19 Torque is calculated by multiplying the
force (1 pound) by the distance from the center of the shaft
to the point (points A, B, and C) where force is exerted.

20 lb.

1 ft.   = 20 ft.-lb.

2 ft.   = 20 ft.-lb.

10 lb.

Tightening a Bolt

FIGURE 2-20 The torque applied to both bolts is 20 foot-pounds.

For example, if you tighten a bolt with a wrench that is 1 foot long and apply a force of 10 lb. to the wrench, you are applying 10 ft.-lbs. of torque to the bolt. Likewise, if you apply a force of 20 lb. to the wrench, you are applying 20 ft.-lbs. of torque. You also could apply 20 ft.-lbs. of torque by applying only 10 lb. of force if the wrench were 2 ft. long (Figure 2-20).

The distance from the center of a circle to its outside edge is called the radius. On a gear, the radius is the distance from the center of the gear to the point on its teeth at which force is applied.

> **AUTHOR'S NOTE:** The term *radius* defines the length of the gear lever. We should be careful when we discuss gear diameter, however. For example, some manufacturers use outside diameter (actual outside measurement from one side of a gear to another) in naming a differential ring gear size. Others use pitch diameter (a smaller dimension measuring the dimension from one pitch point on the gear's surface to the opposite pitch point). The pitch point of a gear is "where the work gets done," or the point of contact where force is applied from one tooth to another. This difference in measurements can make all the difference when ordering differential seals or other drive train components.

If a tooth on the driving gear is pushing against a tooth on the driven gear with a force of 25 lb. and the force is applied at a distance of 1 foot, which is the radius of the driving gear, a torque of 25 ft.-lbs. is applied to the driven gear. The 25 lb. of force from the teeth of the smaller (driving) gear is applied to the teeth of the larger (driven) gear. If that same force were applied at a distance of 2 ft. from the center, the torque on the shaft at the center of the driven gear would be 50 ft.-lbs. The same force is acting at twice the distance from the shaft center (Figure 2-21).

The amount of torque that can be applied from a power source is proportional to the distance from the center at which it is applied. If a fulcrum or pivot point is placed closer to the object being moved, more torque is available to move the object, but the lever must move farther than if the fulcrum was farther away from the object (Figure 2-22). The same principle is used for gears in **mesh**: A small gear will drive a large gear more slowly but with greater torque.

A drivetrain consisting of a driving gear with 24 teeth and a radius of 1 inch and a driven gear with 48 teeth and a radius of 2 inches will have a torque multiplication factor of 2 and a

The **meshing** of gears describes the fit of one tooth of one gear between two teeth of the other gear.

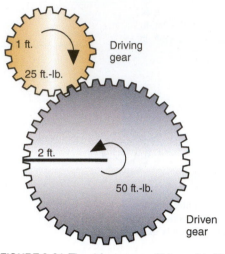

FIGURE 2-21 The driven gear will turn at half the speed but twice the torque because it is two times larger than the driving gear.

speed reduction of one-half. Thus, it doubles the amount of torque applied to it at half the speed (Figure 2-23). The radii between the teeth of a gear act as levers; therefore, a gear that is twice the size of another has twice the lever arm length of the other.

Gear ratios express the mathematical relationship of one gear to another. Gear ratios can be varied by changing the diameter and number of teeth of the gears in mesh. A gear ratio also expresses the amount of torque multiplication between two gears. The ratio is obtained by dividing the diameter or number of teeth of the driven gear by the diameter or teeth of the drive gear. If the smaller driving gear had 10 teeth and the larger gear had 40 teeth, the ratio would be 4:1 (Figure 2-24). The gear ratio tells you how many times the driving gear has to turn to rotate the driven gear once. With a 4:1 ratio, the smaller gear must turn four times to rotate the larger gear once.

Gear ratios are normally expressed in terms of some number to one (1) and use a colon (:) to show the numerical comparison, for example, 3.5:1, 1:1, 0.85:1.

Transmission Gears

Transmissions contain several combinations of large and small gears. In low or first gear, a small gear on the input shaft drives a large gear on another shaft. This reduces the speed of the larger gear but increases its turning force or torque. Connected to the second (counter)

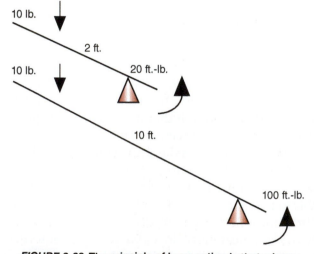

FIGURE 2-22 The principle of lever action is that a long lever is able to perform more work with less force than a short lever.

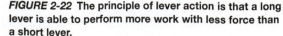

FIGURE 2-23 The 1-inch gear will turn the 2-inch gear at half its speed but twice the torque.

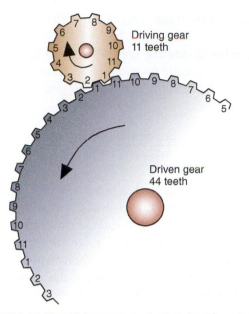

FIGURE 2-24 The driving gear must rotate four times to rotate the driven gear once. The ratio of the gear set is 4:1.

shaft is a small gear, which drives a larger gear, which is connected to the output shaft, which in turn is connected to the drive shaft and driving axle. This reduces the speed and increases the torque still more, giving a higher gear ratio for starting movement or pulling heavy loads. First gear is primarily used to initiate movement. It has the lowest gear ratio of any gear in a transmission. It also allows for the most torque multiplication (Figure 2-25). Recall our discussion in the first chapter on gear terminology where a higher numerical gear ratio is called a "low" gear.

Power moves through the transmission via four gears (two sets of two gears). Speed and torque were altered in steps. To explain how this works, let us put numbers to each of the gears. The small gear on the input shaft has 20 teeth. The gear it meshes with has 40. This provides a gear ratio of 2:1. The output of this gear set moves along the shaft of the 40-tooth gear and rotates other gears. The gear involved with first gear has 15 teeth. This gear rotates with the same speed and with the same torque as the 40-tooth gear. However, the 15-tooth gear is meshed with a larger gear with 35 teeth. The gear ratio of the 15-tooth and the

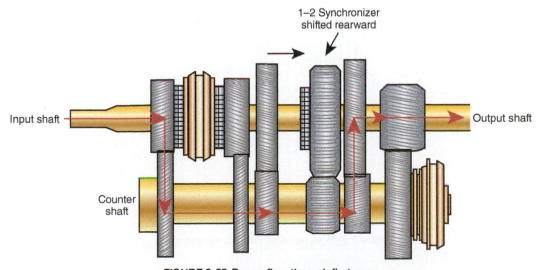

FIGURE 2-25 Power-flow through first gear.

35-tooth gear set is 2.33:1. However, the ratio of the entire gear set (both sets of two gears) is 4.67:1.

To calculate this gear ratio, divide the driven (output) gear of the first set by the drive (input) gear of the first set. Do the same for the second set of gears, then multiply the answer from the first by the second. The result is equal to the gear ratio of the entire gear set. The mathematical formula is as follows:

$$\frac{\text{driven (A)}}{\text{drive (A)}} \times \frac{\text{driven (B)}}{\text{drive (B)}} = \frac{40}{20} \times \frac{35}{15} = 4.67:1$$

Because power is transmitted through the same input gear and larger gear on the countershaft, second gear uses the same first pair of gears as low does. However, the second pair of gears used for first gear is disconnected and the power flows through another gear set. This set consists of a larger drive gear and smaller driven gear. A larger gear on the second shaft drives a smaller gear connected to the drive shaft, which results in less overall speed reduction than in first gear. Second gear is used when the need for torque multiplication is less than the need for vehicle speed and acceleration. Because the car is already in motion, less torque is needed to move the car (Figure 2-26).

Third gear allows for a further decrease in engine speed and torque multiplication, while increasing vehicle speed and encouraging fuel economy (Figure 2-27).

Fourth gear typically provides a **direct drive** (1:1) ratio, so that the amount of torque that enters the transmission is also the amount of torque that passes through and out of the transmission output shaft (Figure 2-28). This gear is used to achieve cruising speeds and promotes fuel economy. When the car is in fourth gear, it lacks the performance characteristics of the second and third gears. To pass slower-moving vehicles, the transmission often must be downshifted to third to take advantage of third gear's slight torque multiplication, resulting in improved acceleration.

Today's transmissions have an overdrive gear that may be fifth or sixth gear. Overdrive gears have ratios of less than 1:1. These ratios are achieved by using a driving gear meshed with a smaller driven gear (Figure 2-29). Output speed is increased and torque is reduced. The purpose of overdrive is to promote fuel economy and reduce operating noise when maintaining highway cruising speed.

Direct drive is characterized by the transmission's output shaft rotating at the same speed as its input shaft.

Overdrive causes the output shaft of the transmission to rotate faster than the input shaft.

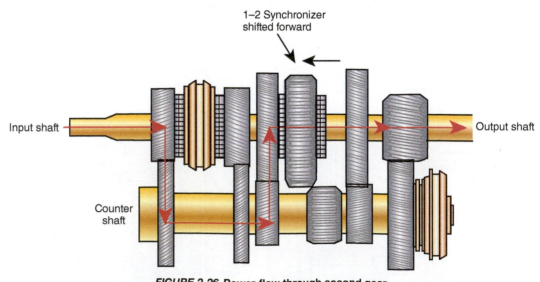

FIGURE 2-26 Power-flow through second gear.

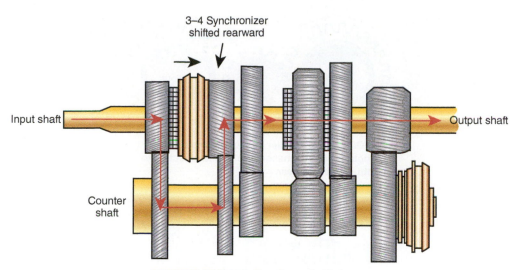

FIGURE 2-27 Power-flow through third gear.

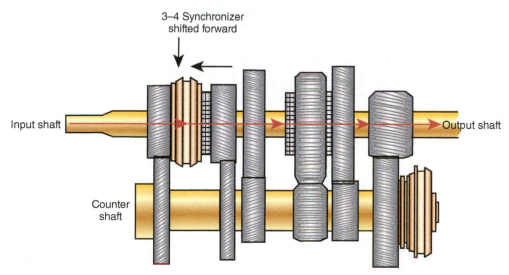

FIGURE 2-28 Power-flow through fourth gear.

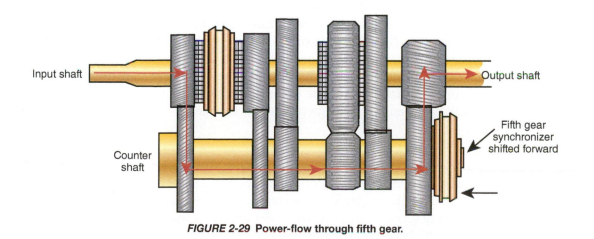

FIGURE 2-29 Power-flow through fifth gear.

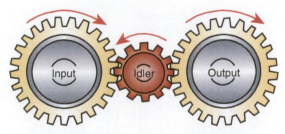

FIGURE 2-30 An idler gear is used to transfer motion without changing rotational direction to obtain reverse gear.

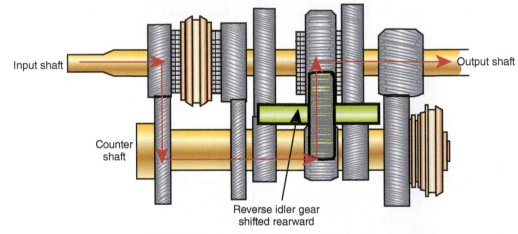

FIGURE 2-31 Power-flow through reverse gear.

Through the use of a reverse idler gear (Figure 2-30), the direction of the incoming torque is reversed and the transmission output shaft rotates in the opposite direction of the forward gears (Figure 2-31). The idler gear merely transfers motion from one gear to another and has no effect on gear ratio. Normally, reverse uses a ratio similar to first gear with the addition of the reverse idler gear. Therefore, only low speeds and high torque can be obtained in reverse.

Gear Spacing. Another aspect of transmission gear ratios is their spacing. Spacing is the "distance" between gear ratios of the various gears. For example, in a four-speed transmission, first gear may have a gear ratio of 3.63:1, second gear a 2.37:1 ratio, third gear a 1.41:1 ratio, and fourth gear a 1:1 ratio, this is a wide ratio transmission. In a close ratio transmission for the same car, the ratios could be 2.57:1 in first, 1.72:1 in second, 1.26:1 in third, and 1:1 in fourth. Fourth gear is the same in both transmissions, but the close ratio gearbox moves the three lower gears closer to fourth. This makes the car more difficult to start from a dead stop but allows for faster acceleration once the car is rolling. Quicker acceleration is possible because there is less loss of engine speed between gear changes.

Final Drive Gears

The transmission's gear ratios are further increased by the gear ratio of the ring and pinion gears in the drive axle assembly. Typical axle ratios are between 2.5 and 4.5:1. The final (overall) drive gear ratio is calculated by multiplying the transmission gear ratio by the final drive ratio. If a transmission is in first gear with a ratio of 3.63:1 and has a final drive ratio of 3.52:1, the overall gear ratio is 12.87:1 (Figure 2-32). If fourth gear has a ratio of 1:1, using the same final drive ratio, the overall gear ratio is 3.52:1.

Final drive ratio is also called the overall gear ratio.

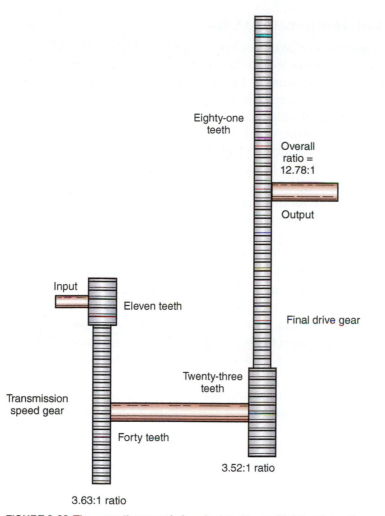

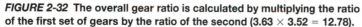

Input

Eleven teeth

Twenty-three teeth

Final drive gear

Eighty-one teeth

Overall ratio = 12.78:1

Output

Transmission speed gear

Forty teeth

3.52:1 ratio

3.63:1 ratio

FIGURE 2-32 The overall gear ratio is calculated by multiplying the ratio of the first set of gears by the ratio of the second (3.63 × 3.52 = 12.78).

HEAT

Heat is a form of energy and is used in many ways. The main sources of heat are the sun, the earth, chemical reactions, electricity, friction, and nuclear energy. Heat is the result of the kinetic energy that is present in all matter; therefore, everything has heat. Cold objects have low kinetic energy because their atoms and molecules are moving very slowly, whereas hot objects have more kinetic energy because their atoms and molecules are moving fast.

Temperature is an indication of an object's intensity (not volume) of kinetic energy. Temperature is measured with a thermometer, which has either a Fahrenheit (F) or Celsius (Centigrade) (C) scale. At absolute zero ($-273°$ C, also referred to as $0°$ Kelvin), particles of matter do not vibrate but, at all other temperatures, particles have motion. The temperature of an object is also a statement of how well the object will transfer heat or kinetic energy to or from another object. Heat and temperature are not the same thing; heat is the volume (commonly measured in British Thermal Units, or BTUs) of kinetic energy something has. Temperature is an indication of the intensity of kinetic energy something has. Energy from something hot will always move to an object that is colder, until both are at the same temperature. The greater the difference in temperature between the two objects, the faster the heat will flow from one to the other.

Heat is measured in BTUs and calories. One BTU is the amount of heat required to heat 1 pound of water by $1°$ F. One calorie is equal to the amount of heat needed to raise the temperature of one gram of water $1°$ C.

The Effects of Temperature Change

Any time the temperature of an object has changed, a transfer of heat has occurred. A transfer of heat also may cause the object to change size or its state of matter. The amount of heat required to raise the temperature of 1 gram of mass 1° C is called the specific heat capacity. Every substance has its own specific heat capacity and this factor is assigned to material based on its difference from water, which has a specific heat capacity of 1. For example, the temperature of 1 gram of water will increase by 10° C if 10 calories of heat were transferred to it. But if 10 calories of heat were added to 1 gram of copper, the temperature would increase by 111° C. This is because copper has a specific heat capacity of only 0.09, as compared to the 1.0 specific heat capacity of water.

As heat moves in and out of a mass, the movement of atoms and molecules in that mass increases or slows down. With an increase in motion, the size of the mass tends to get bigger or expand. This is commonly called thermal **expansion**. Thermal **contraction** takes place when a mass has heat removed from it and the atoms and molecules slow down. All gases and most liquids and solids expand when heated, with gases expanding the most. Water is an exception; it expands when enough heat is removed to turn it into a solid ice. Solids, because they are not fluid, expand and contract at a much lower rate. It is important to realize that all materials do not expand and contract at the same rate. For example, an aluminum component will expand at a faster rate than the same component made of iron. This explains why aluminum cylinder heads have unique service requirements and procedures when compared to iron cylinder heads.

Thermal expansion takes place every time fuel and air are burned in an engine's cylinders. The sudden temperature increase inside the cylinder causes a rapid expansion of the gases, which pushes the piston downward and causes engine rotation.

Typically, when heat is added to a mass, the temperature of the mass increases. This does not always happen, however. In some cases, the additional heat causes no increase in temperature but causes the mass to change its state (solid to liquid or liquid to gas). For example, if we take an ice cube and heat it to 32° F (0° C) and continue to apply heat to it, it will begin to melt (Figure 2-33). As heat is added to the ice cube, the temperature of the ice cube will not increase until it becomes a liquid. The heat added to the ice cube that did not raise its temperature but caused it to melt is called **latent heat** or the heat of fusion. Each gram of ice at 0° C requires 80 calories of heat to melt it to water at 0° C. As more heat is added to the 0° C water, the

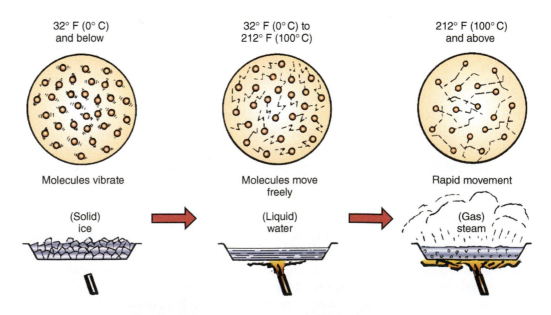

FIGURE 2-33 Water can exist in three different states of matter.

water's temperature will once again increase. This continues until the temperature of the water reaches 212° F (100° C). This is the boiling temperature of water. At this point, any additional heat applied to the water is latent heat causing the water to change its state to that of a gas. This added heat is called the heat of evaporation.

To change the water gas or steam back to liquid water, the same amount of heat required to change the liquid to a gas must be removed from the gas. At that point, the steam begins to condense into a liquid. As additional heat is removed, the temperature will drop until enough heat is removed to bring its temperature back down to freezing (melting in reverse) point. At that time, latent heat must be removed from the liquid before the water turns to ice again.

WAVES AND OSCILLATIONS

An **oscillation** is the back and forth movement of an object between two points. When that motion travels through matter or space, it becomes a **wave**. When something is suspended by a spring, it is acted upon by two forces: gravity and the tension in the spring. When there is no movement (the point of equilibrium), the resultant force is zero. When the mass is given a push, the tension of the spring exceeds the weight of the mass. The spring moves the mass back up toward its original position and its momentum carries it farther upward. When the weight exceeds the spring's tension, the mass moves down again and the oscillation repeats itself until the mass is at equilibrium. As the mass oscillates toward the equilibrium position, the size of the oscillation decreases. As the mass oscillates, the air around it moves and becomes an air wave.

Vibrations

When an object oscillates, it vibrates (Figure 2-34). To prevent the vibration of one object from causing a vibration in other objects, the oscillating mass must be isolated from the other objects. This is often a difficult task. For example, all engines vibrate as they run. To reduce the transfer of engine vibrations to the rest of the vehicle, the engine is held by special mounts. The materials used in the mounts must keep the engine in place and must be elastic enough to absorb the engine's vibrations (Figure 2-35). If the engine was mounted solidly, the vibrations would be felt throughout the vehicle.

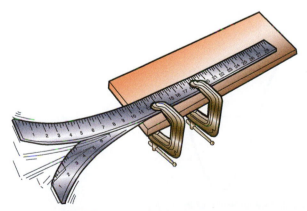

FIGURE 2-34 Vibrations happen in cycles.

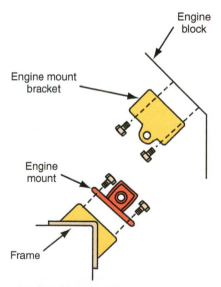

FIGURE 2-35 An engine mount holds the engine in place and isolates engine vibrations from the rest of the vehicle.

Vibration control is important for the reliability of components. If the vibrations are not controlled, the object could shake itself to destruction. Vibration control is the best justification for always mounting parts in the way they were designed to be mounted.

When two waves or vibrations meet, they add up or interfere. This is called the Principle of Superposition and is common to all waves. Making unwanted vibrations tolerable can be done by canceling them with equal and opposite vibrations.

How many times the vibration occurs in 1 second is called **frequency**. Frequency (Figure 2-36) is most often expressed in **hertz (Hz)**. One hertz is equal to one cycle per second. The name is in honor of Heinrich Hertz, a German investigator of radio wave transmission. The **amplitude** of a vibration is its intensity or strength (Figure 2-37). The velocity of a vibration is the result of its amplitude and its frequency. All materials have a unique resonant or natural vibration frequency.

Sound. In air, the vibrations that cause sound are transmitted as a wave between air molecules; many other substances transmit sound in a similar way. A vibrating object causes pressure variations in the surrounding air. Areas of high and low pressure, known as compressions and rarefactions, move through the air as sound waves. Compression makes the sound waves denser, whereas rarefaction makes them less dense. The distance between each compression of a sound wave is called its **wavelength**. Sound waves with a short wavelength have a high frequency and sound high-pitched.

When the rapid variations in pressure occur between about 20 Hz and 20 kHz, sound is audible. Audible sound is the sensation (as detected by the ear) of very small rapid changes in the air pressure above and below atmospheric pressure.

During diagnostics, you often need to listen to the sound of something. You will be paying attention to the type of sound and its intensity and frequency. The tone of the sound usually indicates the type of material that is causing the noise. If there is high pitch, you know the

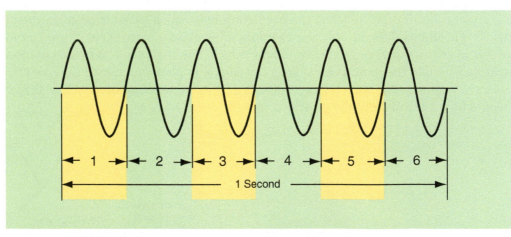

FIGURE 2-36 Frequency is a statement of how many cycles occur in 1 second.

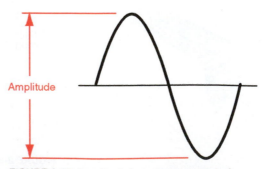

FIGURE 2-37 Amplitude is a measurement of a vibration's intensity.

source of the sound is something that is vibrating quickly. This means the source is less rigid than something that vibrates with a low pitch. Although pitch is dependent on the sound's frequency, the frequency itself can identify the possible sources of the sound.

> **AUTHOR'S NOTE:** Two of the most important acronyms in the technician's vocabulary are NVH and BSR. *Noise, vibration, harshness* and *buzz, squeak, rattle* are standard descriptive terms of frequency-related issues that can drive a customer crazy. Top-level technicians are known for their ability to diagnose these often difficult concerns.

ELECTRICITY AND ELECTROMAGNETISM

All electrical effects are caused by electric charges. There are two types of electric charge: positive and negative. These charges exert electrostatic forces on each other because of the strong attraction of electrons to protons. An electric field is the area on which these forces have an effect. In atoms, protons carry positive charge, whereas electrons carry negative charge. Atoms are normally neutral; they have an equal number of protons and electrons, but an atom can gain or lose electrons, for example, by being rubbed. It then becomes a charged atom, or ion. Electricity has many similarities with magnetism. For example, the lines of the electric fields between charges take the same form as the lines of magnetic force, so magnetic fields can be said to be an equivalent to electric fields. Charges of the same type repel, whereas charges of a different type attract (Figure 2-38).

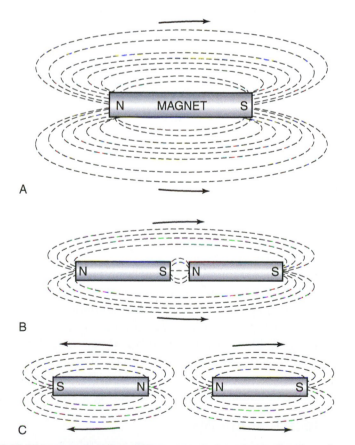

FIGURE 2-38 (A) In a magnet, lines of force emerge from the north pole and travel to the south pole before passing through the magnet back to the north pole. (B) Unlike poles attract, while (C) similar poles repel each other.

Electricity

An electric circuit is simply the path along which an electric current flows. Electrons carry negative charge and can be moved around a circuit by electrostatic forces. A circuit usually consists of a conductive material, such as a metal, where the electrons are held very loosely to their atoms, thus making movement possible. The strength of the electrostatic force is the voltage. The resulting movement of the electric charge is called an electric current. The higher the voltage, the greater the current will be. But the current also depends on the thickness, length, temperature, and nature of the materials that conduct it. The resistance of a material is the extent to which it opposes the flow of electric current. Good conductors have low resistance, which means that a small amount of voltage will produce a large current. In batteries, chemical reactions in the metal electrode cause the freeing of electrons, which results in their movement to another electrode and the formation of a current.

Magnets

Some materials are natural magnets; however, most magnets are produced. The materials typically used to make a permanent magnet are called ferromagnetic materials. These materials consist of mostly iron compounds that are heated. The heat causes the atoms to shift direction. Once they all point in the same direction, the metal becomes a magnet. This sets up two distinct poles, called the north and south poles. The poles are at the ends of the magnet and there is an attraction between the north pole and the south pole. This attraction or force set up by a magnet can be observed but the type of force is not known.

The lines of a magnetic field form closed lines of force, from the north to the south. If another iron or steel object enters into the magnetic field, it is pulled into the magnet. If another magnet is introduced into the magnetic field, it will either move into the field or push away from it. This is the result of the natural attraction of a magnet from north to south. If the north pole of one magnet is introduced to the north pole of another, the two poles will oppose each other and will push away. If the south pole of a magnet is introduced to the north pole of another, the two magnets will join together because the opposite poles are attracted to each other.

The strength of the magnetic force is uniform all around the outside of the magnet. The force is strongest at the surface of the magnet and weakens with distance. If you double the distance from a magnet, the force is reduced by 1/8.

The strength of a magnetic field is typically measured with devices known as magnetometers and in units of Gauss (G).

Electromagnetism

Any electrical current will produce magnetism that affects other objects in the same way as permanent magnets. The arrangement of force lines around a current-carrying conductor, its magnetic field, is circular. The magnetic effect of electrical current is increased by making the current-carrying wire into a coil (Figure 2-39).

When a coil of wire is wrapped around an iron bar, it is called an **electromagnet**. The magnetic field produced by the coil magnetizes the iron bar, strengthening the overall effect. A field like that of a bar magnet is formed by the magnetic fields of the wires in the coil. The strength of the magnetism produced depends on the number of coils and the size of the current flowing in the wires. Electromagnetic coils and permanent magnets are arranged inside an electric motor so that the forces of electromagnetism create rotation of the armature. Electromagnetic principles are also used in the operation of automotive solenoids that are used to shift automated manual transmissions, among many other applications.

Producing Electrical Energy

There are many ways to generate electricity. The most common is to use coils of wire and magnets in a generator. Whenever a wire and magnet are moved relative to each other, a voltage is produced (Figure 2-40). In a generator, the wire is wound into a coil. The more

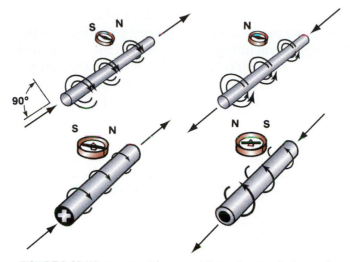

FIGURE 2-39 When current is passed through a conductor such as a wire, magnetic lines of force are generated around the wire at right angles to the direction of the current flow.

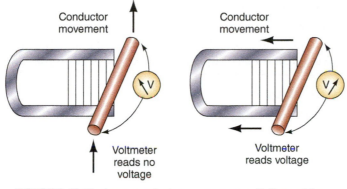

FIGURE 2-40 Moving a conductor across magnetic lines of force induces a voltage in the conductor.

turns in the coil and the faster the coil moves, the greater the voltage. The coils or magnets spin around at high speed, typically turned by steam pressure. The steam is usually generated by burning coal or oil, a process that creates pollution. Renewable sources of electricity, such as hydroelectric power, wind power, solar energy, and geothermal power, produce only heat as a pollutant. In automobiles, the generator is spun by a belt driven by the engine's crankshaft. In a generator, the kinetic energy of a spinning object is converted into electrical energy.

A solar cell converts the energy of sunlight directly into electrical energy, using layers of semiconductors. Electricity is produced by causing electrons to leave the atoms in the semiconductor material. Each electron leaves behind a hole or gap. Other electrons move into the hole, leaving holes in their atoms. This process continues all the way around a circuit. The moving chain of electrons is an electrical current.

CHEMICAL PROPERTIES

The properties of something describe the characteristics of an object. Physical properties are readily observable: color, size, and smell. Chemical properties are only observable during a chemical reaction and describe how one type of matter reacts with another to form a new and different substance.

A solution is a mixture of two or more substances. Most solutions are liquids, but solutions of gases and solids are possible. An example of a gas solution is the air we breathe; it is mostly composed of oxygen and nitrogen. Brass is an example of a solid solution; it is composed of copper and zinc. The liquid in a solution is called the **solvent**, and the substance added to it is the solute. If both are liquids, the one present in the smaller amount is considered the solute. Solutions can vary widely based on how much of the dissolved substance is actually present.

Specific Gravity

Specific gravity is the heaviness or relative **density** of a substance as compared with water. If something is 3.5 times as heavy as an equal volume of water, its specific gravity is 3.5. Its density is 3.5 grams per cubic centimeter, or 3.5 kilograms per liter. Specific gravity checks of a battery's electrolyte are an indication of the battery's state of charge.

Density is a statement of how much mass there is in a particular volume. Water is denser than air; therefore there will be less air in a given container than water in that same container (Figure 2-41). The density of a material changes with temperature (Figure 2-42).

Chemical Reactions

Chemical reactions produce a different substance, such as wood turning into carbon after it has been completely burned. A chemical reaction is always accompanied by a change in energy. This means energy is given off or taken in during the reaction. A reaction takes place when two or more molecules interact and a chemical change occurs, single reactions occur as part of a large series of reactions, or ions, molecules, or pure atoms are formed.

Catalysts and Inhibitors

A **catalyst** lowers the amount of energy required to make a reaction happen. A catalyst is any substance that affects the speed of a chemical reaction without itself being consumed or changed. Catalysts tend to be highly specific, reacting with one substance or a small set of substances. In a catalytic converter, the platinum catalyst converts unburned hydrocarbons and nitrogen compounds into products that are harmless to the environment. Water, especially salt water, catalyzes oxidation and corrosion. An inhibitor is the opposite of a catalyst and stops or slows the rate of a reaction.

Substance	Density in g/cm^3
Air	0.0013
Ice	0.92
Water	1.00
Aluminum	2.70
Steel	7.80
Gold	19.30

FIGURE 2-41 A look at the density of different substances compared with water.

Temp F°	Temp C°	Approx. Change In Density
200°	93°	–21%
180°	82°	–16.8%
160°	71°	–12.6%
140°	60°	–8.4%
120°	49°	–4.2%
100°	38°	–
80°	27°	+4.2%
60°	16°	+8.4%
40°	4°	+12.6%
20°	–7°	+16.8%
0°	–18°	+21%

FIGURE 2-42 The effect that temperature has on the density of air at atmospheric pressure.

Acids/Bases

An ion is an atom or group of atoms with one or more positive or negative electric charges. Ions are formed when electrons are added to or removed from neutral molecules or other ions. Ions are what make something an **acid** or a **base**.

Acids are compounds that break into hydrogen (H^+) ions and another compound when placed in a water solution. They have a sour taste, are corrosive, react with some metals to produce hydrogen, react with carbonates to produce carbon dioxide, change the color of litmus from blue to red, and become less acidic when combined with alkalis. Most acids are slow reacting, especially if they are weak acids. Acids also react with bases to form salts.

Alkalis (bases) are compounds that release hydroxide ions (OH^-) and react with hydrogen ions to produce water, thus neutralizing each other. Most substances are neutral (not an acid or a base). Alkalis feel slippery, change the color of litmus from red to blue, and become less alkaline when they are combined with acids.

A hydroxide is any compound made up of one atom of hydrogen and oxygen, bonded together. An oxide is any chemical compound in which oxygen is combined with another element. Metal oxides typically react with water to form bases or with acids to form salts. Oxides of nonmetallics react with water to form acids or with bases to form salts.

A salt is a compound formed when the hydrogen of an acid is replaced by a metal. Typically, an acid and a base react to form a salt and water.

pH. The **pH scale** is used to measure how acidic or basic a solution is. Its name comes from the fact that pH is the absolute value of the power of the hydrogen ion concentration. The scale goes from 0 to 14. Distilled (pure) water is 7. Acids are found between 0 and 7 and bases are from 7 to 14. When the pH of a substance is low, the substance has many H^+ ions. When the pH is high, the substance has many OH^- ions. The pH value helps define the nature, composition, or extent of reaction of substances.

The pH of something is typically checked with litmus paper. Litmus is a mixture of colored organic compounds obtained from several species of lichen. Lichen is a type of plant that is actually a combination of a fungus and algae. Litmus test strips can be used to check the condition of the engine's coolant (Figure 2-43).

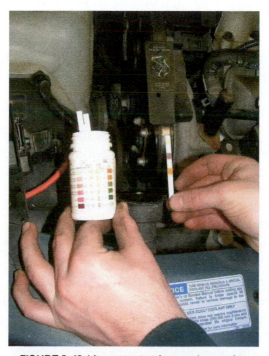

FIGURE 2-43 Litmus test strips can be used to check the condition of an engine's coolant.

Reduction and Oxidation

Oxidation is a chemical reaction in which a substance combines with oxygen. Rapid oxidation produces heat fast enough to cause a flame. When fuel burns, it combines with oxygen to form other compounds. This chemical reaction is combustion, which produces heat and fire.

The addition of hydrogen atoms or electrons is **reduction**. Oxidation and reduction always occur simultaneously: one substance is oxidized by the other, which is reduced. During oxidation, a molecule provides electrons. During reduction, a molecule accepts electrons. Oxidation and reduction reactions are usually called **redox** reactions. Redox is any chemical reaction in which electrons are transferred. Batteries produce an electrical current at a constant voltage through redox reactions.

An oxidizing agent is something that accepts electrons and oxidizes something else while being reduced in the process. A reducing agent is something that provides electrons and reduces something else while being oxidized.

Every atom or ion has an oxidation number. This value compares the number of protons and electrons in that atom. The oxidation number is reduced during reduction by adding electrons. The oxidation number is increased during oxidation by removing electrons. All free, uncombined elements have an oxidation number of zero. Hydrogen, in all its compounds except hydrides, has an oxidation number of $+1$. Oxygen, in all its compounds except peroxides, has an oxidation number of -2.

Metallurgy

Metallurgy is the art and science of extracting metals from their ores and modifying them for a particular purpose. This includes the chemical, physical, and atomic properties and structures of metals and the way metals are combined to form alloys. An **alloy** is a mixture of two or more metals. Steel is an alloy of iron plus carbon and other elements.

Metals have one or more of the following properties:

- Good heat and electric conduction.
- Malleability—can be hammered, pounded, or pressed into a shape without breaking.
- Ductility—can be stretched, drawn, or hammered without breaking.
- High light reflectivity—can make light bounce off its surface.
- Capacity to form positive ions in a solution and hydroxides rather than acids when their oxides meet water.

About three-quarters of the elements are metals. The most abundant metals are aluminum, iron, calcium, sodium, potassium, and magnesium.

> **AUTHOR'S NOTE:** The science of metallurgy has many practical applications in the automotive world. The makeup of differential ring and pinion sets is a perfect example. Ring and pinions designed for road applications must be hard and durable enough to withstand many thousands of miles of service. Drag racing applications call for a more ductile, less brittle alloy to prevent gear fracture under extreme and sudden load.

Rust and Corrosion. The rusting of iron is an example of oxidation. Iron combines with oxygen to form rust. Unlike fire, rusting occurs so slowly that little heat is produced. The rate at which this occurs depends on several factors: the temperature, the exposed surface area, and the presence of catalysts.

Corrosion is the wearing-away of a substance due to chemical reactions. It occurs whenever a gas or liquid chemically attacks an exposed surface. This action is accelerated by heat, acids, and salts. Some materials naturally resist corrosion; others can be protected by painting, coatings, galvanizing, or anodizing.

Galvanizing involves the coating of zinc onto iron or steel to protect them against exposure to the atmosphere. If galvanizing is properly applied, it can protect the metals for 15–30 years or more.

Metals can be anodized for corrosion resistance, electrical insulation, thermal control, abrasion resistance, sealing, improving paint adhesion, and decorative finishing. Anodizing is a process that electrically deposits an oxide film onto the surface of a metal, often aluminum. During the process, dyes can be added to the process to give the material a colored surface.

Hardness. The hardness of something describes its resistance to scratching. **Hardening** is a process that increases the hardness of a metal, deliberately or accidentally, by hammering, rolling, carburizing, heat treating, tempering, or other processes. All of these deform the metal by compacting the atoms or molecules to make the material denser.

Carburizing hardens the surface of steel with heat, while leaving the core relatively soft. The combination of a hard surface and soft interior withstands very high stress. It also has a low cost and offers flexibility for manufacturing. To carburize, the steel parts are placed in a carbonaceous environment (with charcoal, coke, and carbonates, or carbon dioxide, carbon monoxide, methane, or propane) at a high temperature for several hours. The carbon diffuses into the surface of the steel, altering the crystal structure of the metal. Gears, ball and roller bearings, and piston pins are often carburized.

Heat treating changes the properties of a metal by using heat. **Tempering** is the heat treating of metal alloys, particularly steel. For example, raising the temperature of hardened steel to 752° F (4000° C) and keeping it at that temperature for a while before quenching it in oil decreases its hardness and brittleness and produces strong steel.

Solids under Tension

Solids have a greater density than most liquids and gases. The rigidity results from the strong attraction between its atoms. A force pulling on a solid moves these atoms farther apart, creating an opposing force called tension. If a force pushes on a solid, the atoms move closer together, creating compression. These are the principles of how springs function.

An elastic substance is a solid that gets larger under tension, smaller under compression, and returns to its original size when no force is acting on it. Most solids show some elastic behavior, but there is usually a limit to the force that the material can face. When excessive force is applied, the material will not return to its original size and it will be distorted or break. The limit depends on the material's internal structure; for example, steel has a low elastic limit and can only be extended about 1 percent of its length, whereas rubber can be extended to about 1000 percent.

Tensile strength defines the maximum load a material can support without breaking while being stretched. It is dependent on the cross-sectional area of the material. When stresses less than the tensile strength are removed, the material returns to its original size and shape. Greater stresses form a narrow, constricted area in the material, which is easily broken. Tensile strengths are measured in units of force per unit area.

Electrochemistry

Electrochemistry is concerned with the relationship between electricity and chemical change. Many spontaneous chemical reactions release electrical energy, and some of these are used in batteries and fuel cells to produce electric power.

Electrolysis is an electrochemical process. During this process, electric current is passed through a substance, causing a chemical change. This change causes either a gain or loss of electrons. Electrolysis normally takes place in an electrolytic cell made of separated positive and negative electrodes immersed in an electrolyte.

An **electrolyte** is a substance that conducts current as a result of the breaking-down of its molecules into positive and negative ions. The most familiar electrolytes are acids, bases, and salts, which ionize when dissolved in water and alcohol. Ions drift to the electrode of the opposite charge and are the conductors of current in electrolytic cells.

SUMMARY

TERMS TO KNOW

Acceleration

Acid

Alloy

Amplitude

Atom

Base

Carburizing

Catalyst

Centrifugal force

Centripetal force

Contraction

Curb weight

Deceleration

Density

Direct drive

Electrolysis

Electrolyte

Electromagnet

Element

Equilibrium

Evaporate

Expansion

Force

Frequency

Friction

Gear

Gear ratios

Gross weight

Hardening

Heat

Heat treating

Hertz (Hz)

Impermeable

Inertia

Ion

Kinetic energy

- Matter is anything that occupies space, and it exists as a gas, liquid, or solid.
- When a solid dissolves into a liquid, a solution is formed. Not all solids will dissolve in a liquid; rather, the liquid will be either absorbed or adsorbed.
- Materials that *absorb* fluids are permeable substances. Impermeable substances *adsorb* fluids.
- Energy is the ability to do work and all matter has energy.
- The total amount of energy never changes; it can only be transferred from one form to another, not created or destroyed.
- When energy is released to do work, it is called kinetic energy. Stored energy may be called potential energy.
- Mass is the amount of matter in an object. Weight is a force and is measured in pounds or kilograms. Gravitational force gives the mass its weight.
- A force is a push or pull, and can be large or small and can be applied to objects by direct contact or from a distance.
- When an object moves in a circle, its direction is continuously changing and all changes in direction require a force. The forces required to maintain circular motion are called centripetal and centrifugal force.
- Pressure is a force applied against an opposing object and is measured in units of force per unit of surface area (pounds per square inch or kilograms per square centimeter).
- The greater the mass of an object, the greater the force needed to change its motion. Inertia is the tendency of an object at rest to remain at rest, or the tendency of an object in motion to stay in motion.
- When a force overcomes static inertia and moves an object, the object gains momentum. Momentum is the product of an object's weight times its speed.
- Speed is the distance an object travels in a set amount of time. Velocity is the speed of an object in a particular direction. Acceleration, which only occurs when a force is applied, is the rate of increase in speed. Deceleration is the reverse of acceleration, as it is the rate of decrease in speed.
- Newton's laws of motion state that an object at rest tends to remain at rest and an object in motion tends to remain in motion, unless some force acts on it; when a force acts on an object, the motion of the object will change; and for every action there is an equal and opposite reaction.
- Friction is a force that slows or prevents motion of two moving objects that touch.
- Friction can be reduced in two main ways: by lubrication or by the use of rollers.
- When a force moves a certain mass a specific distance, work is done.
- A machine is any device that can be used to transmit a force and, in doing so, change the amount of force or its direction. Examples of simple machines are inclined planes, pulleys, levers, gears, and wheels and axles.
- Torque is a force that tends to rotate or turn things and is measured by the force applied and the distance traveled.
- Gear ratios express the mathematical relationship of one gear to another.

- Gears are used to apply torque to other rotating parts of the drivetrain and to multiply torque.
- Power is a measurement of the rate at which work is done. It is measured in watts.
- Horsepower is the rate at which torque is produced.
- Heat is a form of energy caused by the movement of atoms and molecules and is measured in British Thermal Units (BTUs) and calories.
- Temperature is an indication of an object's kinetic energy and is measured with a thermometer, which has either a Fahrenheit (F) or Celsius (Centigrade) (C) scale.
- As heat moves in and out of a mass, the size of the mass tends to change.
- Any electrical current will produce magnetism. When a coil of wire is wrapped around an iron bar, it is called an electromagnet.
- The most common way to produce electricity is to use coils of wire and magnets in a generator.
- An oscillation is the back and forth movement of an object between two points. When an object oscillates, it vibrates.
- The number of times a vibration occurs in 1 second is called frequency and is most often expressed in hertz.
- A fluid is something that does not have a definite shape; therefore liquids and gases are fluids. A gas will readily expand or compress according to the pressure exerted on it. Liquids are basically incompressible, which gives them the ability to transmit force.
- Pascal's Laws explain how hydraulics can be used to do work.
- The chemical properties of something are only observable during a chemical reaction and describe how one type of matter reacts with another to form a new and different substance.
- Specific gravity is the heaviness or relative density of a substance as compared with water.
- Density is a statement of how much mass there is in a particular volume.
- A catalyst is any substance that affects the speed of a chemical reaction without itself being consumed or changed.
- An inhibitor is the opposite of a catalyst and stops or slows the rate of a reaction.
- Acids are compounds that break into hydrogen (H^+) ions and another compound when placed in an aqueous (water) solution.
- Alkalis (bases) are compounds that release hydroxide ions (OH^-) and react with hydrogen ions to produce water, thus neutralizing each other.
- The pH scale is used to measure how acidic or basic a solution is.
- Oxidation is a chemical reaction in which a substance combines with oxygen.
- The addition of hydrogen atoms or electrons is reduction.

REVIEW QUESTIONS

Short Answer Essays

1. Describe Newton's first law of motion and give an application of this law in automotive theory.

2. Explain Newton's second law of motion and give an example of how this law is used in automotive theory.

3. Describe four different forms of energy.

4. Describe four different types of energy conversion.

5. Why are engines not rigidly mounted in a vehicle?

6. Why does torque increase when a smaller gear drives a larger gear?

7. How are gear ratios calculated?

8. Why does the size of something change with a change in heat?

9. What is torque?

10. What is a solution and how is it formed?

Fill-in-the-Blanks

1. The nucleus of an atom contains _____ and _____.

2. Work is calculated by multiplying _____ × _____.

3. Energy may be defined as the ability to do _____.

4. When one object is moved over another object, the resistance to motion is called _____.

5. Weight is the measurement of the earth's _____ _____ on an object.

6. Torque is a force that does work with a _____ action.

7. Torque is calculated by multiplying the applied force by the _____ from the center of the _____ to the point where the force is exerted.

8. Torque is measured in _____-_____ and _____-_____.

9. Gear ratios are determined by dividing the number of teeth on the _____ gear by the number of teeth on the _____ gear.

10. One horsepower is equal to _____ foot-pounds per minute, or _____ foot-pounds per second.

ASE-Style Review Questions

1. Which of the following statements about friction is not true?
   A. Friction can be reduced by lubrication.
   B. Bearings are a type of roller used to increase the friction between moving parts such as a wheel and its axle.
   C. The presence of oil or another fluid between two surfaces keeps the surfaces apart and thereby reduces friction.
   D. Friction can be reduced by the use of rollers.

2. When discussing different types of energy,
   Technician A says that when energy is released to do work, it is called potential energy.
   Technician B says that stored energy is referred to as kinetic energy.
   Who is correct?
   A. A only   C. Both A and B
   B. B only   D. Neither A nor B

3. When discussing friction in matter,
   Technician A says that friction is a force that slows or prevents motion of two moving objects or surfaces that touch.
   Technician B says that friction occurs in liquids, solids, and gases.
   Who is correct?
   A. A only   C. Both A and B
   B. B only   D. Neither A nor B

4. When discussing mass and weight,
   Technician A says that mass is the measurement of an object's inertia.
   Technician B says that mass and weight may be measured in cubic inches.
   Who is correct?
   A. A only   C. Both A and B
   B. B only   D. Neither A nor B

5. When applying the principles of work and force,
   A. work is accomplished when force is applied to an object that does not move.
   B. in the metric system the measurement for work is cubic centimeters.
   C. no work is accomplished when an object is stopped by mechanical force.
   D. if a 50-pound object is moved 10 feet, 500 ft.-lbs. of work are produced.

6. All these statements about energy and energy conversion are true, *except*:
   A. thermal energy may be defined as light energy.
   B. chemical to thermal energy conversion occurs when gasoline burns.
   C. mechanical energy is defined as the ability to do work.
   D. mechanical to electrical energy conversion occurs when the engine drives the generator.

7. When discussing gear ratios,

Technician A says that they express the mathematical relationship, according to the number of teeth, of one gear to another.

Technician B says that they express the size difference of two gears by stating the ratio of the smaller gear to the larger gear.

Who is correct?

A. A only
B. B only
C. Both A and B
D. Neither A nor B

8. Which of the following statements about mass and weight is true?

A. Mass is the amount of matter in an object.
B. Weight is a measurement of mass expressed in pounds or kilograms.
C. Gravitational force gives an object its mass.
D. Something that weighs much on Earth will weigh nothing once it is lifted from the earth's surface.

9. A screw is a simple machine that operates as a(n):

A. gear.
B. pulley.
C. inclined plane.
D. lever.

10. Which of the following statements is *not* true?

A. Materials that absorb fluids are permeable substances.
B. When a solid dissolves into a liquid, its particles break away from this structure and mix evenly in the liquid, forming a solution.
C. When most liquids are heated, they evaporate.
D. Permeable substances adsorb fluids.

Chapter 3

CLUTCHES

UPON COMPLETION AND REVIEW OF THIS CHAPTER, YOU SHOULD BE ABLE TO:

- Understand and define the purpose of a clutch assembly.

- Understand and describe the major components of a clutch assembly.

- Understand and describe the operation of a clutch.

- Understand and describe the operation of a wet or dry dual clutch.

- Understand and define the role of each major component in a clutch.

- Describe the operation of the various mechanical and cable-type clutch linkages.

- Describe the operation of a hydraulic clutch linkage.

INTRODUCTION

A clutch is a device that transfers rotational motion from one object to another. When engaged, the clutch allows both objects to rotate at the same speed. When disengaged, the objects can rotate independent of each other.

In an automobile, many different types of clutches are used. In automatic transmissions, multiple disc clutch packs are used to drive or hold a member of the transmission's planetary gear set. The clutch for a manual transmission is also a disc or plate clutch. Cone clutches are used in synchronizer assemblies in a manual transmission and in some limited-slip differentials. The gears in a manual transmission are locked to their shafts by a dog clutch. This type of clutch typically has teeth on one part that mesh into a recessed area on the other part. Air conditioning compressors and some 4WD transfer cases are equipped with an electromagnetic clutch. The action of this type of clutch is controlled by current flow to one-half of the clutch assembly.

Some engines have an engine-driven cooling fan with a thermostatically controlled viscous clutch. A viscous clutch is a multiple-disc clutch assembly housed in a sealed container filled with a special fluid. The action of the clutch is controlled by the temperature of the fluid inside the clutch. As the fluid gets hotter, it becomes thicker and the clutch becomes more engaged. Some limited-slip differentials also rely on a viscous clutch.

The manual transmission clutch is a device used to connect and disconnect engine power-flow to the transmission at the will of the driver (Figure 3-1). A driver operates the clutch with a clutch pedal inside the vehicle. This pedal allows engine power-flow to be gradually applied when the vehicle is starting out from a stop and interrupts power-flow to avoid gear clashing when shifting gears. Engagement of the clutch allows for power transfer from the engine to the transmission and eventually to the drive wheels. Disengagement of the clutch provides the necessary halt of power transfer that allows the engine to continue running while no power is supplied to the drive wheels. Engagement and disengagement of the clutch in most vehicles is controlled by a pedal and clutch linkage. Many newer vehicles dispose of the pedal and linkage in favor of an electronically controlled release mechanism. These systems often offer

Approximately 35 percent of all cars sold in North America are equipped with a manually operated clutch and transmission.

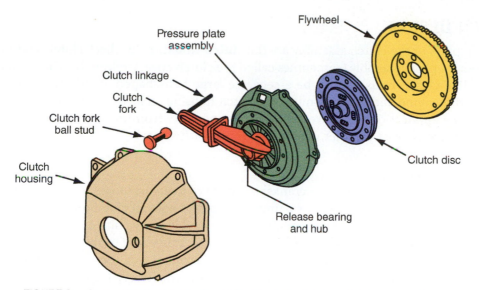

FIGURE 3-1 When the clutch is engaged, the driven member disc is squeezed between the two driving members, the pressure plate and flywheel. The transmission is connected to the driven member.

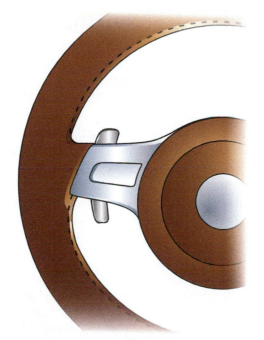

FIGURE 3-2 Basic clutch components. Steering wheel with paddles that control shifts on a car with electronic clutch and shift control. There is no clutch pedal.

driver control of shifts through steering wheel paddles (Figure 3-2). The machined surfaces of the flywheel and pressure plate must be parallel and free of cracks and scores in order to adequately clamp the clutch disc. Clutch slippage, vibration, and noise is minimized by the proper alignment of engine and transmission/transaxle and of the clutch components.

CLUTCH LOCATION

The clutch assembly is placed between the engine and the transmission on nearly all cars and trucks. A flywheel is bolted to the rear of the engine's crankshaft and the clutch assembly is bolted to the flywheel.

Exceptions to this typical location include the late-model Corvette that uses a flywheel and clutch assembly that is placed together with a rear-mounted transaxle. This drivetrain arrangement improves weight distribution and handling.

Normally, a clutch will last over 100,000 miles. However, the life of the clutch depends heavily on the driver, the loads the vehicle carries, and how well the vehicle is maintained.

A clutch housing is often called a bell housing. A high-performance steel clutch housing is often called a scattershield.

Shop Manual
Chapter 3, page 105

A heat sink is a piece of material that absorbs heat to prevent the heat from settling on another component.

A flywheel is a large-diameter heavy disc, usually made of nodular cast iron with a high graphite content.

On aluminum flywheels, a steel plate is bolted onto the flywheel to provide a frictional surface for the clutch.

A pressure plate may be referred to as a clutch cover.

A **flexplate** is also referred to as a drive plate.

CLUTCH DESIGN

The main parts of the clutch assembly are the **clutch housing**, flywheel, clutch shaft, clutch disc, pressure plate assembly (sometimes called the clutch cover), release bearing, and clutch linkage. The clutch housing may be a separate large bell-shaped aluminum casting that connects the engine and the transmission. Clutch housings may be made of steel for high-performance applications to prevent broken components from exiting the housing. Most clutch housings are cast together with the transmission/transaxle.

Flywheel

A flywheel is bolted to the engine's crankshaft to serve many purposes. It acts as a balancer for the engine and it smooths out, or dampens, engine torsional vibrations caused by firing pulses. It adds inertia to the rotating crankshaft. It provides a machined surface from which the clutch can contact and pick up engine torque and transfer it to the transmission. The flywheel also acts as a friction surface and heat sink for one side of the clutch disc.

The flywheel is normally made of nodular cast iron, which has a high graphite content to lubricate the engagement of the clutch. The rear surface of the flywheel is a friction surface, machined very flat to ensure smooth clutch engagement. Performance flywheels are often made of steel, or of aluminum with a steel insert for the friction surface. Lightweight flywheels allow the engine to reach a higher rpm quicker.

The flywheel is bolted to the engine's crankshaft and the clutch assembly's pressure plate is bolted to the flywheel. In the center of the flywheel or crankshaft is the bore for the pilot bearing or bushing to support the front of the input shaft. The teeth around the circumference of the flywheel form a ring gear for the engine starting motor to contact. The ring gear is not actually a part of the flywheel; rather, it is pressed or welded around the outside of the flywheel (Figure 3-3).

Vehicles with automatic transmissions do not have a flywheel. Instead, they use a **flexplate** and the weight of a torque converter and transmission fluid to dampen the engine vibrations and provide inertia for the crankshaft. Flexplates are lightweight, stamped steel discs and are used as the attaching point for the torque converter to the engine's crankshaft. They have no clutch friction surface and will not interchange with manual transmission flywheels. They do have a ring gear for the engine's starter motor. However, the ring gear is normally part of the plate and is not replaceable.

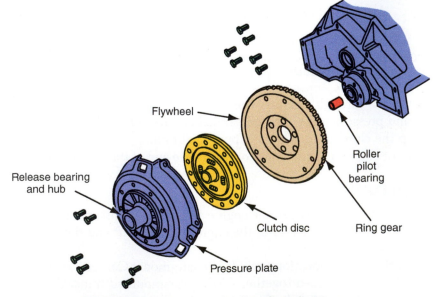

Release bearing and hub

Flywheel

Roller pilot bearing

Clutch disc

Ring gear

Pressure plate

FIGURE 3-3 Location and mounting of a typical flywheel.

Dual-Mass Flywheel. Many late-model cars and light trucks with manual transmissions use a **dual-mass flywheel (DMF)**. These are also used in some semiautomatic transmissions. These flywheels are used to eliminate engine vibrations before they are transmitted to the transmission (Figure 3-4). They also provide for increased fuel economy, longer engine life, smoother shifting, and reduced gear noise. As vehicle manufacturers try to improve fuel economy by keeping engine rpm low at cruising speed, the DMF has played a larger role in reducing NVH (noise, vibration, harshness) issues. The goal is rumble-free driving and to protect the drivetrain from harmful vibrations. A DMF is tuned by the manufacturer to match the engine's torque and horsepower curves, engine harmonics, and vehicle load dynamics. It is also tuned to match the gear ratios in a transmission and the final drive, as well as the rolling diameter of the tires.

Dual-mass flywheels have been used in many pickup trucks with diesel engines. The flywheel reduces the high torsional spikes of the high-compression engines. By eliminating the torsion spikes, the flywheel helps eliminate potential damage to transmission internal components.

The flywheel comprises two rotating plates, connected by a spring and damper assembly (Figure 3-5), called a friction pack. The forward-most portion of the flywheel is bolted to the

Dual-mass flywheels are commonly found on Dodge, Ford, and GM light trucks with diesel engines and on Audis, Volkswagens, Hyundais, BMWs, Corvettes, and Porsches.

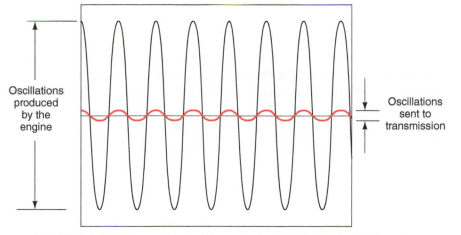

Oscillations produced by the engine

Oscillations sent to transmission

FIGURE 3-4 **A dual-mass flywheel dampens crankshaft vibrations before they enter the transmission.**

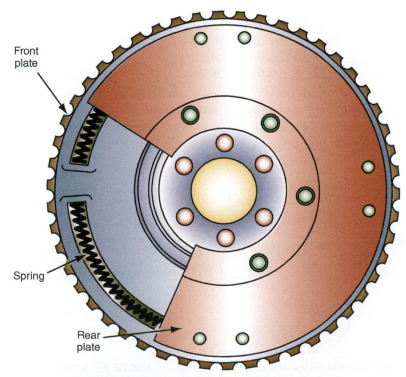

Front plate

Spring

Rear plate

FIGURE 3-5 **The construction of a dual-mass flywheel.**

end of the crankshaft. The pressure plate of the clutch is bolted to the rearward portion of the flywheel. The friction pack is located between the two plates. It allows the plates to rotate out of synch with each other. This action is what dampens the vibrations. Engine torque moves from the front plate through the damper and spring assembly to the rear plate before it enters the transmission. The two halves are rubber-sealed.

Some dual-mass flywheels have a torque-limiting feature that prevents damage to the transmission during peak torque loads. This allows the forward plate to absorb torque spikes and not pass them along through the transmission. When there is too much torque on the flywheel, the flywheel slips and prevents transmission damage.

Dual-mass flywheels are effective and quite durable. If the engine has been modified to produce more power, the flywheel is out of tune and cannot dampen the vibrations. If the vehicle is frequently overloaded, the damper assembly can burn up or its springs destroyed.

Pilot Bearing

The **clutch shaft** projects from the front of the transmission. Most clutch shafts have a smaller shaft or pilot that projects from its outer end. This pilot rides in the pilot bearing in the engine crankshaft flange or flywheel. The pilot bearing or bushing serves as a support for the outer end of the input shaft and it maintains proper alignment of the shaft with the crankshaft (Figure 3-6). A pilot bearing or bushing is normally pressed into the bore of the crankshaft. Some pilot bearings are placed in the center of the flywheel rather than in the back of the crankshaft. Most transaxles are not equipped with a pilot bearing or bushing because their input shaft is supported by two bearings inside the transaxle. The splined area of the clutch shaft allows the clutch disc to move laterally a small amount along the **splines**, while preventing the disc from rocking on the shaft. When the clutch is engaged, the clutch disc drives the transaxle's input shaft through these splines (Figure 3-7).

Clutch Disc

The **clutch (friction) disc** is a steel plate covered with frictional material that fits between the flywheel and the pressure plate. In the center of the disc is the hub, which is splined to fit over the splines of the input shaft. When the clutch is engaged, the disc is firmly squeezed between the flywheel and pressure plate and power from the engine is transmitted by the disc's hub to the transmission's input shaft. The width of the hub prevents the disc from rocking on the shaft while it moves between the flywheel and the pressure plate.

A clutch disc (Figure 3-8) has frictional material riveted or bonded to both sides. Frictional facings are either woven or molded. Molded facings are preferred because they can withstand high-pressure plate loading forces without being damaged. Woven facings are used when additional cushioning during clutch engagement is desired. Grooves are cut across the face of the friction facings to allow for smooth clutch action, increased cooling, and a place for the facing dust to go as the clutch disc wears. Like brake-lining material, the frictional facing wears

The **clutch shaft** is commonly called the transmission's input shaft or main drive shaft; the clutch-driven disc drives this shaft.

Splines resemble a series of keyways cut into a shaft or hole. Parts splined together rotate as a unit, but the parts are free to move somewhat along the shaft's centerline.

FIGURE 3-6 Examples of pilot bushings and bearings.

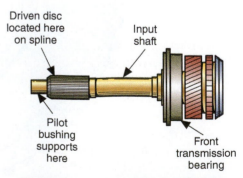

Driven disc located here on spline

Input shaft

Pilot bushing supports here

Front transmission bearing

FIGURE 3-7 A typical clutch (input) shaft.

FIGURE 3-8 A typical clutch disc.

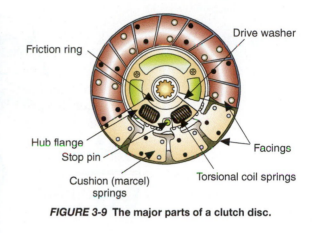

Friction ring

Drive washer

Hub flange

Stop pin

Facings

Cushion (marcel) springs

Torsional coil springs

FIGURE 3-9 The major parts of a clutch disc.

as the clutch is engaged. **Asbestos** wire-woven material was the most common facing for clutch discs. Due to awareness of the health hazards resulting from asbestos, alternative lining materials have been developed and are widely used on newer vehicles. The most commonly used are paper-based and ceramic materials that are strengthened by the addition of cotton and brass particles and wire. Kevlar is another material used in many clutch discs. These increase the torsional strength of the facings and prolong the life of the clutch disc.

The facings are attached to wave springs that cause the contact pressure on the facings to rise gradually as the springs flatten out when the clutch is engaged. These springs eliminate chatter when the clutch is engaged and also help to move the disc away from the flywheel when the clutch is disengaged. The wave springs and facings are attached to the steel disc.

There are two types of clutch discs: rigid and flexible. A **rigid clutch disc** is a solid circular disc fastened directly to a center-splined **hub**. The **flexible clutch disc** is easily recognized by the torsional dampener springs that circle the center hub. The dampener is a shock-absorbing feature built into a flexible clutch disc. A flexible disc absorbs power impulses from the engine that would otherwise be transmitted directly to the gears in the transmission. A flexible clutch disc has torsion springs and friction discs between the plate and hub of the clutch. When the clutch is engaged, the springs cushion the sudden loading by compressing and allowing some twist between the hub and plate. When the "surge" is past, the springs release and the disc transmits power normally. The number and tension of these springs is determined by the amount of engine torque and the weight of the vehicle. Stop pins limit this torsional movement to approximately ⅜ inch (Figure 3-9). Some clutch discs are designed for a specific engine. The torsional springs are tuned to that engine. Like the dual-mass flywheel, tuned clutch discs are ineffective if the engine has been modified.

Asbestos is a mineral fiber composed of a silicate of calcium and magnesium that occurs in long threadlike fibers. It has a high resistance to heat and is known to cause the lung disease asbestosis.

A **hub** is the center part of a wheel or disc.

The facings of a disc are often referred to as the lining of the disc.

The clutch disc's wave springs are also called cushion springs.

Cushion springs are often called marcel springs.

AUTHOR'S NOTE: Because a dual-mass flywheel has a damper assembly, the clutch disc used with it does not have torsional dampener springs. These flywheels are designed to dampen harmonic vibrations at idle, but they also cushion clutch engagement. That allows for the elimination of the clutch disc's dampener springs.

High-Performance Clutches. Standard friction materials will not withstand the high torque demands of many modified engines. Friction materials may have high Kevlar content, or be made of cerametallic or ceramic compounds. Some types of performance clutch discs divide the friction surface into separate "pucks" (Figure 3-10).

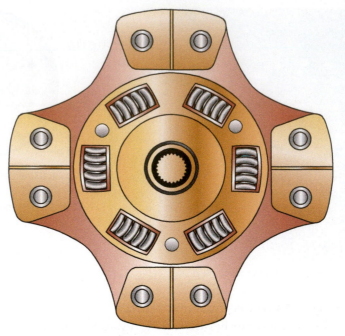

FIGURE 3-10 A puck-type cerametallic clutch disc.

Some high-performance clutch assemblies use multiple clutch discs. An intermediate plate is used in these assemblies to separate the clutch discs. When the clutch is engaged, the first clutch disc is held between the clutch pressure plate and intermediate plate, and the second clutch disc is held between the intermediate plate and the flywheel. When disengaged, the intermediate plate, flywheel, and pressure plate assembly rotate as a unit, while the clutch discs, which are not in contact with the plates, rotate freely within the assembly and do not transmit power to the transmission (Figure 3-11).

Pressure Plate Assembly

The pressure plate (Figure 3-12) squeezes the clutch disc onto the flywheel when the clutch is engaged and moves away from the disc when the clutch pedal is depressed. These actions allow the clutch disc to transmit or not transmit the engine's torque to the transmission. A pressure plate is basically a large spring-loaded clamp that is bolted to and rotates with the flywheel. A pressure plate assembly includes a sheet metal cover, a release spring or springs,

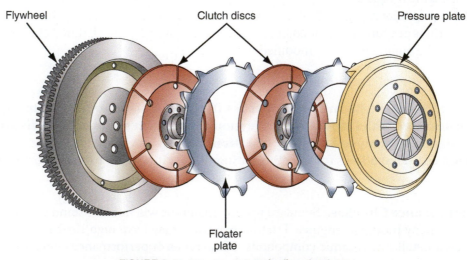

Flywheel   Clutch discs   Pressure plate

Floater plate

FIGURE 3-11 A heavy-duty twin disc clutch kit.

FIGURE 3-12 A typical pressure plate.

a metal pressure ring that provides a friction surface for the clutch disc, a thrust ring or fingers for the release bearing, and release levers. The release levers release the holding force of the springs when the clutch is disengaged. The spring used in most pressure plates is a single **Belleville spring** or **diaphragm spring**; however, a few use multiple coil springs. Some pressure plates are of the semicentrifugal design and use centrifugal weights that increase the clamping force on the thrust springs as engine speed increases.

Diaphragm spring pressure plate (Figure 3-13) assemblies use a cone-shaped diaphragm spring between the pressure plate and the cover to clamp the pressure plate against the clutch disc. This spring is normally secured to the cover by **rivets**. When pressure is exerted on the center of the spring, the outer diameter of the spring tends to straighten out. As soon as the pressure is released, the spring resumes its normal cone shape. The center portion of the spring is slit into numerous fingers that act as release levers. When the clutch is disengaged, these fingers are depressed by the release bearing. The diaphragm spring pivots over the fulcrum ring and its outer rim moves away from the flywheel. The retracting spring pull the pressure plate away from the clutch disc, thereby disengaging the clutch.

When the clutch is engaged, the release bearing is moved away from the diaphragm spring's release fingers. As the spring pivots over the **fulcrum ring**, its outer rim forces the pressure plate tightly against the clutch disc. At this point the clutch disc is clamped between the flywheel and pressure plate. At the outer rim, the spring is moved outward and the pressure plate is forced against the clutch disc. Diaphragm pressure plates have another distinguishing characteristic. As the clutch disc wears, the pressure plate load increases. For all other types of pressure plates, the load decreases. Diaphragm pressure plates are preferred because they are compact, lightweight, require less pedal effort, and have few moving parts to wear.

Coil spring pressure plate assemblies (Figure 3-14) use helical springs that are evenly spaced around the inside of the pressure plate cover. These springs exert pressure to hold the pressure plate tightly against the flywheel. During clutch disengagement, release levers release the holding force of the springs and the clutch disc no longer rotates with the pressure plate and flywheel. Normally these pressure plates are equipped with three release levers and each

A **Belleville spring** is a **diaphragm spring** made from a thin sheet of metal formed into a cone shape.

Some diaphragm-type springs are fitted with a thrust pad at the ends of the release fingers.

A **fulcrum** is the support that provides a pivoting point for a lever.

The fulcrum ring is also referred to as the pivot ring.

A coil spring pressure plate is often called a Borg and Beck.

The holding power of a clutch assembly is determined by its surface area, coefficient of friction, spring pressure, and the number of clutch discs in the assembly.

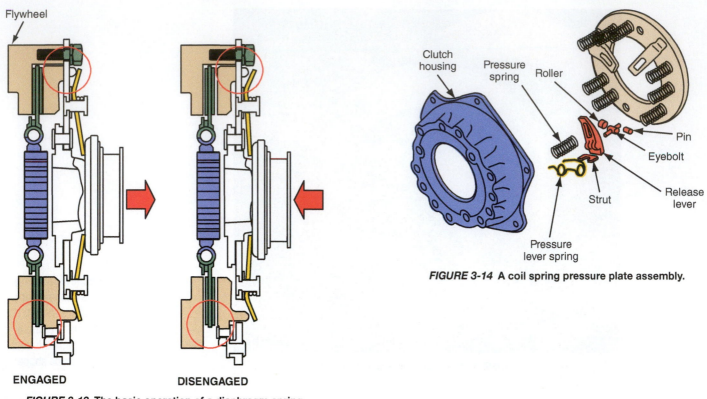

Flywheel

ENGAGED   DISENGAGED

FIGURE 3-13 The basic operation of a diaphragm spring
clutch assembly.

Clutch housing Pressure spring Roller

Pin

Eyebolt

Release lever

Strut

Pressure lever spring

FIGURE 3-14 A coil spring pressure plate assembly.

lever has two pivot points. One pivot point attaches the lever to a pedestal cast into the pressure plate and the other attaches the lever to a release yoke that is bolted to the cover. The levers pivot on the pedestals and release lever yokes to move the pressure plate through its engagement and disengagement operations (Figure 3-15).

To disengage the clutch, the release bearing pushes the inner ends of the release levers toward the flywheel. The release levers act as a fulcrum for the levers and the outer ends of the release levers move to pull the pressure plate away from the clutch disc. This action compresses the coil springs and disengages the clutch. When the clutch is engaged, the release bearing moves and allows the springs to exert pressure to hold the pressure plate against the clutch disc. This forces the disc against the flywheel and the engine's power is transmitted to the transmission through the clutch disc.

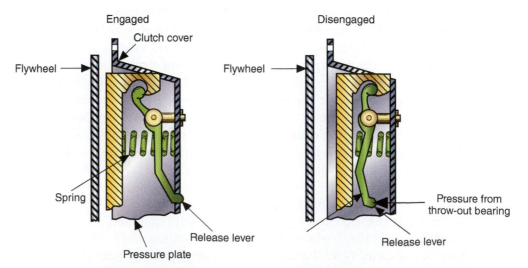

Engaged   Disengaged

Clutch cover

Flywheel   Flywheel

Spring   Pressure from throw-out bearing

Release lever   Release lever

Pressure plate

FIGURE 3-15 The action of a coil-spring pressure plate's release levers.

FIGURE 3-16 A diaphragm spring pressure plate with centrifugal assist weights.

A **semicentrifugal pressure plate** is a design variation that alters the holding force on the clutch disc according to engine speed. Pressure plates utilizing centrifugal assist may be either a coil or diaphragm spring design (Figure 3-16). As engine speed increases, so does the clamping force. A weighted end on the release levers acts to increase centrifugal force as the rotational speed of the pressure plate increases. The centrifugal force adds to the spring pressure to produce greater holding force against the clutch disc (Figure 3-17). This design allows for the use of diaphragm or coil springs with less tension, as the centrifugal force compensates for the decrease in spring tension. As a result, less pedal effort is needed to apply the clutch at low speeds. However, at high speeds, more pedal pressure is needed due to the need to overcome the centrifugal force and spring tension.

The individual parts of a pressure plate assembly are contained in the cover. Some covers are vented to allow heat to escape and air to enter. Other covers are designed to provide a fan action to force air circulation around the clutch assembly. The effectiveness of the clutch is affected by heat; therefore, by allowing the assembly to cool, it is able to work better.

Centrifugal force is an outward pull from the center of a rotating axis.

Some manufacturers use a modular clutch assembly. The flywheel, clutch disc, and pressure plate come as a single unit and cannot be separated. The assembly is bolted to a thin plate similar to an automatic transmission flexplate.

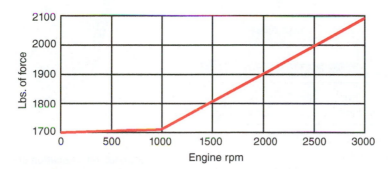

FIGURE 3-17 A graph showing how a semicentrifugal pressure plate changes plate loading according to engine speed.

Release Bearing

Shop Manual
Chapter 3, page 114

Release bearings are commonly referred to as throw-out bearings.

The hollow shaft on the front of the front bearing retainer is often referred to as the *quill shaft*.

The **clutch release bearing** (Figure 3-18) is a ball-type bearing located in the bell housing and operated by the clutch linkage (Figure 3-19). Release bearings are usually sealed and pre-lubricated to provide smooth and quiet operation as they move against the pressure plate to disengage the clutch. A clutch release bearing is often referred to as a throw-out bearing. When the clutch pedal is depressed to disengage the clutch, the release bearing moves toward the flywheel, depressing the pressure plate's release fingers or thrust pad and moving the pressure plate fingers or levers against pressure plate spring force. This action moves the pressure plate away from the clutch disc, thus interrupting power flow. A few vehicles have release bearings that pull on the springs of the pressure plate rather than push them.

Release bearings are mounted on a sleeve, called a hub, which slides back and forth on a hollow shaft at the front of the transmission housing. This hollow shaft is part of the transmission's front bearing retainer (Figure 3-20). The sleeve is grooved or has raised flat surfaces and retaining springs that hold the inner ends of the release fork in place on the release bearing assembly. The fork and connecting linkage convert the movement of the clutch pedal to the back-and-forth movement of the release bearing.

To disengage a clutch, the release bearing is moved toward the flywheel by the clutch fork. As the bearing contacts the release levers or fingers, it begins to rotate with the pressure plate assembly. As the release bearing continues to move forward, the pressure on the release levers or fingers causes the force of the pressure plate's spring to move away from the clutch disc.

To engage the clutch, the clutch pedal is released and the release bearing moves away from the pressure plate. This action allows the pressure plate's springs to force against the clutch disc, engaging the clutch to the flywheel. Once the clutch is fully engaged, the release bearing is normally stationary and does not rotate with the pressure plate.

Some release bearings must be adjusted so they do not touch the release fingers, beveled springs, or thrust pad when the pedal is released. However, most release bearings are designed to ride lightly against the pressure plate assembly; these are called constant running release bearings (Figure 3-21) and are used on transmissions equipped with self-adjusting cable or hydraulically operated clutch linkages.

The clutch linkage connects the driver-operated clutch pedal to a bell housing-mounted release fork that acts directly on the release bearing. The release bearing slides back and forth

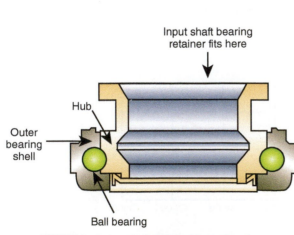

FIGURE 3-18 A typical clutch release bearing.

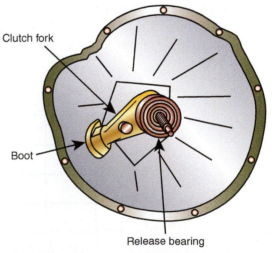

FIGURE 3-19 Location of clutch fork and release bearing.

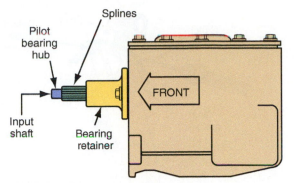

FIGURE 3-20 The release bearing slides on the hollow shaft nose of the transmission's front bearing retainer.

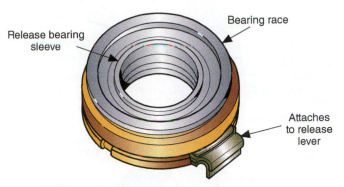

FIGURE 3-21 A typical constant running release bearing.

on the transmission's front bearing retainer in response to clutch pedal movement. On vehicles with a concentric slave cylinder inside the bell housing the release bearing is mounted on the slave cylinder.

Clutch Release Systems

Clutches are normally operated by either mechanical, hydraulic, or electronically controlled systems. Two types of mechanical linkages are used: the cable type and the shaft and lever type. The shaft and lever clutch linkage has many parts and pivot points and transfers the movement of the clutch pedal to the release bearing via shafts, levers, and bell cranks. In older vehicles, the pivot points were equipped with grease fittings. Improved systems pivot on low-friction plastic grommets and bushings. As the pivot points wear, the extra play in the linkage makes precise clutch pedal free-play adjustments difficult.

A typical shaft and lever clutch control assembly (Figure 3-22) includes a release lever and rod, an equalizer or cross shaft, a pedal-to-equalizer rod, an assist or Over-center spring, and the pedal assembly. Depressing the pedal moves the equalizer, which, in turn, moves the release rod. When the pedal is released, the assist spring returns the linkage to its normal

Shop Manual
Chapter 3, page 101

Clutch pedal free-play is the amount the pedal can move without applying pressure on the pressure plate.

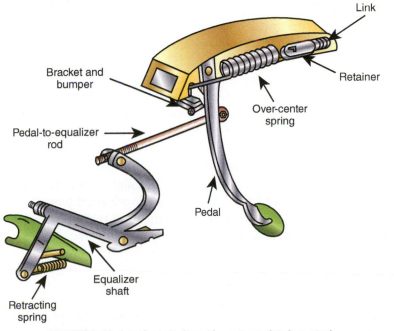

FIGURE 3-22 A typical shaft and lever type clutch control.

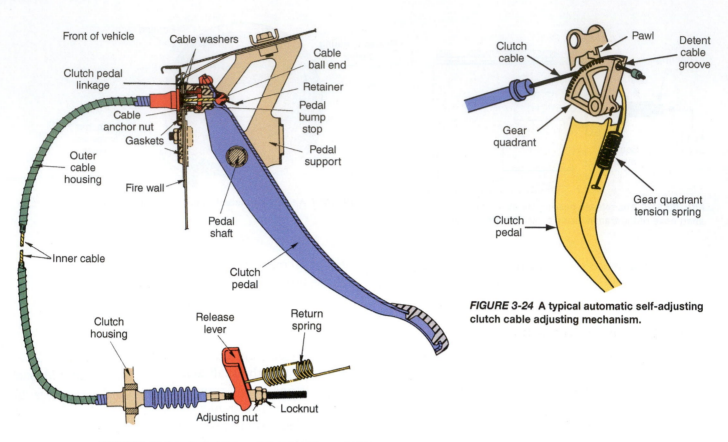

FIGURE 3-24 A typical automatic self-adjusting clutch cable adjusting mechanism.

FIGURE 3-23 A typical clutch pedal and cable assembly.

position and removes the pressure on the release rod. This action causes the release bearing to move away from the pressure plate.

A cable-type clutch linkage is simple and lightweight. Normally the cable connects the pivot of the clutch pedal directly to the release fork (Figure 3-23). This simple setup is compact, flexible, and eliminates the wearing pivot points of a shaft and lever linkage. However, cables will gradually stretch and can break due to electrolysis.

Typically, one end of the cable is connected to the pedal assembly. At the assembly is a spring that keeps the pedal in the up position. The cable is held under tension by the spring and a cable stop located on the fire wall. The other end of the cable is connected to the outer end of the clutch release fork. This end is threaded and fitted with an adjusting nut and locknut that allows for pedal free-play adjustments. When the clutch pedal is depressed, the cable pulls on the clutch fork, which causes the release bearing to move against the pressure plate.

On many vehicles, the cable is self-adjusting. At the pedal pivot, the cable is wrapped around and attached to a toothed wheel (Figure 3-24). Slight contact of the release bearing on the pressure plate is maintained by a ratcheting, spring-loaded pawl that is engaged to the toothed wheel. When the clutch pedal is released, the pawl takes any slack out of the cable by engaging the next tooth of the wheel. Self-adjusting clutches use a constant running release bearing and have no built-in free play.

Hydraulic Systems

Another method used to move the clutch fork is with a hydraulic system. A hydraulic system multiplies force, which reduces the force required to depress the clutch pedal. Most newer vehicles have hydraulic clutch linkages.

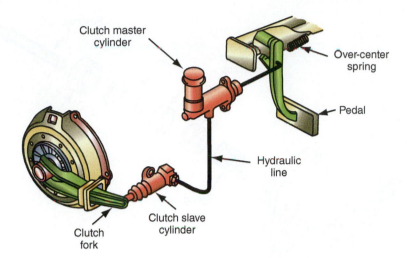

FIGURE 3-25 A typical hydraulic clutch linkage.

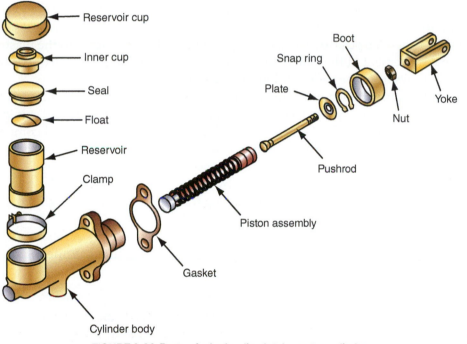

FIGURE 3-26 Parts of a hydraulic clutch master cylinder.

A hydraulic clutch linkage, like a brake system, consists of a master cylinder, hydraulic tubing, and a slave cylinder (Figure 3-25). The **master cylinder** is attached to and activated by the clutch pedal through the use of an actuator rod. The master cylinder is mounted on the firewall and receives its fluid from an attached or remote reservoir filled with brake fluid. The **slave cylinder** is connected to the master cylinder by a flexible pressure hose or metal tubing. The slave cylinder is positioned so that it can work directly on the clutch release yoke lever.

Depressing the clutch pedal pushes the actuator rod into the bore of the master cylinder. This action forces a plunger or piston up the bore of the master cylinder (Figure 3-26). During the initial 1/32-inch of pedal travel, the valve seal at the end of the master cylinder bore closes the port to the fluid reservoir. As the pedal is further depressed, the movement of the plunger forces fluid from the master cylinder through a line to the slave cylinder (Figure 3-27). This fluid is under pressure and causes the piston of the slave cylinder to move. Moving the slave cylinder's piston causes its pushrod to move against the release fork and bearing, thus

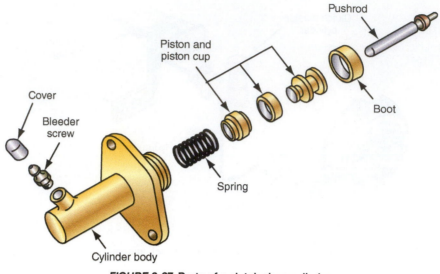

FIGURE 3-27 Parts of a clutch slave cylinder.

disengaging the clutch. When the clutch pedal is released, the springs of the pressure plate push the slave cylinder's pushrod back, forcing the hydraulic fluid back into the master cylinder. The final movement of the clutch pedal and the master cylinder's piston opens the valve seal and fluid flows from the reservoir into the master cylinder.

A clutch master cylinder performs the following functions: moves fluid through the hydraulic line to the slave cylinder, compensates for temperature change and minimal fluid loss to maintain the correct fluid volume through the use of a bleed port and compensation port, and compensates for a worn clutch disc and pressure plate by displacing fluid through the reservoir bleed port, thereby eliminating the need for periodic adjustment.

The slave cylinder disengages the clutch by extending the slave cylinder pushrod and makes sure the clutch release bearing is contacting the pressure plate through the use of the slave cylinder preload spring.

Some hydraulic clutch systems have a clutch damper (Figure 3-28). A flexible diaphragm in the damper absorbs hydraulic pulses during engagement and disengagement of the clutch

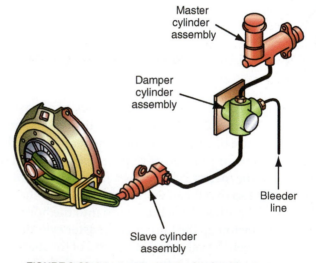

FIGURE 3-28 A hydraulic clutch circuit with a damper cylinder assembly. Note the extra hydraulic line from the damper. This line is not connected to any other component and is used to bleed the system.

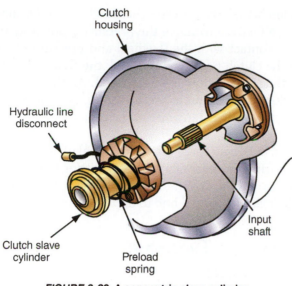

FIGURE 3-29 A concentric slave cylinder.

and eases clutch engagement. Other systems may use a **clutch delay valve** that provides a restriction in the line between the master cylinder and slave cylinder to prevent abrupt clutch engagement.

Concentric Slave Cylinders. A **concentric slave cylinder** is found on many cars and light trucks. These units (Figure 3-29) replace the conventional release fork, pivot ball, linkage, and throw-out bearing. By having the slave cylinder directly behind the release bearing, the movement of the release bearing is linear. The use of this slave cylinder also eliminates packaging problems caused by the placement of the bell housing and the required height of the release lever.

A concentric slave cylinder is a doughnut-shaped unit that mounts to the front of the transmission and the transmission's input shaft passes through it. The concentric slave cylinder is either bolted to the transmission's front bearing cover or is secured by a pressed pin.

> **AUTHOR'S NOTE:** I have seen many students search and search for the clutch slave cylinder on Ford trucks and other vehicles. Many vehicles have concentric slave cylinders that are located inside the bell housing. You can only see them when you separate the transmission from the engine. Always refer to the service manual when servicing a clutch assembly.

CLUTCH OPERATION

The clutch disc is sandwiched between the flywheel and pressure plate assembly to become the driven member of the clutch assembly. The flywheel drives the front side while the pressure plate drives the rear side of the clutch disc. In the center of the clutch disc is a splined hub that meshes with the splines on the clutch shaft. The pressure plate is held in contact with the rear friction facing of the clutch disc by spring tension. When the clutch is disengaged, the pressure plate is released by a release bearing that utilizes a lever action to pull the pressure plate away from the clutch disc. The release yoke, with the release bearing clipped to it, is mounted on a pivot located inside the clutch housing and is operated by the clutch linkage and pedal assembly.

The sudden release of the clutch pedal, typically done by sliding the foot off the side of the pedal, is often called "dropping the clutch." The practice risks damage to the vehicle's drivetrain components.

To disengage the clutch, the driver presses the clutch pedal. The linkage forces the release bearing and release yoke forward to move the pressure plate away from the disc. Because the disc is no longer in contact with the flywheel and pressure plate, the clutch is said to be disengaged. When the clutch pedal is depressed, the flywheel, clutch disc, and pressure plate are disengaged, thus power-flow is interrupted. As the clutch pedal is released, the pressure plate moves closer to the clutch disc, clamping the disc between the pressure plate and the flywheel. Therefore, if the transmission is in gear, the drive wheels will turn when the clutch disc turns.

To engage the clutch, the driver eases the clutch pedal up from the floor. The control linkage moves the release bearing and lever rearward, permitting the pressure plate spring tension to force the pressure plate and the driven disc against the flywheel. Engine torque again acts on the disc's friction facings and splined hub to drive the transmission input shaft.

SELF-SHIFTING MANUAL TRANSMISSIONS

One of the latest trends in manual transmission design is that of a self-shifting manual unit. Some of these units use a dual-clutch arrangement. Dual clutches are used in these units to improve the shifting quality and driveability in self-shifting manual transmissions. Fuel economy is improved because the engine speed can be better maintained at its most efficient point and there is less time between shifts when power is not being transmitted to the drive wheels. These transmissions have two separate gear set shafts, each with its own corresponding clutch, to engage and disengage the gears. For example, in a six-speed manual transmission, one shaft takes care of the one-, two-, five-, and reverse speed gears, and the second shaft handles the two-, four-, and six-speed gears (Figure 3-30). By clutching the shaft for the outgoing gear when simultaneously engaging the shaft for the next desired gear, the unit greatly reduces the time required for the gearshift, but, more important, drastically improves shift quality by nearly eliminating shift shock and harsh response to throttle inputs.

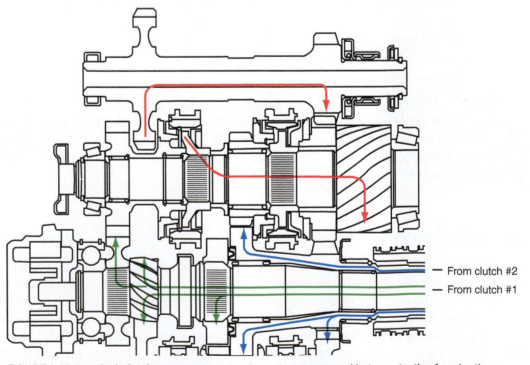

— From clutch #2
— From clutch #1

FIGURE 3-30 In a dual-clutch arrangement, the disc sets are engaged independently of each other through hydraulic pressure of pistons. When a clutch is engaged, torque is transmitted to the allocated input shaft via the respective clutch assembly.

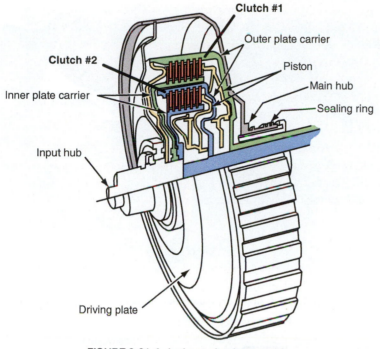

FIGURE 3-31 A dual wet clutch assembly.

The electronic control unit for the dual-clutch transmission relies on inputs from both the engine control system and the driver. The driver can select from different modes to adjust shift timing and clutch engagement feel. There is no need for a clutch pedal or shift lever, and most manufacturers use steering wheel paddles to allow the driver to control upshifts and downshifts.

The dual clutches can be dry or wet (similar to those used in automatic transmissions) design. Dry dual clutches are used primarily in vehicles with low engine torque. Dual **wet clutch** systems are used on more powerful engines. A wet clutch assembly has multiple plates immersed in fluid (Figure 3-31). The fluid cleans and cools the friction surfaces, allows for smoother and quieter operation, and extends the life of the clutch. The clutch is applied by hydraulic pressure against an apply piston. The two clutches in the dual-clutch setup can be engaged independently from each other. When a clutch is engaged, torque is transmitted to the allocated input shaft via the respective clutch assembly. These systems typically include a torsional damper, hydraulic control unit (HCU) for actuating and cooling the dual clutch, and transmission actuators with an electronic transmission and system control unit (TCU).

ELECTROMECHANICAL CLUTCH SYSTEMS

Some manufacturers have maintained the use of a single dry clutch plate/pressure plate system but have chosen to actuate the clutch with an electronically controlled solenoid (Figure 3-32). Other manufacturers use a dual dry clutch setup with two solenoids, one for each clutch disc (Figure 3-33). These systems have many advantages for the driver. The rate of clutch release can be controlled by placing the system in different driving modes. The clutch and transmission can be left in "automatic" mode, freeing the driver from having to think about the clutch or transmission at all. In fact, some makers advertise these electromechanical transmissions/transaxles as automatic transmissions even though the transmission is no different internally than a typical manual unit.

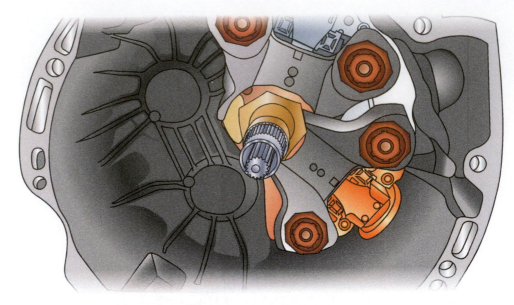

FIGURE 3-32 A single dry clutch electromechanical system.

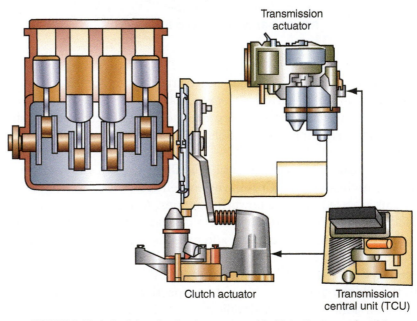

Transmission actuator

Clutch actuator

Transmission central unit (TCU)

FIGURE 3-33 A dual dry clutch release assembly. Note the two solenoid-activated release forks and two input shafts.

SUMMARY

- The main parts of the clutch assembly are the clutch housing, flywheel, input shaft, disc, pressure plate assembly, release bearing, and linkage.
- The flywheel acts as a balancer and smoothens out, or dampens, engine vibrations caused by firing pulses and adds inertia to the rotating crankshaft of the engine.
- The flywheel also provides a machined surface for the clutch friction disc.
- A dual-mass flywheel incorporates dampener springs to smooth out engine torsional vibrations.
- Vehicles with an automatic transmission are equipped with a drive plate or flexplate rather than a heavy flywheel.

- The clutch disc is splined to the input shaft, which allows the disc to move without rocking on the shaft.
- The clutch disc is a steel plate with friction material bonded to both sides that fits between the flywheel and the pressure plate.
- Some friction discs contain asbestos fibers. Always follow safety precautions if it is possible that asbestos is present.
- A rigid or solid clutch disc is a solid disc fastened directly to a center splined hub.
- A flexible clutch disc has torsional dampener springs in its center hub.
- The primary purpose of a flexible disc is to absorb power impulses from the engine that would otherwise be transmitted directly to the transmission.
- The pressure plate is a large spring-loaded plate that engages the clutch by pressing the disc against the flywheel surface.
- The pressure plate moves away from the flywheel when the clutch pedal is depressed, releasing the clamping force and stopping engine torque from reaching the transmission.
- The clutch release bearing is operated by the clutch release mechanism.
- When the clutch pedal is depressed, the bearing typically moves toward the flywheel, depressing the pressure plate fingers or thrust pad and moving the pressure plate away from the clutch disc.
- The clutch linkage connects the clutch pedal to a release fork that acts on the release bearing.
- The clutch is usually located between the engine and the transmission.
- Clutches are operated by either mechanical, hydraulic, or electromechanical systems.
- A mechanical clutch linkage transfers the clutch pedal movement to the release bearing via shafts, levers, and bell cranks, or by a cable.
- A hydraulic clutch system consists of a master cylinder, hydraulic tubing, and a slave cylinder.
- Dual-clutch systems engage two separate transmission input shafts, allowing for the preselection of the next expected gear.

TERMS TO KNOW

Asbestos
Belleville spring
Clutch delay valve
Clutch (friction) disc
Clutch housing
Clutch release bearing
Clutch shaft
Concentric slave cylinder
Diaphragm spring
Dual-mass flywheel (DMF)
Flexible clutch disc
Flexplate
Fulcrum ring
Hub
Master cylinder
Rigid clutch disc
Rivets
Semicentrifugal pressure plate
Slave cylinder
Splines
Wet clutch

REVIEW QUESTIONS

Short Answer Essays

1. Define the purpose of a clutch assembly.

2. List and describe the major components of a clutch assembly.

3. Describe the operation of a clutch.

4. Compare and contrast the operation of a coil spring pressure plate and a diaphragm spring pressure plate.

5. Define the role of each major component in a clutch assembly.

6. Describe the operation of a cable-type clutch linkage.

7. Describe the operation of a hydraulic clutch linkage.

8. Describe the construction of a flexible clutch disc and state why it is different than a rigid disc.

9. Explain why some vehicles are fitted with a dual-mass flywheel and briefly explain how they work.

10. What is the commonly used name for the clutch release bearing?

Fill-in-the-Blanks

1. A clutch housing is also called a _____ _____.

2. The teeth around the outside of the flywheel form a ring gear for the _____-_____.

3. The clutch disc slides over the splines of the _____ shaft.

4. There are two types of dry clutch discs: _____ and _____.

5. The pressure plate engages and disengages the
   _____.

6. A flexible clutch disc has _____ springs in
   its center hub.

7. The pressure plate moves away from the flywheel
   when the clutch pedal is _____.

8. When the clutch is disengaged, the pressure plate is
   released by a _____ _____.

9. A hydraulic clutch linkage consists of a
   _____ _____, _____
   _____, and a _____.

10. Tension is maintained in a self-adjusting cable-type
    clutch linkage by the use of a _____ and a
    toothed _____.

ASE-Style Review Questions

1. When discussing where a clutch disc is normally
   mounted,
   Technician A says that it is bolted to the flywheel.
   Technician B says that it is splined to the transmission
   input shaft.
   Who is correct?
   A. A only   C. Both A and B
   B. B only   D. Neither A nor B

2. Which of the following statements about the torsional
   springs in a clutch disc are not true?
   A. They dampen vibrations coming from the
       transmission and drive shaft.
   B. They may be tuned to a particular engine and
       driveline.
   C. They cushion the sudden load on the disc when it is
       quickly engaged.
   D. They are only found on flexible clutch discs.

3. When discussing the purpose of the flywheel,
   Technician A says that it serves as a heat sink for the
   clutch disc.
   Technician B says that it serves as a vibration
   dampener for the engine.
   Who is correct?
   A. A only   C. Both A and B
   B. B only   D. Neither A nor B

4. When discussing the operation of the clutch assembly,
   Technician A says that when the pedal is depressed, the
   release bearing is pushed against the release fingers of
   the pressure plate, which releases the plate's pressure
   on the clutch disc.
   Technician B says that the clutch disc is pressed against
   the flywheel by the pressure plate.
   Who is correct?
   A. A only   C. Both A and B
   B. B only   D. Neither A nor B

5. When discussing the major components of a hydraulic
   clutch linkage,
   Technician A says that the master cylinder for the car's
   brakes is also used for the clutch.
   Technician B says that the slave cylinder is directly
   connected to the clutch pedal and increases hydraulic
   pressure as the pedal is depressed.
   Who is correct?
   A. A only   C. Both A and B
   B. B only   D. Neither A nor B

6. When discussing release bearings,
   Technician A says that constant running release bear-
   ings are used with hydraulically controlled clutches.
   Technician B says that the release bearing moves the
   pressure plate fingers to disengage the clutch.
   Who is correct?
   A. A only   C. Both A and B
   B. B only   D. Neither A nor B

7. When discussing the different types of clutches,
   Technician A says that rigid clutch discs are tuned to
   dampen engine firing impulses.
   Technician B says that flexible clutch discs are
   necessary when the vehicle has a dual-mass flywheel.
   Who is correct?
   A. A only   C. Both A and B
   B. B only   D. Neither A nor B

8. When discussing pilot bearings or bushings,
   Technician A says that they are often eliminated in
   front-wheel drive vehicles.
   Technician B says that they serve as a low-friction
   pivot point for the clutch linkage.
   Who is correct?
   A. A only   C. Both A and B
   B. B only   D. Neither A nor B

9. While discussing hydraulic clutch systems,

   Technician A says the slave cylinder responds directly to the clutch pedal and sends hydraulic fluid to the clutch master cylinder.

   Technician B says the clutch delay valve between the master and slave cylinder allows for immediate clutch release.

   Who is correct?

   A. A only
   B. B only
   C. Both A and B
   D. Neither A nor B

10. While discussing semicentrifugal pressure plates,

    Technician A says they alter the holding force of the clutch assembly in response to engine speed.

    Technician B says because they require high pedal effort at low engine speeds.

    Who is correct?

    A. A only
    B. B only
    C. Both A and B
    D. Neither A nor B

MANUAL TRANSMISSIONS/ TRANSAXLES

UPON COMPLETION AND REVIEW OF THIS CHAPTER, YOU SHOULD BE ABLE TO:

- Understand and discuss the purpose and operation of typical manual transmissions.

- Understand and describe the purpose, design, and operation of a synchronizer assembly.

- Understand and discuss the flow of power through a manual transmission.

- Compare and contrast the design and operation of a transmission and a transaxle.

- Understand and discuss the flow of power through a manual transaxle.

- Understand and describe how gears are shifted in a manual transmission/transaxle.

- Discuss the importance of correct lubrication in a manual transmission/transaxle.

INTRODUCTION

The purpose of a transmission is to use various-sized gears to keep the engine in its best operating range in a variety of vehicle speeds and loads.

Some manufacturers call the output shaft the main shaft.

A gear transfers power and motion from one shaft to another shaft.

Transmissions and transaxles serve basically the same purpose and operate by the same basic principles. Although the assembly of a transaxle is different than a transmission, the fundamentals and basic components are the same. In fact, a transaxle is basically a transmission with other driveline components housed within the assembly. All basics covered in this chapter refer to both a transaxle and a transmission unless otherwise noted.

A transmission (Figure 4-1) is a system of gears that transfers the engine's power to the drive wheels of the car. The transmission receives torque from the engine through its **input shaft** when the clutch is engaged. The torque is then transferred through a set of gears that either multiply torque, transfer it directly, or reduce it. The resultant torque turns the transmission's **output shaft**, which is indirectly connected to the drive wheels. All transmissions have two primary purposes: (1) select different speed ratios for a variety of conditions and (2) provide a way to reverse the movement of the vehicle.

A manual transaxle is a single unit composed of a manual transmission, final drive, and differential. Most front-wheel-drive (FWD) cars are equipped with a transaxle. Transaxles are also found on some front-engined and rear-wheel-drive (RWD) and four-wheel-drive (4WD) cars and on rear-engined and rear-wheel-drive cars (Figure 4-2).

With a transaxle, the engine is normally mounted transversely across the front of FWD cars (Figure 4-3). The compactness of a transaxle allows designers to offer increased passenger room, reduce vehicle weight, and reduce vibration and alignment problems normally caused by the long drive shafts of rear-wheel-drive cars.

Shifting is often accomplished by a mechanical linkage from inside the car to the transmission/transaxle. The linkage may be a cable or cables, rod-type linkage, or go directly from the shift lever to the transmission. On some later vehicles shifting is done through electronically

FIGURE 4-1 A late-model transaxle.

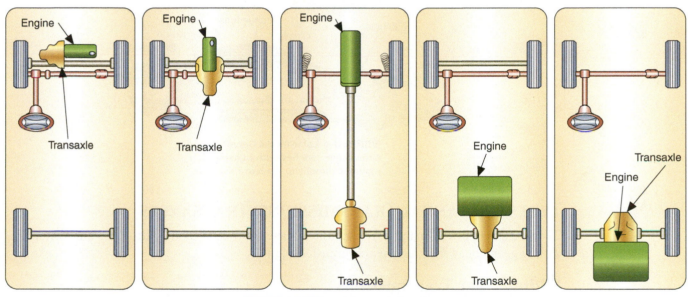

FIGURE 4-2 Different transaxle locations.

controlled solenoids. This setup eliminates the mechanical linkage and makes for more precise shifting. Increased driver control of shift timing is available through the selection of different shift modes.

Remote-Mounted Transmissions

Late-model Corvettes, as well as other cars, have a driveline setup that is somewhat unique. These vehicles are front-engined, RWD vehicles. The six-speed transmission is mounted directly to the rear axle assembly (Figure 4-4) instead of the engine. The clutch assembly is still at the rear of the engine and a drive shaft (torque tube) connects the output from the clutch to the transmission. The primary advantage of moving the transmission to the rear of the car is to shift the weight of the unit to the rear. This gives the car a near balance of weight on the front and rear wheels. The weight balance provides for improved handling and braking.

Transmission gears are normally helical-type gears; however, some reverse speed gears and many racing gears are spur-type gears.

A transaxle is a final drive unit and transmission housed in a single unit.

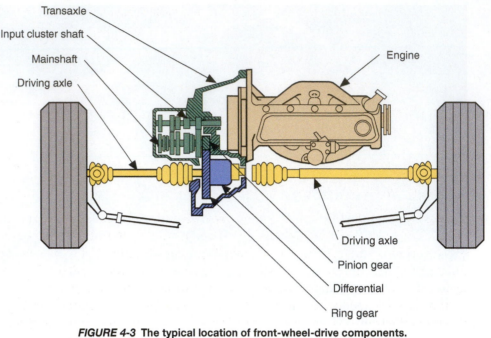

FIGURE 4-3 The typical location of front-wheel-drive components.

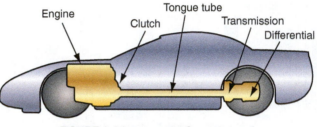

FIGURE 4-4 Later-model Corvettes and a few other cars have the transmission mounted directly to the rear drive axle.

TYPES OF MANUAL TRANSMISSIONS/TRANSAXLES

Three major types of manual transmissions have been used by the automotive industry: the sliding gear, collar shift, and synchromesh designs.

Sliding Gear Transmissions

A **sliding gear transmission** uses two or more shafts mounted in parallel, with sliding spur gears arranged to mesh with each other to provide a change in speed or direction. The driver moves the shifter, which in turn moves the appropriate gear into mesh with its mate. If either of the gears are rotating, the changing of gears is very difficult and will normally grind into mesh. This type of transmission is currently used only on farm and industrial machines. However, some transmissions use a sliding gear mechanism for the engagement of reverse gear.

Collar Shift Transmissions

A **collar shift transmission** has parallel shafts fitted with gears in **constant mesh**. The change of gear ratios is accomplished by locking the free-running gears to their shafts using sliding collars. The sliding collars connect and lock the appropriate gear to its shaft. One side of the gear has short splines into which the internal splines of the sliding collar mesh. Because the collar is splined to the shaft, the connection of the collar to the gear connects the gear to the rotating shaft. Although the splines on the gears and collars have rounded ends for easier shifting, gear clashing will occur when the gears are changed or engaged, unless speeds are matched by the driver by double-clutching.

AUTHOR'S NOTE: If you don't know what double-clutching is, ask your instructor to explain.

Synchromesh Transmissions

A **synchromesh transmission** is a constant mesh (Figure 4-5), collar shift transmission equipped with synchronizers that equalize the speed of the shaft and gear before they are engaged. The action of the synchronizer eliminates gear clashing and allows for smooth changing of gears. Synchromesh transmissions are used on all current models of cars and are commonly found in other machines wherever shifting while moving is required.

Engine torque is applied to the transmission's input shaft when the clutch is engaged. The input shaft enters the transmission case, where it is supported by a bearing and fitted with a gear (Figure 4-6). The gear on the input shaft is called the input gear or **clutch gear**. In RWD transmissions, the output shaft is inserted into, but rotates independently of, the input shaft. The main shaft is supported by a bearing inside the input shaft and a bearing at the rear of the transmission case. The various speed gears rotate on the main shaft. Located below or to the side of the input and main shaft assembly is a countershaft that is fitted with several-sized gears. All of these gears, except for one, are in constant mesh with the gears on the main shaft. The remaining gear is in constant mesh with the input gear.

One of the primary differences between a RWD transmission and a transaxle is that the counter (cluster) gear assembly is eliminated. The input shaft's gears drive the output shaft gears directly. The output shaft rotates according to each synchro-activated gear's operating ratio. The shafts usually ride on roller, tapered roller, or ball type bearings.

Normally a transaxle has two separate shafts: an input shaft and an output shaft. The engine's torque is applied to the input shaft and the revised torque (due to the transaxle's gearing) rotates the output shaft. Usually the input shaft is located above and parallel to the output shaft. The main gears freewheel around the output shaft unless they are locked to the shaft by the synchronizers. The main speed gears are in constant mesh with their mating gears on the input shaft and rotate whenever the input shaft rotates. Because the input shaft in most transversely mounted transaxles are supported by bearings in the housing, these units do not need a pilot bearing or bushing to support the portion of the input shaft that extends into the clutch assembly.

> Free-running gears are gears that rotate independently on their shafts.

FIGURE 4-5 The gears in a transmission transmit rotating power (torque) from the engine.

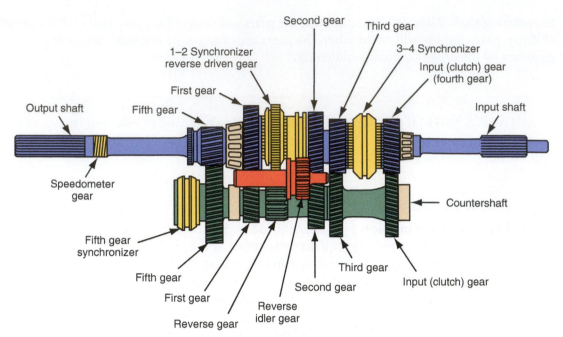

FIGURE 4-6 The arrangement of the gears and shafts in a typical five-speed transmission.

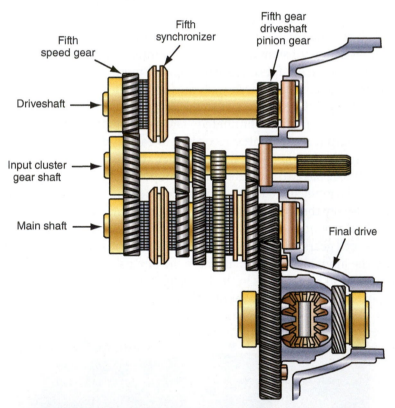

FIGURE 4-7 A typical five-speed transaxle with three gear shafts.

Some transaxles are equipped with an additional shaft designed to offset the power-flow on the output shaft (Figure 4-7). This shaft is placed in parallel with the input and output shafts. Power is transferred from the output shaft to the third shaft by helical gears. The third shaft is added only when an extremely compact transaxle installation is required. Therefore, most transaxles use two shafts: the output shaft and the input shaft. Some transaxles with a third shaft use an offset input shaft, which receives the engine's power and transmits it to a main shaft, which serves the same functions as an input shaft.

FIGURE 4-8 The shift forks and rails are assembled in the cover of this transmission with a direct shift linkage.

Gear changes occur when a gear is selected by the driver and is locked or connected to the main shaft. This is accomplished by the movement of a collar that connects the gear to the shaft (Figure 4-8). Smooth and quiet shifting can only be possible when the gears and shaft are rotating at the same speed. This is the primary function of the synchronizers.

Forward Speeds

All transmissions are equipped with a varied number of forward speed gears and one reverse speed. Transmissions can be classified by the number of forward gears (speeds) it has. For many years, the three-speed manual transmission was the most commonly used and four-speed transmissions were found only in heavy-duty vehicles and high-performance cars. The concern for improved fuel mileage led to smaller engines with four-speed transmissions. The additional gear allowed the smaller engines to perform better by matching the engine's torque curve with vehicle speeds.

> **AUTHOR'S NOTE:** Although four-speed transmissions are no longer used by manufacturers, maybe the most well-known transmission in modern history was the "Muncie Four Speed." This General Motors transmission was used in many high-performance applications and proved to be a very durable unit. Many versions of this transmission were produced from 1963 until 1974, with the most famous being the "M22 rockcrusher." This transmission had gear sets of higher nickel alloy, and the helical gears were of a straighter angle to reduce side thrust forces. This straighter angle created more gear noise, hence the nickname "rockcrusher."

Five-speed and six-speed transmissions and transaxles are now the most commonly used units. Some earlier units were actually four speeds with an add-on fifth or overdrive gear. Many late-model transmissions incorporate fifth gear in their main assemblies. This is also true of late-model six-speed transmissions and transaxles. The fifth and sixth gears are part of the main gear assembly. Most often, the two high gears in a six-speed provide overdrive gear ratios.

The latest developments in transmission/transaxle design include seven- and even eight-speed versions. Increased fuel economy and lowered emissions are achieved because the engine can be maintained at more efficient engine speeds without large rpm drops between shifts. These units are dual-clutch, fully electronically controlled units and are considered in Chapter 11.

SYNCHRONIZERS

Shop Manual
Chapter 4, page 168

A synchronizer's primary purpose is to bring components that are rotating at different speeds to one synchronized speed. It also serves to lock these parts together. The forward gears of all current automotive transmissions are synchronized. Some older transmissions and truck transmissions were not equipped with synchronizers on first or reverse gears. These gears could only be easily engaged when the vehicle was stopped. Reverse gear on some late-model transmissions and transaxles is also synchronized. A single synchronizer is placed between two different speed gears, therefore transmissions have two or three synchronizer assemblies.

In the past, reverse gear was not normally synchronized. Today, most late-model transmissions have a synchronized reverse gear, or at least a blocking ring in the design to stop the shaft from turning during reverse engagement. This assembly is often called a reverse brake. Many reverse-gear arrangements use spur gears, but manufacturers have gone to helical gearing in most units to eliminate noise concerns from the customer.

The term *synchro* is a commonly used slang word for synchronizer.

> **AUTHOR'S NOTE:** Sometimes the best way to really understand how a synchronizer works is to hold a complete synchro assembly in your hands and move each part around and back and forth. Notice what else moves when the parts are in their various positions, then think about the rest of the transmission parts and their movements.

Synchronizer Designs

There are four types of synchronizers used in synchromesh transmissions: block, disc and plate, plain, and pin. The most commonly used type on current transmissions is the block type. All synchronizers use friction to synchronize the speed of the gear and shaft before the connection is made.

Block synchronizers consist of a hub (called a **clutch hub**), sleeve, blocking ring, and inserts or spring-and-ball detent devices (Figure 4-9). The synchronizer sleeve surrounds the synchronizer assembly and meshes with the external splines of the hub. The hub is internally splined to the transmission's main shaft. The outside of the sleeve is grooved to accept the shifting fork. Three slots are equally spaced around the outside of the hub and are fitted with the synchronizer's inserts or spring-and-ball detent assemblies.

Block synchronizers are commonly referred to as cone synchronizers.

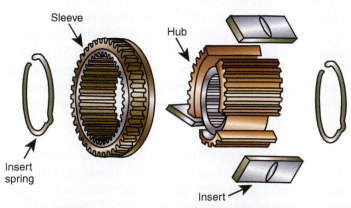

FIGURE 4-9 A typical block synchronizer assembly.

These inserts are able to freely slide back and forth in the slots. The inserts are designed with a ridge on their outer surface and insert springs hold this ridge in contact with an internal groove in the synchronizer sleeve. When the transmission is in the "neutral" position, the inserts keep the sleeve lightly locked into position on the hub. If the synchronizer assembly uses spring-and-ball detents, the balls are held in this groove by their spring. The sleeve is machined to allow it to slide smoothly on the hub.

Blocking rings are positioned at the front and rear of each synchronizer assembly. Blocking rings are made of often made of brass or bronze because of the high heat produced by friction. The use of these metals minimizes the wear on the hardened steel gear's cone. A common trend in the automotive industry today is to reduce frictional losses as much as possible. The materials used to manufacture blocking rings have not been overlooked. Powdered metal, Kevlar, and organic frictional materials are currently being used and/or being tested by some manufacturers. Blocking rings have notches to accept the insert keys that cause them to rotate at the same speed as the hub (Figure 4-10). Around the outside of the blocking ring is a set of beveled **dog teeth**. The inside of the blocking ring is shaped like a cone, the surface of which has many sharp grooves. The inner surfaces of the blocking rings match the conical shape of the shoulders of the driven gear. These cone-shaped surfaces serve as the frictional surfaces for the synchronizer. The shoulder of the gear also has a ring of beveled dog teeth designed to align with the dog teeth on the blocking ring (Figure 4-11).

A synchronizer hub is also called a **clutch hub**.

Synchronizer inserts are often referred to as keys.

Dog teeth are small teeth that surround the side of a gear. The term is used to describe a set of gear teeth that has gaps between the teeth and are not close together.

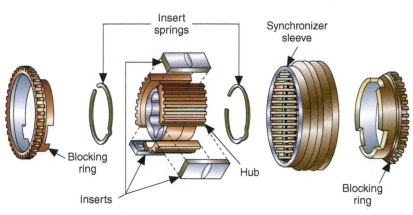

FIGURE 4-10 The notches in the blocking ring correspond with the spacing of the inserts.

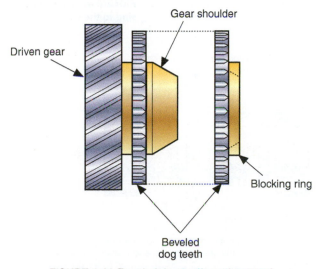

FIGURE 4-11 Beveled dog teeth on the speed gear and the blocking ring.

Synchronizer Operation

A fully synchronized transmission is a transmission in which all of its forward gears are equipped with a synchronizer assembly.

When the transmission is in neutral (Figure 4-12), the synchronizers are in their neutral position and are rotating with the main shaft. The main shaft's gears are meshed with the counter gears and are rotating with the countershaft. However, they turn freely, at various speeds, on the main shaft and do not cause the shaft to rotate because they are not connected to it.

When a gear is selected, the shifting fork forces the sleeve toward the selected speed gear. As the sleeve moves, so do the inserts since they are pressed by insert springs into the sleeve's internal groove. The movement of the inserts pushes the blocking ring into contact with the shoulder of the driven gear. When this contact is made, the grooves on the blocking ring's cone cut through the film of lubrication on the gear's shoulder. Contact between the two is then made and the speed synchronization of the two parts begins (Figure 4-13).

The resultant friction between the two causes the blocker ring to rotate tightly against the inserts, which places the teeth directly in line with the splines or engagement teeth of the synchronizer sleeve. The pressure from the shift fork presses the synchronizer spring and blocker ring toward the gear. This brings the blocker ring to nearly the same speed as the gear.

The friction created by the contact of the blocker ring and gear cone causes them to rotate at the same speed. When this occurs, the dog teeth on the blocker ring can become aligned with the teeth on the synchronizer sleeve. This allows the sleeve to slide over the blocker ring and over the dog teeth on the speed gear shoulder (Figure 4-14). This action locks the speed gear to the shaft and the gear is fully engaged. The blocking ring will only allow the sleeve to

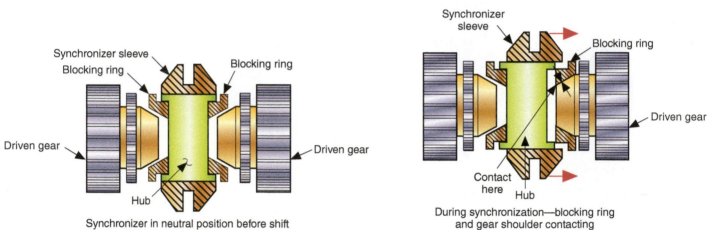

Synchronizer in neutral position before shift

FIGURE 4-12 A block synchronizer in its neutral position.

During synchronization—blocking ring and gear shoulder contacting

FIGURE 4-13 Initial contact of the blocking ring to the gear's cone.

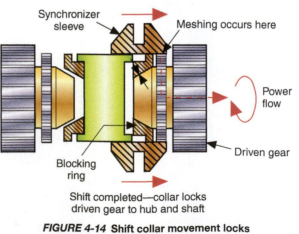

Shift completed—collar locks driven gear to hub and shaft

FIGURE 4-14 Shift collar movement locks the gear to the synchronizer hub.

mesh with the gear when its teeth are lined up with the dog teeth of the gear. If speeds do not match because of synchronizer problems or incorrect gear selection, the shift is "blocked."

Power now flows from input gear to the counter gear, back up to the speed gear locked by the synchronizer. Power then flows from the gear through the locking teeth to the sleeve, then to the hub, and finally to the main shaft.

To disengage a gear, the shifter is moved to the neutral position, which causes the synchronizer sleeve to move away from the previous gear, thereby disconnecting it from the shaft.

In summary, synchronization occurs in three stages. In the first stage, the movement of the sleeve causes the inserts to place light pressure on the blocking ring onto the cone of the gear. In the second stage, the sleeve contacts the blocking ring teeth and this causes additional friction between the ring and the speed gear cone. In the third stage, the synchronizer ring completes its fit over the gear cone and the gear is brought up to the same speed as the synchronizer assembly. The sleeve slides onto the gear's teeth and locks the gear to its synchronizer assembly and to the main shaft.

The locking teeth on a speed gear are also called the gear's dog or clutch teeth.

Advanced Synchronizer Designs

Synchronizers act as clutches to speed up or slow down the gear sets that are being shifted to, and greater friction area results in easier shifting for the driver. This is the reason manufacturers use multiple cone-type synchronizers in their transmissions. Many transmissions are fitted with single-cone, double-cone, or triple-cone synchronizers on their forward gears. For example, first and second gears may have triple-cone synchronization, third and fourth may have double-cone, and fifth and sixth have single-cone.

Double-cone synchronizers (Figure 4-15) have friction material on both sides of the synchronizer rings. The extra friction surfaces spread the workload of accelerating and decelerating driven gears to the desired speed over a larger area. This results in decreased shift effort and greater synchronizer durability. Triple-cone synchronizers provide a third surface in the synchronizer assembly lined with friction material.

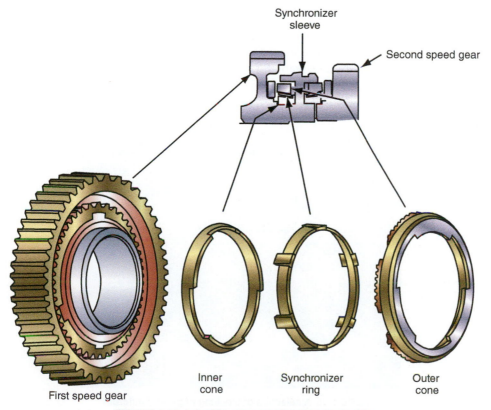

FIGURE 4-15 A double-cone synchronizer assembly.

With these multiple-cone synchronizers, the size of the transmission can be reduced. The multiple-cone synchronizers offer a high synchronizer capacity in a smaller package. To obtain the same results in shifting, a single-cone synchronizer would need to have a larger diameter, which would increase the overall size and weight of the transmission.

Stepped Clutch Teeth. Stepped clutch teeth on the synchronizer sleeves engage the gears sequentially in some transmissions. This makes gear engagement feel lighter and more positive.

Lengthened Sleeves. Synchronizer sleeves can be lengthened, which will reduce the distance the synchronizers must travel when engaging a gear. This reduces the shift throw distance and provides for quicker gear changes.

Materials. A variety of materials are used as the friction surface on today's blocking rings. Most use sintered bronze, whereas others use carbon and molybdenum. The latter two materials are extremely durable friction surfaces that remain stable even under extreme heat. Some synchronizers are made of a paper friction material, which offers more durability and clash resistance than brass. It is common to find different materials used for the different synchronizers in a transmission (Figure 4-16). For example, in the commonly used five-speed truck transmission, the Aisin AR-5, there are sintered bronze double-cone blocker rings on the synchronizers for first and second gears, whereas third and fourth gears use carbon fiber blocker rings, and fifth and reverse gears use molybdenum on their synchronizers.

The development of different blocking ring friction surfaces has been accompanied by new requirements for transmission lubricant. Many manufacturers specify fluids with designed friction characteristics that help the synchros do their job better. The use of incorrect transmission fluid can have a negative effect on shift quality.

Zeroshift. Zeroshift is the trade name for a new gear change system from a company in the United Kingdom. The system replaces normal synchronizers to provide a change in gear ratios without interrupting the flow of engine torque to the wheels. The system relies on two drive rings that are moved by small shift forks. The shift forks can be moved by electrical solenoids or by hydraulics or pneumatics. The rest of the transmission is conventional.

The system replaces synchromesh components with two drive rings with alternating faces sharing a common hub. Each of the drive rings has a set of "bullets" for shifting. Each bullet is sprung and has a ramp face and a retention face. When the transmission is in first gear and the vehicle is moving, one set of bullets transmits torque to the dog teeth of first gear through the retention faces. When second gear is selected, the second set of bullets is moved into position on the second gear and, for an instant, both first and second gears are engaged. Second gear

FIGURE 4-16 A selection of blocking rings with different friction surfaces.

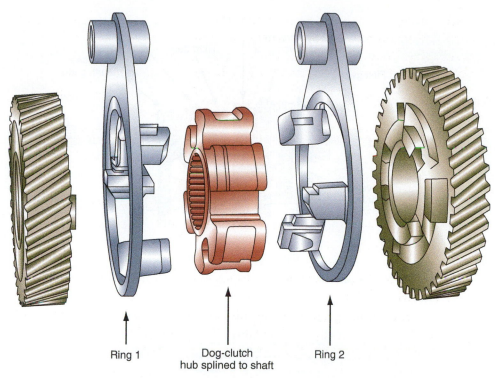

Ring 1   Dog-clutch   Ring 2
          hub splined to shaft

FIGURE 4-17 The major components in a Zeroshift gear change system.

then immediately overdrives the hub, which pushes the first set of bullets to disengage via the ramp faces (Figure 4-17).

This action provides a seamless transition when changing gears. It also improves overall performance, shift quality, and fuel economy. In addition, it reduces CO_2 emissions. All of these result from the fact that no fuel is wasted while gear changes are made since there is no time when the engine is not driving the wheels.

Dog Ring Shifters. Some racing transmissions do not require synchronization between shifts. Instead, dog rings, or face rings, connect the individual speed gears to hubs on the output shaft. One dog ring is connected to, or part of, the speed gear. The matching ring is splined to a hub on the output shaft and is moved into position by a shift fork. This shifting arrangement allows for shifting without the clutch if desired, and shift times are dramatically reduced without a synchronizer assembly to slow things up. Dog ring shifters are not designed for street use.

> **AUTHOR'S NOTE:** Many racing transmission components, including dog ring shifter parts, may be cryogenically treated to increase durability and strength. In this process components are placed in a deep dry freezer that uses flashed liquid nitrogen. Temperatures reach negative 300°F, and the process can take up to 72 hours. The molecular structure of the part is changed during the process.

TRANSMISSION DESIGNS

Nearly all automotive transmissions are equipped with a varied number of forward speed gears and one reverse speed. Transmissions can be divided into groupings according to the number of forward gears (speeds) it has.

Four-speed transmissions were common until the early 1980s (Figure 4-18). Five-speed and six-speed transmissions and transaxles are the most commonly used units. Some of the earlier units were actually four-speeds with an add-on fifth or overdrive gear. Many late-model

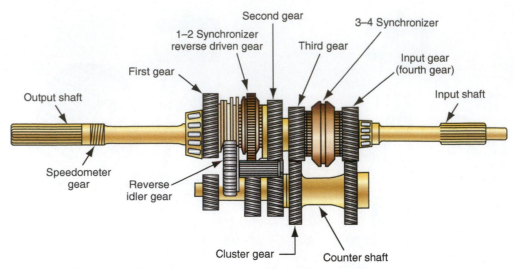

FIGURE 4-18 The gear and shaft layout for a typical four-speed transmission.

GEAR	GEAR RATIO	FINAL DRIVE	OVERALL RATIO
1st	3.66:1	4.06:1	14.86:1
2nd	2.37:1	4.06:1	9.62:1
3rd	1.75: 1	4.06:1	7.11:1
4th	1.32:1	4.06:1	5.36:1
5th	0.98:1	4.06:1	3.98:1
6th	0.73:1	4.0 6:1	2.96:1

FIGURE 4-19 The speed gears, final drive, and overall gear ratios for a Mitsubishi Lancer Ralliart six-speed transmission.

units incorporate fifth gear in their main assemblies. This is also true of six-speed transmissions and transaxles. The fifth and sixth gears of these units are part of the main gear assembly. Most often the two high gears in a six-speed provide overdrive gear ratios (Figure 4-19). Again, overdrive reduces engine speed at a given speed. This typically improves fuel economy and lowers engine wear and noise.

AUTHOR'S NOTE: Currently the only available vehicle that does not have multiple forward speeds is the Tesla, which is a pure electric vehicle. The manufacturer has chosen to use a single-speed transmission. This is possible because the torque from electric motors is at its highest level when its speed is at its lowest level. Therefore, its behavior accomplishes the same thing as a change in gear ratios, as speed increases there is a reduction in torque. Think about this while trying to understand it. This is a fact that may be very important in the future.

AUTHOR'S NOTE: The top gear in a transmission has been called the "high" gear, simply because it achieves the highest vehicle speed. This is not necessarily true today. Many high-performance vehicles achieve their top speed while running in the gear below the top gear. This results from the air pushing against the vehicle, at speed, and the need for torque to push the vehicle. The top gear is mostly used to decrease fuel consumption and emissions.

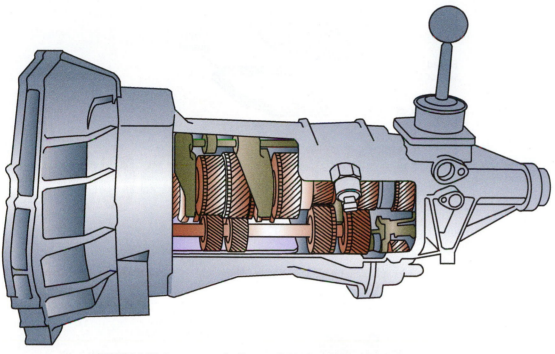

FIGURE 4-20 A cutaway of a Tremec T-56 six-speed manual transmission.

A five-speed transmission is usually a four-speed plus an overdrive or fifth gear (see Figure 4-6). The 1:1 ratio of fourth gear is usually retained, and fifth gear is made into an overdrive. Typical fifth-gear overdrive ratios range from 0.70 to 0.90:1. Such ratios greatly reduce engine rpm at freeway speeds and increase fuel mileage and engine life.

Six-speed transmissions are becoming more popular. Previously they were only found in performance cars (Figure 4-20), but with environmental and fuel economy concerns they can now be found in many vehicles. These units are sometimes based on a five-speed transmission and have an extra gear set to provide for a sixth gear.

BASIC OPERATION OF MANUAL TRANSMISSIONS

All manual transmissions function in much the same way and have similar parts. (Although transaxles operate in basically the same way as transmissions, there are enough differences to warrant a separate discussion. The operation of transaxles is explained after this section on transmissions.) Before you can completely understand how a transmission works, you must be familiar with the names, purposes, and descriptions of the major components of all transmissions.

The transmission case normally is a cast aluminum case that houses most of the gears in the transmission. The housing is shaped to accommodate the gears within the transmission. It also has machined surfaces for a cover, rear extension housing, and mounting to the clutch's bell housing (Figure 4-21). On transmissions, the clutch bell housing may be a separate stamped or cast metal part, which houses the clutch and connects the transmission to the back of the engine. Most bell housings have a pivot point mounted to it where the clutch throw-out bearing lever arm is mounted. On many transmissions and most transaxles, the bell housing is part of the transmission housing.

The front bearing retainer is an aluminum or cast-iron piece bolted to the front of the transmission case. The retainer serves several functions: It houses an oil seal that prevents oil from leaking out the input shaft and into the clutch disc area, it holds the input shaft bearing rigid, and because the input shaft is pressed into the bearing, it prevents undesired axial movement of the input shaft (Figure 4-22).

A BIT OF HISTORY

The first American car to use a four-speed gearbox was the Locomobile in 1903. In 1934, Chrysler introduced the first transmission with overdrive.

Shop Manual
Chapter 4, page 138

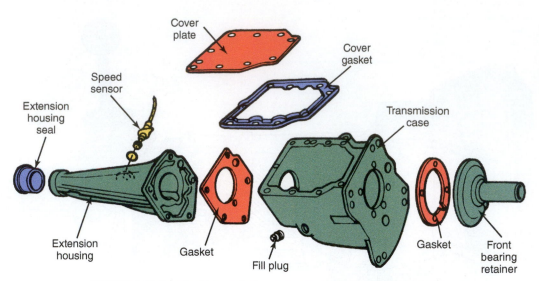

FIGURE 4-21 A typical transmission case component with attachments.

The **counter gear assembly** is a one-piece machined unit, containing first, second, third, and fourth counter gears. The fifth-speed counter gear is often a separate assembly that is splined to the countershaft.

The counter gear assembly is often referred to as the **cluster gear** assembly.

Reverse idler gear in a transmission is an additional gear that must be meshed to obtain reverse gear; it is a gear used only in reverse that does not transmit power when the transmission is in any other position.

Shift forks move the synchronizer sleeves to engage and disengage gears.

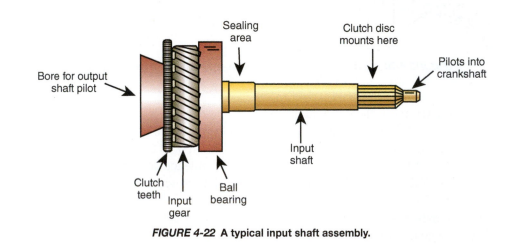

FIGURE 4-22 A typical input shaft assembly.

The **counter gear assembly** is normally located in the lower portion of the transmission case and is constantly in mesh with the input gear. The counter gear assembly has several different-sized gears on it that rotate as one solid assembly. Normally the counter gear assembly is supported in the case by ball, roller, or tapered roller bearings on each end. The counter gear on some older transmissions rotates on a countershaft using roller or needle bearings (Figure 4-23).

The speed gears are located on the main shaft (Figure 4-24). These gears are not fastened to the main shaft and rotate freely on the main shaft journals. The gears are constantly in mesh with the gears on the counter gear assembly. On two-wheel-drive vehicles a tone ring or worm gear is mounted on the output shaft to allow for speed sensor operation. These components will be on the transfer case output shaft on four-wheel-drive vehicles. The outer end of the main shaft has splines for the slip-joint yoke of the drive shaft.

A reverse gear is not meshed with the counter gear like the forward gears; rather, the **reverse idler gear** is meshed with the counter gear (Figure 4-25). Normally, reverse gear is engaged by sliding it into mesh with the reverse idler gear. The addition of this third gear causes the reverse gear to rotate in the opposite direction as the forward gears.

Most **shift forks** have two fingers that ride in the groove on the outside of the synchronizer sleeve. The forks are bored to fit over shift rails. Roll pins are commonly used to fasten the shift forks to the rails (Figure 4-26). Each of the shift rails have shift lugs that the shift lever fits into. These lugs are also fastened to the rails by roll pins. The shift rails slide back and forth

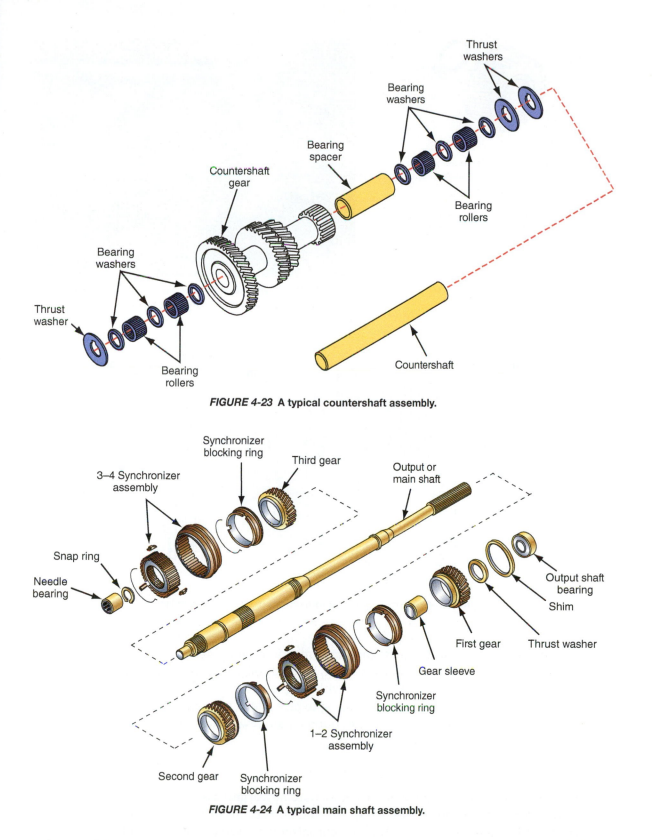

FIGURE 4-23 A typical countershaft assembly.

FIGURE 4-24 A typical main shaft assembly.

in bores of the transmission case. Each shift rail has three notches, in which a spring-loaded ball or bullet rides to give a detent feel to the shift lever and precisely locate the proper position of the shift fork during gear changes. The shift rails also have notches cut in their sides for interlock plates or pins to fit in. Interlock mechanisms prevent the engagement of two gears at the same time. The lower portion of the shifter assembly fits into the shift rail lugs and moves the shift rails for gear selection.

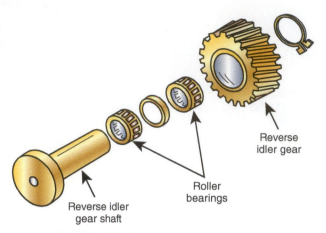

FIGURE 4-25 A reverse idler gear assembly.

FIGURE 4-26 The shift forks, riding on shift rails, fit into the grooves of the synchronizer sleeve.

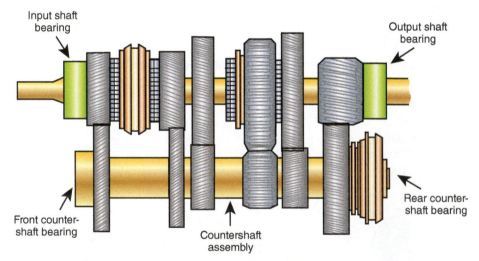

FIGURE 4-27 Location of various bearings in a typical transmission.

The arrangement of the shafts in a transmission require bearings to support them. The ends of the shafts are fitted with large roller, tapered roller, or ball bearings pressed onto the shaft (Figure 4-27). Some transmissions with long shafts use an intermediate bearing that is fitted into the intermediate bearing housing to give added strength to the shaft. Small roller or needle bearings are often used on the countershaft, reverse idler gear shaft, and at the connection of the output shaft to the input shaft.

Operation of a Five-Speed Transmission

In a typical five-speed transmission, all five forward helical gear assemblies are in constant mesh. They are activated by the first/second speed synchronizer, the third/fourth speed synchronizer, and the fifth speed synchronizer. Each synchronizer is activated by its own shift fork. All three shift forks may slide along a transmission's single shift rail. Other transmissions may have a separate shift rail for each synchronizer assembly. The reverse idler gear slides along another rail to engage the reverse spur gear. The transmission's floor-mounted shift lever is spring loaded, so the driver must push down or pull up on the lever in order to engage reverse gear.

With the gears in constant mesh, all the gears will rotate when the input shaft is supplying input power. However, the gears will not transfer power to the main shaft until one of the synchronizers engages with a gear. If the gears are not engaged by a synchronizer, they are free-wheeling on the main shaft. The individual output shaft gears are mechanically locked to the output shaft only when the synchronizers are activated. At other times they rotate independently of the output shaft.

Power-Flow in Neutral

In the neutral position, the input shaft drives the countershaft gears but no power is transferred out of the transmission. Because they are in mesh, all of the gears on the main shaft rotate, but power is not transferred to the output shaft because the synchronizers are not engaged with any of the gears.

Power-Flow in First Gear

In the first gear (Figure 4-28), power enters the transmission through the input shaft and rotates the countershaft gear. The first/second synchronizer sleeve is engaged with the dog teeth on the first-speed gear, locking the gear to the main shaft. The power coming in the input shaft transfers through the counter gear and up into the first gear. The gear rotates the synchronizer sleeve, which rotates the hub and main shaft for output power. All other gears mounted on the main shaft rotate freely.

Power-Flow in Second Gear

In the second-gear position (Figure 4-29), the input shaft again drives the countershaft gear. The first/second synchronizer sleeve is moved to engage with the dog teeth of the second gear, locking it to the output shaft. Power comes in through the input shaft, down to the counter gear, and up to the second gear. The dog teeth of the second gear rotate the synchronizer sleeve, which rotates the hub and main shaft for output power.

Shop Manual
Chapter 4, page 165

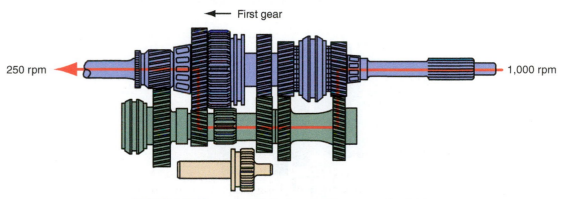

FIGURE 4-28 Power-flow in first gear with a gear ratio of 4:1.

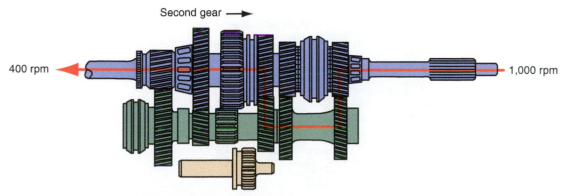

FIGURE 4-29 Power-flow in second gear with a gear ratio of 2.5:1.

Power-Flow in Third Gear

The third-gear position causes the countershaft gear, which is driven by the input shaft, to be mechanically connected to the third gear on the output shaft (Figure 4-30). The third/fourth synchronizer sleeve is moved to engage with the dog teeth of the third gear. The power comes from the input shaft to the counter gear and then to the third gear. The dog teeth on the third gear rotate the synchronizer sleeve, which rotates the hub and main shaft for output power.

Power-Flow in Fourth Gear

The fourth-gear position mechanically locks the output shaft to the input shaft (Figure 4-31). The third/fourth synchronizer sleeve is moved to engage with the dog teeth of the input gear. The power flows in from the input shaft, through the synchronizer sleeve and hub, and then through the main shaft for output power. This directly links the two shafts, and the output shaft rotates at the same speed as the input shaft to provide for direct drive. The counter gear is still driven by the input gear, but is not transferring power to the main shaft.

Power-Flow in Fifth Gear (Overdrive)

The fifth-gear position causes the countershaft gear, which is driven by the input shaft, to rotate the fifth gear (Figure 4-32). The fifth-gear synchronizer sleeve is moved to engage with the dog teeth on the fifth gear. The power on the input gear transfers to the counter gear and then to the fifth gear. Power transfers through the synchronizer sleeve and hub to the main shaft for output power. As a result, the output shaft rotates at a higher speed than the input shaft.

Power-Flow in Reverse Gear

In the reverse-gear position, if the transmission has a synchronized reverse gear, the reverse synchronizer sleeve moves to engage with the reverse gear. Power comes in through the input shaft, into the counter gear, through the reverse idler gear, and into the reverse gear on the

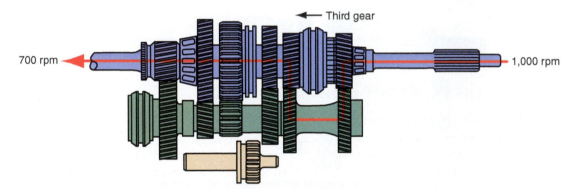

FIGURE 4-30 Power-flow in third gear with a gear ratio of 1.43:1.

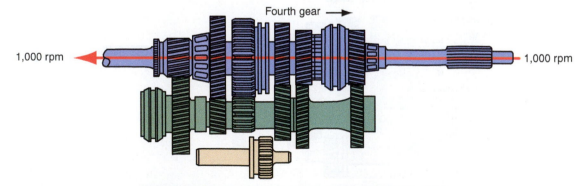

FIGURE 4-31 Power-flow in fourth gear with a gear ratio of 1:1.

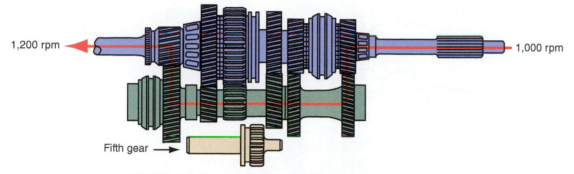

1,200 rpm

1,000 rpm

Fifth gear →

FIGURE 4-32 Power-flow in fifth gear with a gear ratio of 0.83:1.

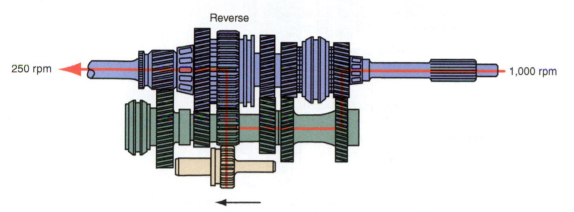

Reverse

250 rpm

1,000 rpm

FIGURE 4-33 Power-flow in reverse gear with a gear ratio of 4:1.

main shaft. The reverse gear rotates the synchronizer sleeve, which rotates the hub and main shaft in a reverse direction.

If reverse is a nonsynchronized gear, a reverse-gear shift relay lever slides the reverse idler gear into contact with the countershaft reverse gear and the reverse gear on the output shaft (Figure 4-33). The reverse idler gear causes the reverse output shaft gear to rotate counterclockwise.

BASIC TRANSAXLE OPERATION

The transmission section of a transaxle is practically identical to RWD transmissions. It provides for torque multiplication, allows for gear shifting, and is synchronized. They also use many of the design and operating principles found in transmissions. However, a transaxle also contains the differential gear sets and the connections for the drive axles (Figure 4-34).

Transaxles normally use fully synchronized, constant mesh helical gears for all forward speeds and either helical or spur gears for reverse. To keep a transaxle compact, many designs use pressed-fit synchronizer hubs and narrower gears.

In a transaxle, the transmission and differential are both located in a lightweight housing bolted to the engine. A pinion gear is machined onto the end of the transaxle's output shaft and is in constant mesh with the differential ring gear (Figure 4-35). When the output shaft rotates, the pinion gear causes the differential ring gear to rotate at the desired final drive ratio. The ring gear is attached to the differential case, which drives the differential pinion shaft, differential pinion gears, and side gears. The side gears drive the two half-shafts.

Reverse is usually engaged by moving a sliding idler gear arrangement rather than a sliding synchronizer collar (Figure 4-36), although some five-speed units have synchronized gears in all forward gears and in reverse. The sliding gears have splines through their center bore and rotate at the speed of the shaft. There is no speed-matching action, so nonsynchronized gears tend to clash when shifted into mesh.

Shop Manual
Chapter 4, page 166

A BIT OF HISTORY

Years ago, a new convenience for the driver was introduced: a column-mounted gear shifter. For many years the shifter was mounted to the floor and took up valuable space in the interior of cars. By mounting the shifter on the column, that floor space was reclaimed. This trend was reversed in the 1960s and today column-shifted manual transmission vehicles are a rarity.

The input shaft transmits the engine's power to the transmission or transaxle's gears.

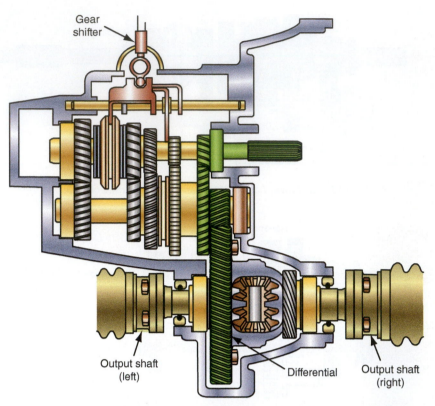

FIGURE 4-34 A typical transaxle assembly.

Gear shifter

Output shaft (left)

Differential

Output shaft (right)

FIGURE 4-35 A transaxle output shaft with integral pinion gear.

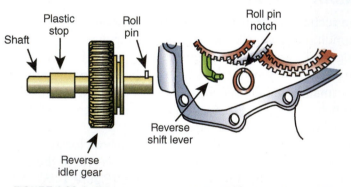

Shaft

Plastic stop

Roll pin

Roll pin notch

Reverse shift lever

Reverse idler gear

FIGURE 4-36 A slider-type reverse gear used in many transaxles.

The output shaft transmits torque from the transmission to the drive axles via a drive shaft.

Power-Flow in Neutral

When a transaxle is in its "neutral" position, no power is applied to the differential. Because the synchronizer collars are centered between their gear positions, the meshed drive gears are not locked to the output shaft. Therefore, the gears spin freely on the shaft and the output shaft does not rotate.

104

Power-Flow through Forward Gears

When first gear is selected (Figure 4-37), the first and second gear synchronizer engages with first gear. Because the synchronizer hub is splined to the output shaft, first gear on the input shaft drives its mating gear (first gear) on the output shaft. This causes the output shaft to rotate at the ratio of first gear. The pinion gear machined at the end of the output shaft drives the differential ring gear and case assembly.

As the other forward gears are selected, the appropriate shift fork moves to engage the synchronizer with the gear. Because the synchronizer's hub is splined to the output shaft, the desired gear on the input shaft drives its mating gear on the output shaft. This causes the output shaft to rotate at the ratio of the selected gear. The pinion gear is machined on the output shaft drives the differential ring gear and case assembly (Figure 4-38).

Power-Flow in Reverse

When reverse gear is selected on transaxles that use a sliding reverse gear (Figure 4-39), the shifting fork forces the gear into mesh with the input and output shafts. The addition of this third gear reverses the normal rotation of first gear and allows the car to change direction.

Differential Action

The final drive ring gear is driven by the transaxle's output shaft. The ring gear then transfers the power to the differential case. One major difference between the differential in a RWD car and the differential in a transaxle is power-flow. In a RWD differential, the power-flow changes 90 degrees between the drive pinion gear and the ring gear. This change in direction is not needed with most FWD cars. The transverse engine position places the crankshaft so that it already is rotating in the correct direction. Therefore, the purpose of the differential is only to provide torque multiplication and divide the torque between the drive axle shafts so that they can rotate at different speeds during cornering (Figure 4-40).

The ring gear and differential case are driven by the output shaft of the transmission. The differential pinion shaft is pinned to the case. Two differential pinion gears are free to rotate

Some transaxles need the 90-degree power-flow change in the differential. These units are used in rear-engine with RWD applications or in longitudinally positioned engines with FWD. Audi, Subaru, Volkswagen, and Porsche are good examples of these types of transaxles.

Constant velocity (CV) joints are much like the universal joints used in the drive shaft of RWD cars. However, the speed of the driven side of the U-joint may vary with relation to the driving side, depending on the angle of the shaft. A CV joint maintains an equal speed on both sides of the joint, which helps reduce vibration and wear.

Shop Manual
Chapter 4, page 172

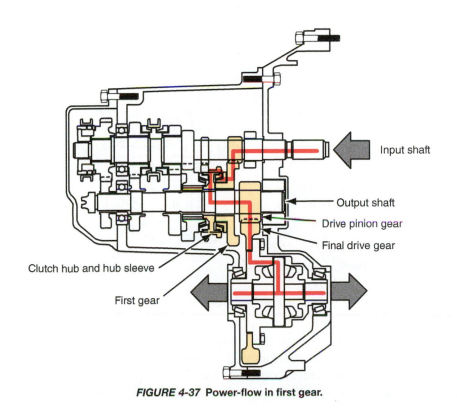

FIGURE 4-37 Power-flow in first gear.

Input shaft

Output shaft

Drive pinion gear

Final drive gear

Clutch hub and hub sleeve

First gear

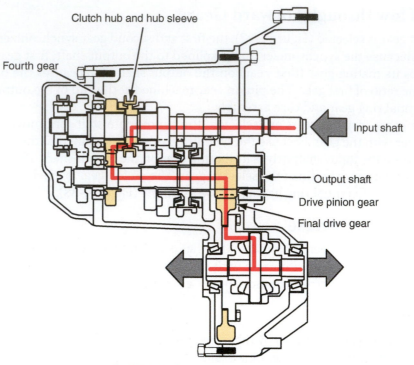

FIGURE 4-38 Power-flow in fourth gear.

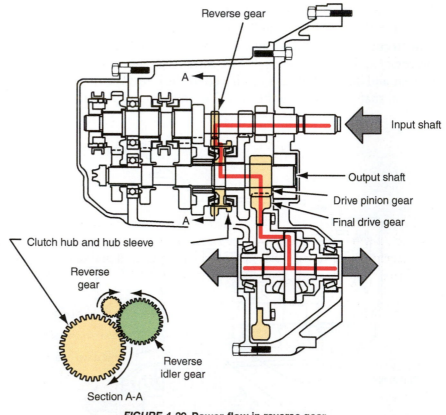

FIGURE 4-39 Power-flow in reverse gear.

on the pinion shaft and transmit their motion to the two side gears. The differential side gears are connected to inboard constant velocity (CV) joints by splines and are secured in the case with circlips. The drive axles extend out from each side of the differential to rotate the car's wheels. The axles are made up of three pieces connected together to allow the wheels to turn

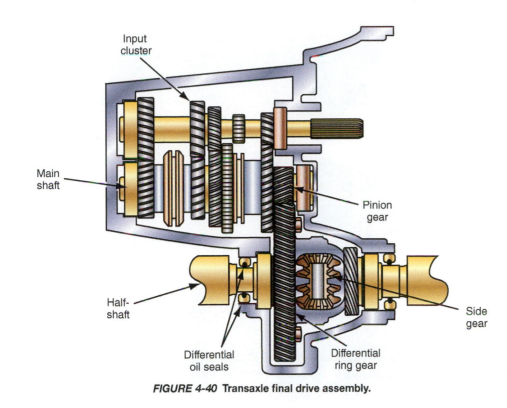

Input cluster

Main shaft

Pinion gear

Half-shaft

Side gear

Differential oil seals

Differential ring gear

FIGURE 4-40 Transaxle final drive assembly.

Shift rails are machined with **interlock** and **detent** notches. The interlock notches prevent the selection of more than one gear during operation. The detent notches give the driver a positive shift feel when engaging the speed gears.

for steering and to move up and down with the suspension. In some models a short stub shaft extends from the differential to the inner CV joint. An axle shaft connects the inner CV joint and the outer CV joint. A short spindle shaft that fits into the hub of the wheels extends from the outer CV joint. To keep dirt and moisture out of the CV joints, a neoprene boot is installed over each CV joint assembly.

GEARSHIFT LINKAGES

There are two basic designs of gearshift linkages, some internal and others external to the transmission. Movement of the shifter moves a **shift rail** and shift fork toward the desired gear and moves the synchronizer sleeve to lock the speed gear to the output shaft.

As the rail moves, a **detent** ball or plunger moves out of its detent notch and drops into the notch for the selected gear as the rail moves. The detent assembly precisely locates the rail and shift fork and provides feel to the driver that a shift has been made (Figure 4-41). At the same time, an **interlock** pin moves out of its interlock notch and prevents the other shift rails from moving so that the transmission cannot be in two gears at once (Figure 4-42).

External Linkages

External linkages (Figure 4-43) are mounted to the outside of the transmission housing. The shifter protrudes from the back of the housing into the vehicle's interior. Linkage arms or rods connect externally to levers connected to the internal shift rails of the transmission.

In a transaxle, the external shift assembly normally consists of a floor-mounted shift lever that pivots through the shifter boot and is held in place by a stabilizer assembly and bushing (Figure 4-44). Shift lever motion is transmitted to the internal shift mechanism by a shift rod that is connected to and operates the transaxle input shift shaft.

Some transaxles use a two-cable assembly to shift the gears. One cable is the transmission selector cable and the other is a shifter cable (Figure 4-45). The selector cable activates the desired shift fork and the shifter cable moves the rail and shift fork to achieve engagement of the desired gear.

Shop Manual
Chapter 4, page 149

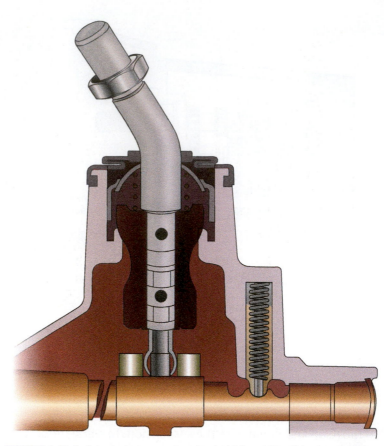

FIGURE 4-41 A detent plunger and spring hold the shift rail in the correct position and give the driver a positive shift feel when engaging a speed gear.

The right interlock pin is moved by the 1–2 shift rail into the 3–4 shift rail slot.

The 3–4 shift rail pushes both the interlock pins outward into the slots of the 5-R and 1–2 shift rails.

The right interlock pin is moved by the lower tab of the left interlock pin into the 1–2 shift rail.

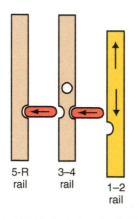

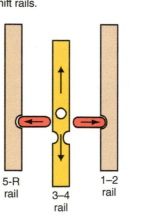

The left interlock pin is moved by the lower tab of the right interlock pin into the 5-R shift rail slot.

The left interlock pin is moved by the 5-R shift rail into the 3–4 shift rail slot.

FIGURE 4-42 Interlock pins prevent the selection of more than one gear at a time.

Remote Transmissions. Vehicles that have a transaxle mounted directly to the rear differential use a shift shaft located at the front of the transmission. This shaft extends from the transmission to the shifter inside the car. On some later-model vehicles, shifting is done through electronically controlled solenoids. This setup reduces the bulk of the shift rod and offers precise shifting.

FIGURE 4-43 An external shifter assembly mounted to a transmission.

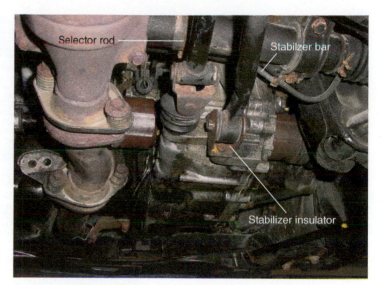

Selector rod

Stabilzer bar

Stabilizer insulator

FIGURE 4-44 The external linkage for a transaxle.

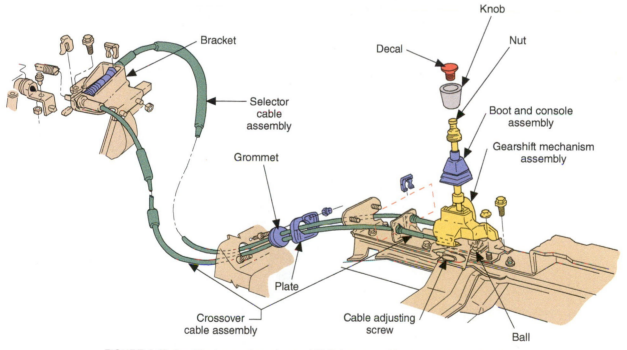

Bracket

Knob

Nut

Decal

Selector cable assembly

Boot and console assembly

Grommet

Gearshift mechanism assembly

Plate

Cable adjusting screw

Ball

Crossover cable assembly

FIGURE 4-45 A cable-type external gearshift linkage used in a transaxle application.

Internal Linkages

Internal linkages typically are connected to the transmission at the top of the unit through a shift tower, and to a transaxle at the top, rear, or side of the unit. The control end of the shifter is mounted inside the transmission, as are all of the shift controls. In some transmission designs the shifter selects individual shift rails mounted in the transmission cover assembly (Figure 4-46). Other designs may have the shifter contact the shift rails in the rear of the transmission.

The driver changes gears on a typical transaxle by moving the shift lever to move the internal shift mechanisms of the transaxle. Movement of the shift lever is transmitted to the main shift control shaft and forks to select and engage the forward gears. An internal shift

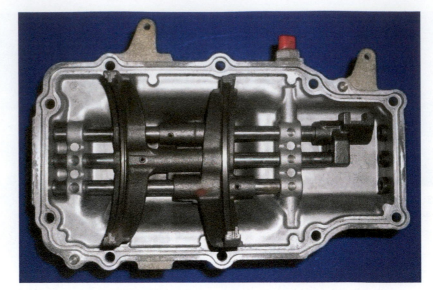

FIGURE 4-46 A cover-mounted internal shift system.

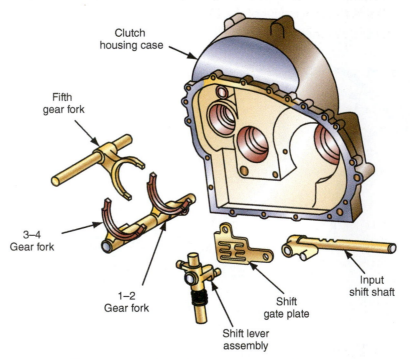

FIGURE 4-47 The internal shift mechanism for a typical transaxle.

mechanism assembly transfers shift lever movements through the main shift control shaft assembly and the shifting forks. A reverse relay lever assembly is commonly used to engage the reverse idler gear with the reverse sliding gear and the input shaft (Figure 4-47).

SUMMARY

- The purpose of a transmission is to use various-sized gears to keep the engine in an efficient operating range through all vehicle speeds and loads.
- A transmission is a system of gears that transfers the engine's power to the drive wheels of the car.
- A manual transaxle is a single unit composed of a transmission, differential, and final drive.

- Three major types of manual transmissions have been used by the automotive industry: the sliding gear, collar shift, and synchromesh designs.
- A synchromesh transmission is a constant mesh, collar shift transmission equipped with synchronizers.
- A synchronizer's primary purpose is to bring components that are rotating at different speeds to one synchronized speed.
- Block synchronizers consist of a hub, sleeve, blocking ring, and inserts.
- Synchronization occurs in three stages. In the first stage, the movement of the sleeve causes the inserts to press the blocking ring onto the cone of the gear. In the second stage, the sleeve makes direct contact with the blocking ring teeth for increased friction on the cone of the gear. In the third stage, the synchronizer ring completes its friction fit over the gear cone and the gear is brought up to the same speed as the synchronizer assembly. The sleeve slides onto the gear's teeth and locks the gear through its synchronizer assembly to the main shaft.
- Most current transmissions have five or six forward speeds.
- The front bearing retainer of a transmission serves to house an oil seal for the input shaft, holds the input shaft rigid, serves as a centering and holding fixture for the clutch release bearing, and limits the movement of the input shaft.
- The counter gear assembly in a RWD transmission is in constant mesh with the speed gears.
- The speed gears are located on the main shaft.
- Reverse gear is not meshed with the counter gear and is normally engaged by sliding the reverse gear into the reverse idler gear.
- Many FWD transaxles have three separate shafts.
- Shift forks are moved by the shift mechanism and move the synchronizer sleeves to engage a gear.
- There are two basic types of gear shift linkages: internal and external.
- One major difference between the differential of a RWD car and the differential of a transverse FWD car is power-flow. In a RWD differential, power-flow changes 90 degrees between the drive pinion gear and the ring gear.

TERMS TO KNOW

Block synchronizers
Cluster gear
Clutch gear
Clutch hub
Collar shift
Constant mesh
Counter gear assembly
Detent
Dog teeth
Input shaft
Interlock
Output shaft
Reverse idler gear
Shift forks
Shift rails
Sliding gear
Synchromesh

REVIEW QUESTIONS

Short Answer Essays

1. Define the purpose of a transmission.
2. What is the primary purpose of a synchronizer?
3. Describe the three stages of synchronization.
4. List the functions of a transmission's front bearing retainer.
5. How is reverse speed obtained in most transmissions?
6. Define the primary differences between a transmission and a transaxle.
7. Describe and explain the major difference between the differential of a RWD car and the differential of a FWD car.

8. What is the primary purpose of the clutch gear in a transmission?
9. Describe the power-flow through a transaxle when it is in first gear.
10. What is the purpose of fifth gear in most transaxles and transmissions?

Fill-in-the-Blanks

1. A transmission is a system of _____ that transfers the _____ to the _____ of the car.
2. A manual transaxle is a single unit composed of a _____, _____, and _____.

3. Block synchronizers consist of a _____, _____, _____ _____, and _____ .

4. _____ _____ are moved by the gear shift and move the synchronizer sleeves to engage a gear.

5. The speed gears are located on the _____.

6. A synchromesh transmission is a _____ _____, _____ _____ transmission equipped with synchronizers.

7. Smooth and quiet shifting can only be possible when the _____ and the _____ are rotating at the same speed.

8. The output shaft is also called the _____ shaft.

9. All synchronizers use _____ to synchronize the speed of the gear and shaft before the connection is made.

10. If the speed gears are not engaged by a synchronizer, they _____ on the main shaft.

ASE-Style Review Questions

1. When discussing the operation of a synchronizer,
   Technician A says that the synchronizer inserts force the blocking ring against the conical face of the gear, which slows the speed of the gear and allows for engagement.
   Technician B says that by matching the speed of the blocking ring and the gear, the synchronizer sleeve is able to engage with the gear dog teeth.
   Who is correct?
   A. A only
   B. B only
   C. Both A and B
   D. Neither A nor B

2. When discussing shift mechanisms,
   Technician A says that the detent system prevents the engagement of more than one gear at a time.
   Technician B says that the interlock system places the shift rail and fork in an exact position.
   Who is correct?
   A. A only
   B. B only
   C. Both A and B
   D. Neither A nor B

3. When discussing current trends in manual transmissions,
   Technician A says that six- and seven-speed transmissions reduce fuel economy because of the weight of their extra parts.
   Technician B says that some manual transmissions are shifted by internal solenoids and do not require a shift lever.
   Who is correct?
   A. A only
   B. B only
   C. Both A and B
   D. Neither A nor B

4. While discussing synchronizers,
   Technician A says that in automotive transmissions, the commonly used type of synchronizer is the block type.
   Technician B says that triple-cone synchronizers are not as effective because they are smaller than single-cone synchronizers.
   Who is correct?
   A. A only
   B. B only
   C. Both A and B
   D. Neither A nor B

5. When discussing overdrive gears,
   Technician A says that the typical ratio is 0.70 to 0.90:1.
   Technician B says that the typical ratio is 1.0:1.
   Who is correct?
   A. A only
   B. B only
   C. Both A and B
   D. Neither A nor B

6. When discussing the power-flow through a five-speed transmission when it is in first gear,
   Technician A says that power enters in on the input shaft, which rotates the countershaft that is engaged with first gear.
   Technician B says that the first gear synchronizer engages with the clutching teeth of first gear and locks the gear to the main shaft, allowing power to flow from the input gear through the countershaft and to first gear and the main shaft.
   Who is correct?
   A. A only
   B. B only
   C. Both A and B
   D. Neither A nor B

7. While discussing power-flow through the forward gears of a typical transaxle,

   Technician A says the main speed gears freewheel around the output shaft unless they are locked to the shaft by the synchronizers.

   Technician B says the main speed gears are in constant mesh with their mating gears on the input shaft and rotate whenever the input shaft rotates.

   Who is correct?

A.	A only	C.	Both A and B
B.	B only	D.	Neither A nor B

8. While discussing reverse gear,

   Technician A says if the transmission has a synchronized reverse gear, the reverse synchronizer sleeve moves to engage with reverse gear. Power comes in through the input shaft, into the counter gear, through the reverse idler gear, and into the reverse gear on the main shaft. The reverse gear rotates the synchronizer sleeve, which rotates the hub and main shaft in a reverse direction.

   Technician B says if reverse is a nonsynchronized gear, a reverse-gear shift relay lever slides the reverse idler gear into contact with the countershaft reverse gear and the reverse gear on the output shaft. The reverse idler gear causes the reverse output shaft gear to rotate counterclockwise.

   Who is correct?

A.	A only	C.	Both A and B
B.	B only	D.	Neither A nor B

9. While calculating the first gear ratio of a five-speed transmission,

   Technician A says to first divide the number of teeth on the front of the counter gear by the number of teeth on the clutch gear.

   Technician B says to next divide the number of teeth on the first speed gear by the number of teeth on the matching counter gear and then multiply the answer from the first set by the second.

   Who is correct?

A.	A only	C.	Both A and B
B.	B only	D.	Neither A nor B

10. While discussing shift forks,

    Technician A says most shift forks have two fingers that ride in the grooves on the outside of the speed gears.

    Technician B says each shift rail has three notches, which a spring-loaded ball or plunger rides in to give a detent feel to the shift lever and locates the proper position of the shift fork.

    Who is correct?

A.	A only	C.	Both A and B
B.	B only	D.	Neither A nor B

FRONT DRIVE AXLES

UPON COMPLETION AND REVIEW OF THIS CHAPTER, YOU SHOULD BE ABLE TO:

- Explain the purposes of a FWD car's drive axles and joints.
- Understand and describe the different methods used by manufacturers to offset torque steer.
- Name and describe the different types of CV joints currently being used.

- Name and describe the different designs of CV joints currently being used.
- Explain how a ball-type CV joint functions.
- Explain how a tripod-type CV joint functions.

INTRODUCTION

Between 1975 and 1979, only 4–5 percent of the total vehicle population had front wheel drive. Today, FWD and all-wheel drive (AWD) accounts for over 80 percent of the vehicle population.

The drive axle assembly transmits torque from the engine and transmission to drive the vehicle's wheels. Front-wheel-drive (FWD) axles transfer engine torque from the transaxle differential to the front wheels. With the engine mounted transversely in the car, the differential does not need to turn the power flow 90 degrees to the drive wheels. However, on a few older FWD cars, the differential unit is a separate unit (like rear-wheel-drive [RWD] cars) and the power-flow must be turned. Both arrangements use the same type of drive axle (Figure 5-1). Some FWD cars have a longitudinally mounted transaxle.

Many AWD vehicles are based on FWD platforms. In these cases, the front drive axles transmit power through the differential to the wheels in the same manner as a FWD vehicle, and the principles in this chapter apply to these vehicles as well.

DRIVE AXLE CONSTRUCTION

The basic FWD driveline consists of two drive shafts. The shafts extend out from each side of the differential to supply power to the drive wheels. Steel tubes or solid bars of steel are used as the drive axles on the front of four-wheel-drive and front-wheel-drive cars. In all FWD

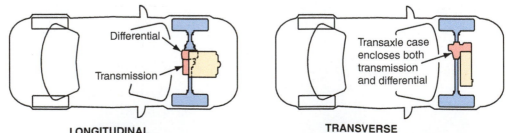

LONGITUDINAL
(front to rear mounted engine)

TRANSVERSE
(sideways mounted engine)

FIGURE 5-1 The two basic FWD engine and transaxle configurations.

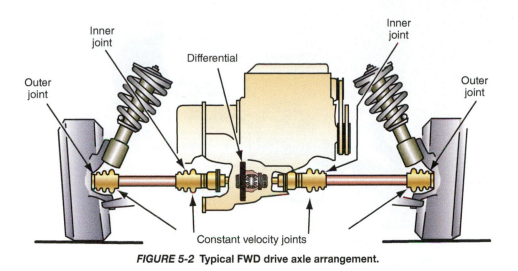

FIGURE 5-2 Typical FWD drive axle arrangement.

and some 4WD/AWD systems, the transaxle is bolted to the engine and the axles pivot on **CV joints** (Figure 5-2). The outer parts of the axles are supported by the steering knuckles that house the axle bearings. Steering knuckles serve as suspension components and as the attachment points for the steering gear, brakes, and other suspension parts.

The drive axles used with FWD systems are actually made up of three pieces attached together in such a way as to allow the wheels to turn and move with the suspension. The **axle shaft** is connected to the differential by the inboard CV joint. The axle shaft then extends to the outer CV joint. A short spindle shaft runs from the outer CV joint to mate with the wheel assembly (Figure 5-3). The hub and wheel bearing assembly connects to the spindle shaft of the outer CV joint.

The drive shafts, on a FWD vehicle operate at angles as high as 40 degrees for turning and 20 degrees for suspension travel (Figure 5-4). Each shaft has two CV joints, an inboard joint that connects to the differential, and an outboard joint that connects to the splines to a hub mounted with the wheel bearing in the steering knuckle. As a front wheel is turned during

A **CV joint** is a constant-velocity joint used to transfer a uniform torque and a constant speed while operating through a wide range of angles.

FWD drive axles are also called **axle shafts**, drive shafts, and half shafts.

The complete drive axle, including the inner and outer CV joints, is typically called a half shaft.

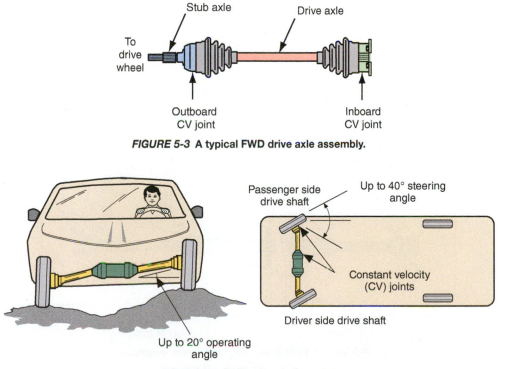

FIGURE 5-3 A typical FWD drive axle assembly.

FIGURE 5-4 FWD drive shaft angles.

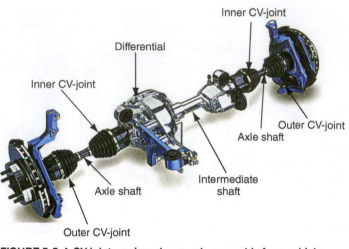

FIGURE 5-5 A CV joint-equipped rear axle assembly for a vehicle with independent rear suspension.

steering, the outboard CV joint moves with it around a fixed center. Up-and-down movements of the suspension system force the inboard joint to slide in and out.

CV joints are used on the front axles of many 4WD vehicles as well. They have also been used on RWD buses and cars that have the engine mounted in the rear, such as Porsches. Mid-engine cars with RWD, such as the Pontiac Fiero, Toyota MR2, Porsche Boxster, and Lotus Evora, also are equipped with CV joints. Some RWD cars with independent rear suspension (IRS), such as BMW, Nissan 300ZX and 370Z, Ford Thunderbird, and Chevrolet Camaro, use CV joints on their drive axles (Figure 5-5). Off-road vehicles and all-terrain vehicles also commonly use CV joints on their drive axles.

DRIVE AXLES

The drive axles from a transaxle extend from the sides out to the drive wheels. With the engine mounted transversely, the transaxle sits to one side of the engine compartment, typically to the left, or driver's side. This means the position of the differential gears is also off-center, which requires that one of the drive axles must be longer than the other (Figure 5-6). Unequal-length drive axles, however, have been found to deliver more torque to the wheel with the shorter shaft. Under light loads, the difference is insignificant. However, when the vehicle is

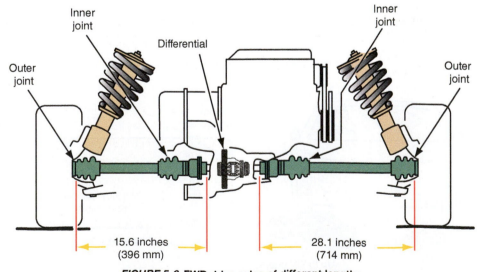

FIGURE 5-6 FWD drive axles of different lengths.

accelerated, the wheel with the shorter shaft has more torque, causing that wheel to pull harder than the other wheel. The stronger wheel tends to pull ahead of the other wheel, which creates the induced steering pull toward the opposite side. This pull is called **torque steer**.

Torque Steer

Due to the placement of the differential, the direction of torque steer in a vehicle with a left-mounted transaxle is usually toward the right. Torque steer describes a condition in which a vehicle tends to steer or pull in one direction as engine power is applied to the drive wheels. This condition increases with the power output of the engine. Therefore, it is more prominent on FWD vehicles with high-output engines. Basically, torque steer occurs when the CV joints of one half shaft operate at different angles than those on the other shaft. Torque steer is caused by many factors:

■ Front wheel alignment angles
■ Tire sidewall construction
■ Unequal tire pressure
■ Unequal amounts of road traction
■ Movement of the engine under loads
■ Unequal drive axle lengths
■ Faulty wheel bearings at one side
■ Dragging brakes on one side
■ Worn suspension components

Front Wheel Alignment Angles. Torque steer becomes more evident when the vehicle's wheels are not properly aligned. If one of the tires has less traction due to its contact with the road surface, it will receive more torque from the differential during hard acceleration. This will cause the vehicle to pull to one side. Also, if the alignment angles are different from one side to the other, the vehicle will tend to pull. Although this may seem to be a service item, sometimes it is just the way the vehicle needs to be to provide good and safe handling during conditions other than hard acceleration. Most vehicles have staggered alignment angles to compensate for road crown.

Tire Sidewall Construction. The flexibility of a tire's sidewalls determines how well it will respond to the road's surface. If the sidewalls are stiff or short, the tires will tend to move more easily when uneven surfaces are encountered. Soft, tall sidewalls have a better chance of staying in a straight-ahead position, in spite of the road conditions. Remember, torque steer will not occur if both front wheels are headed in the same direction and at the same speed.

Unequal Amounts of Road Traction. If one tire of the vehicle has more traction than the other, the vehicle will tend to pull to the other side. This is something that cannot be corrected and occurs on all vehicles. With open differentials, the tire that has the least grip gets the most power. Therefore, when a FWD vehicle is accelerated hard on slippery surfaces, the vehicle will tend to steer toward the side with the least bit of traction. This is one of the advantages of having a FWD vehicle with a limited-slip differential. Not many FWD vehicles are equipped with this type of differential. This is due to cost. However, manufacturers are very sensitive to the adverse effects of torque steer and limit the horsepower output of their engine to avoid severe torque steer. The current limit is approximately 175–200 horsepower. Engines producing more power have high degrees of torque steer or have a limited-slip differential.

Movement of the Engine. If the engine is able to twist on its mounts during heavy loads, the geometric relationship of the axles will also change. This can cause torque steer. Some manufacturers have changed the slant of the engine in the chassis to minimize torque steer. This reduces the effects of the torsional effects on the location of the engine. Some vehicles have shims under the right motor mount to raise the right side of the engine slightly.

The spindle shaft is commonly called the stub axle. **Torque steer** is a term used to describe a condition in which the car tends to steer or pull in one direction as engine power is applied to the drive wheels.

With the engine mounted transversely, the transaxle sits to one side of the engine compartment. Because of this offset, one of the axles must be longer than the other (Figure 5-6). The unequal lengths result in torque steer.

Torque steer also occurs when the CV joints on one drive shaft operate at different angles from those on the other shaft. The joints on the longer shaft almost always operate at less of an angle than those on the shorter drive shaft. Therefore, when the differential in a transaxle sends power out to the wheels, the longer shaft will have less resistance, and most of the engine's torque will be sent to that wheel. This is the same thing that happens when one tire has more traction than the other. The tire with the least traction will get the most power. This, of course, doesn't happen if the differential is equipped with limited-slip.

Equal-Length Shafts Combatting Torque Steer

As engine output increases with advances in technology, manufacturers have gone to great lengths to reduce the effects of torque steer. One method that has been used for many years is to keep axle length the same on both sides. Equal-length shafts are used on most FWD cars to reduce torque steer. To provide for equal-length shafts, manufacturers divide the longer shaft into two pieces (Figure 5-7). One piece comes out of the transaxle and is supported by a bearing. The other piece is made the same length as the shorter axle on the other side of the car. The **intermediate drive shaft** is located between the right and left drive shafts and equalizes drive shaft length.

Vibration Dampers

A small damper weight is sometimes attached to the long half shaft to dampen harmonic vibrations in the driveline and to stabilize the shaft as it spins, not to balance the shaft (Figure 5-8). This torsional vibration absorber is splined to the axle shaft and is often held in place by a snapring.

Unlike the drive shaft of RWD cars, FWD axle shaft balance is not very important because of the relatively slow rotational speeds. In fact, FWD drive shafts operate at about ⅓ the rotational speed of a RWD drive shaft. This is because the shafts drive the wheels directly, with

> The inner piece of the axle shaft is normally called the **intermediate drive shaft**, which serves to equalize the lengths of the two outer drive shafts.

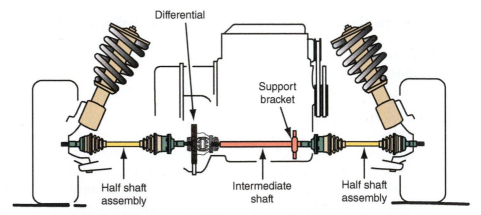

FIGURE 5-7 Equal-length FWD half shafts with an intermediate shaft.

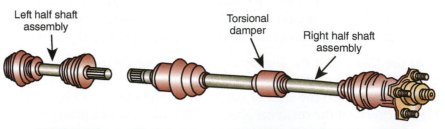

FIGURE 5-8 The long half shaft is sometimes fitted with a torsional damper.

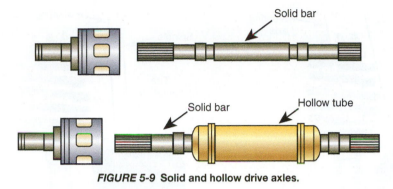

FIGURE 5-9 Solid and hollow drive axles.

the final gear reduction already having taken place inside the transaxle's differential. The lower rotational speed has the advantage of eliminating vibrations that sometimes result from high rotational speeds.

Unequal-Length Half Shafts

If the **half shafts** are not equal in length, the longer one is usually made thicker than the shorter one or one axle may be solid and the other tubular (Figure 5-9). These combinations would allow both axles to twist in the same amount while under engine power. If they twist unequal amounts, the car may experience torque steer.

High-Horsepower Solutions

With horsepower in some FWD vehicles approaching 300-plus, manufacturers have turned to technology to help reduce the effects of torque steer. Many electronic steering systems are programmed to compensate for or counter torque steer sensed by the system. These systems have been labeled Torque Steer Compensation (TSC) programs. Some manufacturers have had to cut back on power or turbocharger boost in the lower gears so drivers can maintain control under hard acceleration. While many traditional limited-slip differential (LSD) designs do not work well in a front differential, the helical gear LSD design has been used on some models to maintain traction and reduce torque steer (Figure 5-10). Other manufacturers have adopted a more costly solution of electronically controlled LSDs. These designs are considered in later chapters. In addition, many traction control systems are programmed to reduce torque steer.

A BIT OF HISTORY

New-design front struts created for better handling and reduced torque steer may not be so new after all. Toyota developed what has been called a "super strut" for some of its high-performance models in the early 1990s that incorporates many of the design features now found on Ford and GM units.

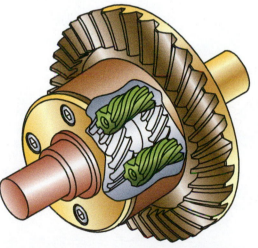

FIGURE 5-10 A spiral gear limited-slip differential.

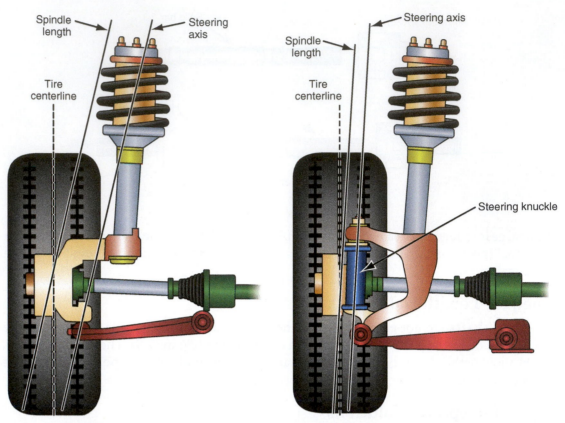

FIGURE 5-11 (A) Typical FWD strut design. (B) New strut design to minimize torque steer.

Improved strut designs have come onto the market that have allowed for a shorter spindle length and smaller tire scrub radius, both key aids to controlling torque steer (Figure 5-11). Ford developed the Revoknuckle strut for its 300 HP Focus RS, and GM uses the HiPer Strut in some high-torque applications. In these designs, the knuckle attaches to the suspension through pivot bearings (Figure 5-12).

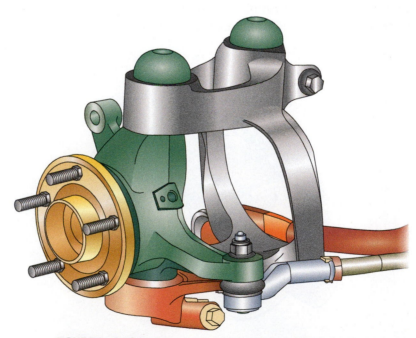

FIGURE 5-12 "Super Strut" detail with upper and lower bearings.

Drive Axle Supports

The drive axles used with transaxles are the full-floating type. This type of axle does not support the weight of the vehicle; rather, all of the vehicle's weight is supported by the suspension. As the car goes over bumps, the axles move up and down, which changes their length. The inner CV joints let the axles slide in and out, changing their length, as they move up and down. The outer joints allow the steering system to turn the wheels, as well as allow for the up-and-down movement of the suspension.

Shop Manual
Chapter 5, page 213

CV JOINTS

One of the most important components of a FWD drive axle is the constant-velocity joint (Figure 5-13). These joints are used to transfer a uniform torque and a constant speed, while operating through a wide range of angles. On FWD and 4WD/AWD cars, operating angles of as much as 40 degrees for the outer steering CV joint are common. The drive axles must transmit power from the engine to the front wheels that must also drive, steer, and cope with the severe angles caused by the up-and-down movement of the vehicle's suspension. CV joints are compact joints that allow the drive axles to rotate at a constant velocity, regardless of the operating angle.

CV joints do the same job as the Universal joints of front-engined RWD cars. The drive shaft of these cars is fitted with Universal joints at each end of the shaft. The joints allow the drive shaft to move with the suspension as it transfers power to the drive wheels. As the shaft rotates on an angle, the first U-joint sets up an oscillation in the drive shaft and then a second U-joint, at the other end of the shaft, cancels the oscillation before it reaches the axle. The ability of the joints to cancel the inherent oscillations lessens as the angle of the shaft increases. U-joints only work well if the shaft angle is 3–6 degrees. Two sets of U-joints are often used on drive shafts with greater operating angles. When the U-joints respond to the changes in the angle of the drive shaft, the speed of the shaft changes during each revolution. This change in speed causes the drive shaft to vibrate or pulse as it rotates. Constant-velocity joints turn at the same speed during all operating angles and therefore can smoothly deliver power to the wheels.

Universal joints are commonly referred to as U-joints.

CV Boots

Bellows-type neoprene **boots** are installed over each CV joint to retain lubricant and to keep out moisture and dirt (Figure 5-14). These boots must be maintained in good condition. Each end of the boot is sealed tightly against the shaft or housing by a retaining clamp or strap (Figure 5-15). These straps may be metal or plastic and are available in many sizes and designs (Figure 5-16).

In some applications, the inboard CV joints operate very close to the engine's exhaust system. Special rubber boots of silicone or thermoplastic materials are required to withstand

Bellows-type boots are rubber or neoprene protective covers with accordion-like pleats used to contain lubricants and exclude contaminating dirt or water.

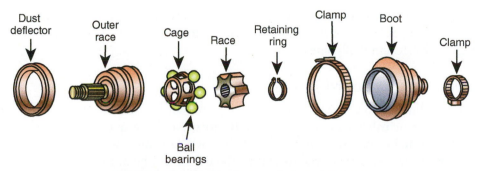

Dust deflector   Outer race   Cage   Race   Retaining ring   Clamp   Boot   Clamp

Ball bearings

FIGURE 5-13 An exploded view of a typical CV joint.

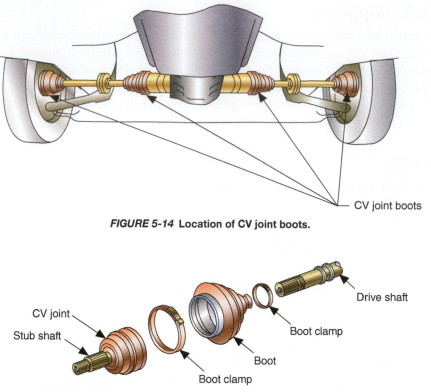

FIGURE 5-14 Location of CV joint boots.

FIGURE 5-15 Clamps seal the boot against the CV joint and drive axle.

If the CV joint boot keeps the joint properly sealed, the joint can last more than 100,000 miles.

the temperatures. In these extremes, a special high-temperature lubricant will be specified by the manufacturer. Silicone boots are also specified for the extreme cold conditions found in northern climates.

> **AUTHOR'S NOTE:** Extreme off-road conditions call for special equipment throughout the truck, including the CV shafts. Standard boot materials would be easily destroyed, so some aftermarket manufacturers have designed special materials that can put up with the punishment. Some shafts use a separate inside boot with a replaceable Kevlar outer boot for extra protection (Figure 5-17). These off-road shafts often use a six-ball inner CV joint that can operate at angles well beyond 25 degrees.

The outer CV joints of most Asian import cars use a modified single retention method for securing the CV joints to the shaft.

TYPES OF CV JOINTS

To satisfy the needs of different applications, CV joints come in a variety of styles. The different types of joints can be referred to by position (inboard or outboard), by function (fixed or plunging), or by design (ball-type or tripod).

Inboard and Outboard Joints

The **inboard constant velocity joint** is often called the plunge joint, and the **outboard constant velocity joint** is called the fixed joint.

In FWD drivelines, two CV joints are used on each half shaft. The joint nearest the transaxle is the inner or **inboard constant velocity joint**, and the one nearest the wheel is the outer or **outboard constant velocity joint**. In a RWD vehicle with IRS, the joint nearest the differential can also be referred to as the inboard joint. The one closer to the wheel is the outboard joint.

DESCRIPTION	APPEARANCE	TYPE
Large ladder clamp		A.C.I. and G.K.N.
Small ladder clamp		A.C.I. and G.K.N.
Small strap and buckle clamp		Citroën
Large strap and buckle clamp		Citroën
Large spring clamp		A.C.I. and G.K.N.
Small rubber clamp		A.C.I. and G.K.N.

FIGURE 5-16 Various sizes and designs of boot clamps.

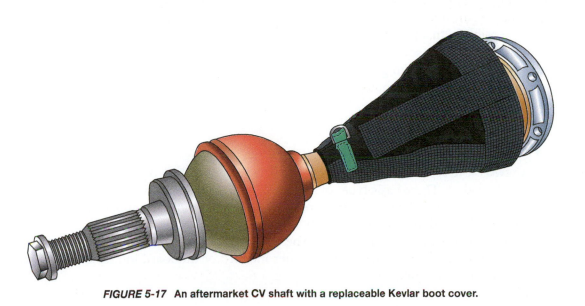

FIGURE 5-17 An aftermarket CV shaft with a replaceable Kevlar boot cover.

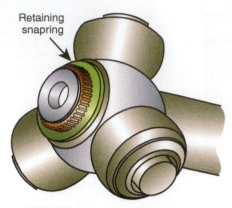

Retaining snapring

FIGURE 5-18 A snapring is often used to positively retain a CV joint to the drive shaft.

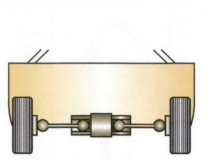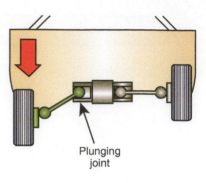

Plunging joint

FIGURE 5-19 Suspension movement and the resulting plunging action of an inboard CV joint.

CV joints are held onto the axle shafts by three different methods: nonpositive, positive, and single retention. Most inner joints use positive retention, whereas outboard joints may be held by any one of the three methods. Nonpositive retention is accomplished by the slight interference fit of the joint onto the shaft. Positively retained joints use a snapring to secure the joint to the shaft (Figure 5-18) or are splined to the shaft and retained with a nut. Single retention is accomplished by a very tight press-fit connection of the joint onto the shaft. Often, joints retained in this way cannot be removed without destroying the joint or the shaft.

Fixed and Plunging Joints

CV joints can also be categorized by function. They are either **fixed-type joints** (meaning they do not plunge in and out to compensate for changes in length) or **plunging constant velocity joints** (one that is capable of in-and-out movement).

In response to the suspension of the car, the drive axles' effective length changes as the distance from the inboard to the outboard CV joint changes. The inboard CV joint must allow the drive shaft to freely move in and out of the joint housing as the front wheels go up and down (Figure 5-19).

The outboard joint is a fixed joint. Both joints do not need to plunge if one can do the job. The outboard joint must be able to handle much greater operating angles for steering than would be possible with a plunging joint.

In RWD applications with IRS, one joint on each axle shaft can be fixed and the other plunging, or both can be plunging joints. The operating angles are not as great because the wheels do not have to steer, thus plunging joints can be used at either or both ends of the axle shafts.

CV JOINT DESIGNS

There are two basic types of outboard CV joints: the Rzeppa or ball-type fixed CV joint and the fixed tripod joint. Three basic types of inboard CV joints are used: double-offset CV joint (DOJ), plunging tripod CV joint, and the cross-groove plunge joint. The applications of these vary with car make and model (Figure 5-20).

CV joints are also classified by design. The two basic varieties are the ball-type and the tripod-type joints. Both types are used as either inboard or outboard joints, and both are available in fixed or plunging designs.

Nonpositive retention is used on the outer CV joints of all models of Ford and Chrysler cars, as well as some German-made cars.

Positive retention is used on most inner CV joints and on the outer joints of some European cars.

A BIT OF HISTORY

In the 1920s an engineer from Ford, Alfred Rzeppa, developed a compact constant-velocity joint using ball bearings between two bearing races. Through the years, this joint was further developed and improved. The improved design, similar to what is being used today, appeared on British and German FWD cars in 1959.

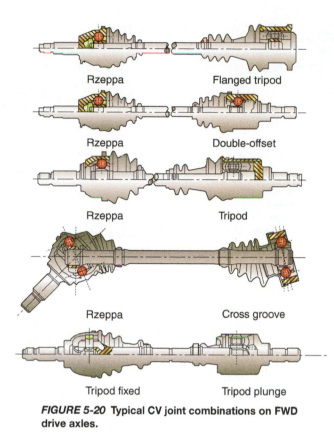

FIGURE 5-20 Typical CV joint combinations on FWD drive axles.

OUTBOARD CV JOINT DESIGNS

Ball-Type Joints

The most commonly used type of CV joint was named after its original designer, A. H. Rzeppa, and is based on a ball-and-socket principle (Figure 5-21). The **Rzeppa constant velocity joint** has its inner race attached to the axle. The inner race has several precisely machined grooves spaced around its outside diameter. The number of grooves equals the number of ball bearings used by the joint. These joints are typically designed with a minimum of three to a maximum of six ball bearings. The bearing cage is pressed into or is part of the outer housing and serves to keep the joint's ball bearings in place as they ride in the groove of the inner race.

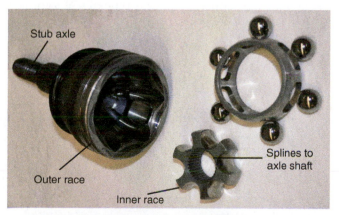

FIGURE 5-21 A Rzeppa ball-type fixed CV joint.
Courtesy of Federal-Mogul Corporation.

Named after its inventor, Alfred Rzeppa, **Rzeppa constant velocity joints** are ball-type CV joints and are usually the outer joints on most FWD cars.

The term *bisect* means to divide by two.

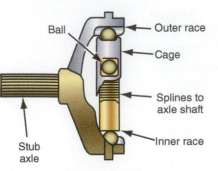

FIGURE 5-22 The inner race is splined to the axle shaft. The balls, placed between the ball groove and cage window, move the cage with the axle.

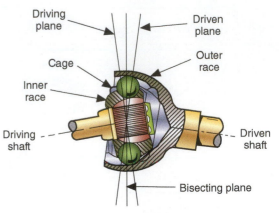

FIGURE 5-23 In a Rzeppa CV joint, the balls bisect the angle of the joint.

When the axle rotates, the inner bearing race and the balls turn with it. The balls, in turn, cause the cage and the outer housing to turn with them (Figure 5-22). The grooves machined in the inner race and outer housing allow the joint to flex. The balls serve both as bearings between the races and the means of transferring torque from one to the other. This type of CV joint is used on almost every make and model of FWD car, except for most French designs.

If viewed from the side, the balls within the joint always bisect the angle formed by the shafts on either side of the joint regardless of the operating angle (Figure 5-23). This reduces the effective operating angle of the joint by a half and virtually eliminates all vibration problems. The cage helps to maintain this alignment by holding the balls snugly in their windows. If the cage windows become worn or deformed over time, the resulting play between the ball and window typically results in a clicking noise when turning. It is important to note that the opposing balls in a Rzeppa CV joint always work together as a pair. Heavy wear in the grooves of one ball almost always results in identical wear in the grooves of the opposing ball.

Another ball-type joint is the disc-style CV joint, which is used predominantly by Volkswagen as well as many German RWD cars. Its design is very similar to the Rzeppa joint.

Tripod-Type Joints

The fixed **tripod constant velocity joint** uses a central hub or tripod that has three trunnions fitted with spherical rollers on needle bearings (Figure 5-24). These spherical rollers or balls ride in the grooves of an outer housing that is attached to the front wheels (Figure 5-25). Because the balls are not held in a set position in the hub, they are free to move back and forth within the hub. This allows for a constant velocity regardless of the movement of the hub as it responds to the steering or suspension system. This type of CV joint is used on most French

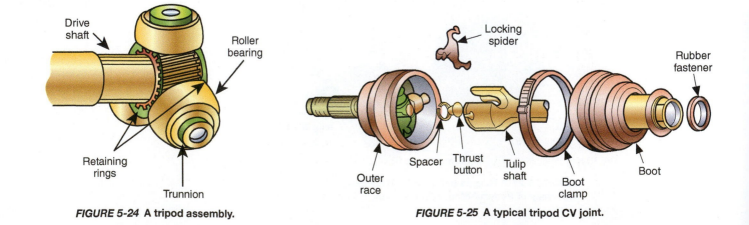

FIGURE 5-24 A tripod assembly.

FIGURE 5-25 A typical tripod CV joint.

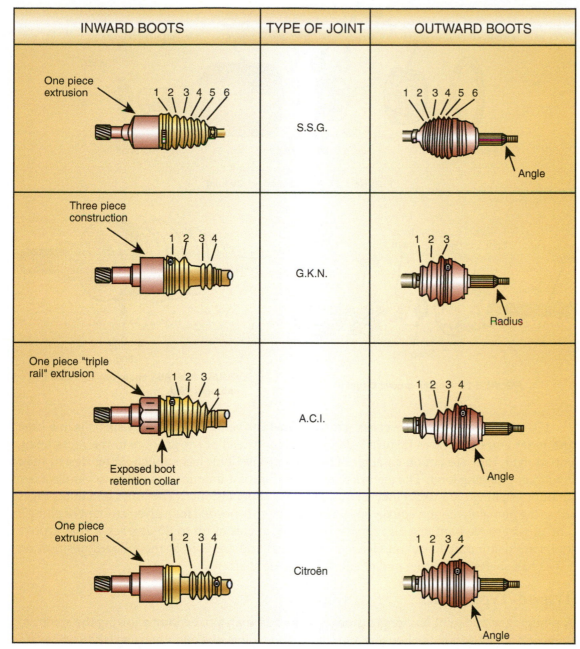

INWARD BOOTS	TYPE OF JOINT	OUTWARD BOOTS
One piece extrusion 1 2 3 4 5 6	S.S.G.	1 2 3 4 5 6 Angle
Three piece construction 1 2 3 4	G.K.N.	1 2 3 Radius
One piece "triple rail" extrusion 1 2 3 4 Exposed boot retention collar	A.C.I.	1 2 3 4 Angle
One piece extrusion 1 2 3 4	Citroën	1 2 3 4 Angle

FIGURE 5-26 The appearance of the joint's boot can identify the design of the joint.

cars and has great angular capability and is known by three subcategory types that define how they are retained: the Citroen, GKN, and ACI joints (Figure 5-26).

INBOARD CV JOINT DESIGNS

Ball-Type Joints

A Rzeppa CV joint can be modified to become a plunging joint, simply by making the grooves in the inner race longer (Figure 5-27). Longer grooves in the inner race of a Rzeppa joint allow the bearing cage to slide in and out. This type of joint is called a **double-offset constant velocity joint** and is typically used in applications that require higher operating angles (up to 25 degrees) and greater plunge depth (up to 2.4 inches). This type of joint can be found at the inboard position on some FWD half shafts as well as on the drive shaft of some 4WD vehicles.

The **double-offset constant velocity joint** is often listed as a DOJ, and is another name for a plunging, inner CV joint.

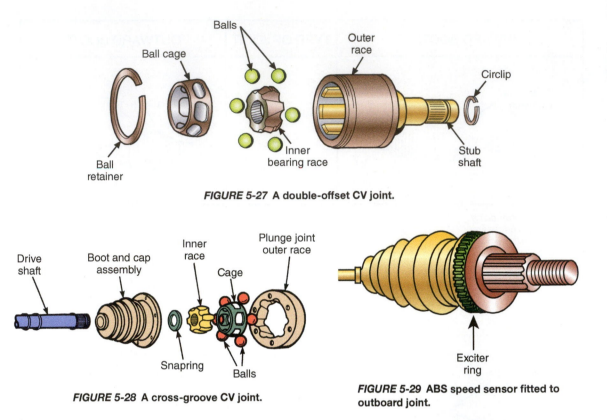

FIGURE 5-27 A double-offset CV joint.

FIGURE 5-28 A cross-groove CV joint.

FIGURE 5-29 ABS speed sensor fitted to outboard joint.

Like the Rzeppa joint, the **cross-groove constant velocity joint** uses six balls in a cage and inner and outer races (Figure 5-28). However, instead of the grooves in the races being cut straight, they are cut on an angle. The cross groove joint has a much flatter design than any other plunging joint. The feature that makes this joint unique is its ability to handle a fair amount of plunge (up to 1.8 inches) in a relatively short distance. The inner and outer races share the plunging motion equally so less overall depth is needed for a given amount of plunge. The cross groove joint can also handle operating angles of up to 22 degrees. Cross groove joints are commonly found on German-made cars and are used at the inboard position on FWD half shafts or at either end of a RWD IRS axle shaft.

Tripod-Type Joints

A plunging tripod joint has longer grooves in its hub than a fixed joint allowing the spider to move in and out within the housing. On some tripod joints, the outer housing is closed, meaning the roller tracks are totally enclosed within it. On others, the tulip is open and the roller tracks are machined out of the housing (Figure 5-30). Plunging tripod-type joints are used on many American and European cars, including some Fords, Chryslers, General Motors, and French cars.

New CV Joint Designs

Manufacturer requirements for increased fuel economy and reduced noise, vibration, harshness (NVH) concerns have resulted in new CV joint designs that have made their way into production. GKN developed the eight-ball Countertrack CV joint that is lighter, stronger, and produces less friction while allowing up to 50 degrees turning angle in fixed joints and 30-degree angle capability in plunging joints (Figure 5-31). GKN's Crosstrack plunging joint combines the benefits of straight ball tracks with angled ball tracks in a single unit that reduces "clunk" on engagement and significantly reduces rotating imbalance (Figure 5-32). The company has also developed the SIO plus Ballspline plunging joint for use in premium SUVs (Figure 5-33). This CV joint provides a low-profile design with advantages in space requirements.

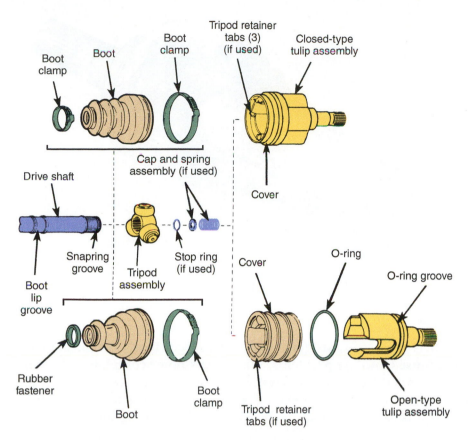

FIGURE 5-30 Inner tripod plunge-type joints, with closed housing and open housing.

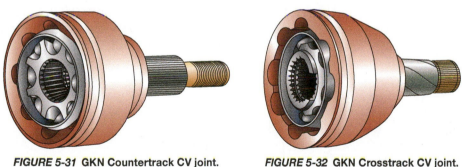

FIGURE 5-31 GKN Countertrack CV joint.   FIGURE 5-32 GKN Crosstrack CV joint.

CV Joint Wear

Regardless of the application, outer joints typically wear at a higher rate than inner joints because of the increased range of operating angles to which they are subjected. Inner joint angles may change only 10–20 degrees as the suspension travels through **jounce** and **rebound**. Outer joints can undergo changes of up to 40 degrees in addition to jounce and rebound as the wheels are steered. That, combined with more flexing of the outer boots, is why outer joints have a higher failure rate. On an average, nine outer joints are replaced for every inner joint. That does not mean you should overlook the inner joints. They wear, too. Every time the suspension travels through jounce and rebound, the inner joints must plunge in and out to accommodate the different arcs between the drive shafts and the suspension. Tripod inner joints tend to develop unique wear patterns on each of the three rollers and their respective tracks in the housing, which can lead to noise and vibration problems.

Jounce is the upward movement of the car in response to road surfaces.

The ABS speed sensors are often called reluctors.

129

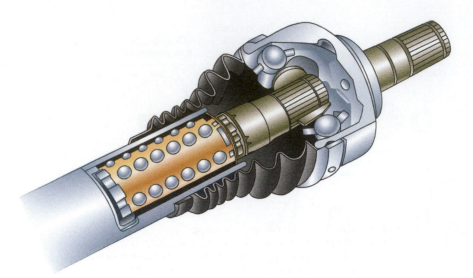

FIGURE 5-33 GKN SIO plus Ballspline plunging joint.

AUTHOR'S NOTE: Many students and techs forget that worn or damaged CV joints will cause handling problems. When a customer complains of a vibration, noise, or pulling, the first reaction is to check the tires, brakes, steering, and suspension systems. To save time and frustration, I suggest you include an inspection of the CV joints as part of your diagnosis of handling problems.

FWD WHEEL BEARINGS

The drive axles are supported in the steering knuckle by wheel bearings (Figure 5-34). These bearings allow the axle to rotate evenly and smoothly and keep the axle in the center of the steering knuckle's hub. Basically there are two types of FWD wheel bearings. Most GM,

Shop Manual
Chapter 5, page 227

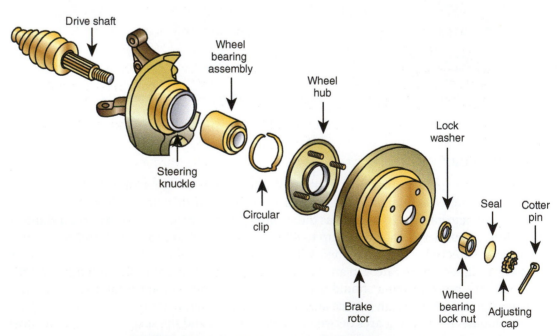

FIGURE 5-34 A typical FWD wheel bearing and hub assembly.

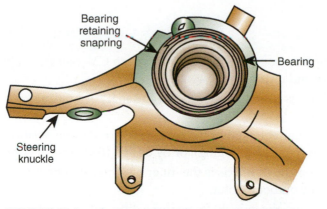

FIGURE 5-35 A sealed front wheel bearing that is held in the wheel hub/steering knuckle assembly by a snapring. On some cars, the bearing is part of the wheel hub assembly; on others, the bearing unit is pressed into the hub; and on a few, the bearings are retained in much the same way as the front wheel bearings on a RWD car.

Chrysler, and European cars use a double-row, angular-contact bearing. These units are simply two rows of ball bearings that are located next to each other. The races for these bearings are slightly offset to control radial loads during cornering. Ford and most Asian cars use opposed tapered roller bearings. Most FWD front-wheel bearings are one-piece sealed units. These units are typically pressed into the wheel hub assembly and held by a snapring (Figure 5-35).

SUMMARY

- The complete drive axle, including the inner and outer CV joints, is typically called a half shaft.
- A FWD car has two short drive shafts, one on each side of the engine, that drive the wheels and adjust for steering and suspension changes.
- CV joints are not only used on FWD cars; they are also used on the front axles of many AWD and 4WD vehicles. They have also been used on IRS rear drive axles and RWD buses and cars that have the engine mounted in the rear.
- *Torque steer* is a term used to describe a condition in which the car tends to steer or pull in one direction as engine power is applied to the drive wheels.
- Equal-length shafts are used in some vehicles to help reduce torque steer. This is accomplished by making the longer side into two pieces. One piece comes out of the transaxle and is supported by a bearing assembly attached to the engine. The other piece is made to the same length as the shorter side axle.
- A small damper weight is sometimes attached to one half shaft to dampen harmonic vibrations in the driveline and to stabilize the shaft as it spins.
- If the half shafts are not equal in length, the longer one is usually made thicker than the shorter one, or one axle may be solid and the other tubular. These combinations would allow both axles to twist the same amount when under engine power.
- Constant-velocity joints are used to transfer a uniform torque and a constant speed when operating through a wide range of angles.

SUMMARY

- Bellows-type neoprene boots are installed over each CV joint to retain lubricant and to keep out moisture and dirt.
- Each end of the boot is sealed tightly against the shaft or housing by a retaining clamp or strap.
- CV joints come in a variety of types that can be referred to by position (inboard or outboard), by function (fixed or plunging), or by design (ball-type or tripod).
- The CV joint nearest the transaxle is the inner or inboard joint, and the one nearest the wheel is the outer or outboard joint.
- CV joints are held onto the axle shafts by three different methods: nonpositive, positive, and single retention.
- CV joints are either fixed (meaning they do not plunge in and out to compensate for changes in length) or plunging (one that is capable of in-and-out movement) joints.
- The most commonly used type of CV joint is a Rzeppa, which has its inner race attached to the axle. The inner race has several precisely machined grooves spaced around its outside diameter. The number of grooves equals the number of ball bearings used by the joint. These joints are designed with a minimum of three to a maximum of six ball bearings. The bearing cage serves to keep the joint's ball bearings in place as they ride in the groove of the inner race.
- The fixed tripod CV joint uses a central hub or tripod that has three trunnions fitted with spherical rollers on needle bearings. These spherical rollers or balls ride in the grooves of an outer housing that is attached to the front wheels.
- A Rzeppa CV joint can be modified to become a plunging joint by making the grooves in the inner race longer. Longer grooves in the inner race of a Rzeppa joint allow the bearing cage to slide in and out. This type is called a double-offset joint.
- Like the Rzeppa joint, the cross-groove CV joint uses six balls in a cage and inner and outer races. However, the grooves in the races are cut on an angle rather than straight. The cross-groove joint has a much flatter design than any other plunging joint.
- A plunging tripod joint has longer grooves in its hub than a fixed joint, allowing the spider to move in and out within the housing.
- Regardless of the application, outer joints typically wear at a higher rate than inner joints because of the increased range of operating angles to which they are subjected.
- The drive axles are supported in the steering knuckle by wheel bearings. These bearings allow the axle to rotate evenly and smoothly and keep the axle in the center of the steering knuckle's hub.

TERMS TO KNOW

Axle shaft

Boots

Constant velocity joint

Cross-groove constant velocity joint

Double-offset constant velocity joint

Fixed-type velocity joint

Half shaft

Inboard constant velocity joint

Intermediate drive shaft

Jounce

Outboard constant velocity joint

Plunging joint

Rebound

Rzeppa constant velocity joint

Torque steer

Tripod constant velocity joint

REVIEW QUESTIONS

Short Answer Essays

1. Define the purpose of a CV joint.

2. Describe the difference between fixed and plunging CV joints.

3. Explain why CV joints are preferred over conventional universal joints.

4. Explain the different methods used by automobile manufacturers to offset torque steer.

5. Describe the purpose of the boot on a CV joint.

6. Describe purposes of the wheel bearing of a FWD car.

7. Describe the major differences between a Rzeppa and a tripod CV joint.

8. Explain why a fixed CV joint tends to wear much faster than a plunging joint.

9. Explain why some half shafts are fitted with a vibration damper.

10. Explain how a ball-type CV joint is constructed.

Fill-in-the-Blanks

1. CV joints come in a variety of types that can be referred to by position (_____ or _____), by function (_____ or _____), or by design (_____ or _____).

2. If the half shafts are not equal in length, the longer one is usually made _____ than the other, or one may be _____ and the other _____.

3. The major components of a Rzeppa joint are three to six _____, an inner _____, and an outer _____.

4. The major components of a tripod CV joint are the _____, three _____, and an outer _____.

5. _____ CV joints typically wear at a higher rate than the _____ ones because of the _____ range of operating _____.

6. The type of joint that allows for changes in axle length is the _____ joint.

7. Half shafts are also called _____ _____ and _____ _____.

8. CV joints are used to transfer _____ _____ and a _____ _____.

9. CV joints are held onto the axle shafts by three different methods: _____, _____, and _____.

10. The use of an intermediate shaft allows for half shafts of _____.

ASE-Style Review Questions

1. When discussing CV joints,
   Technician A says that they are called constant-velocity joints because their rotational speed does not change with their operational angle.
   Technician B says that conventional universal joints cannot operate properly at the angles needed in a FWD axle arrangement.
   Who is correct?
   A. A only
   B. B only
   C. Both A and B
   D. Neither A nor B

2. When trying to decide which is the most commonly used type of CV joint,
   Technician A says that the ball-and-socket type is the most common.
   Technician B says that the Rzeppa type is the most common.
   Who is correct?
   A. A only
   B. B only
   C. Both A and B
   D. Neither A nor B

3. When discussing the differences between RWD and FWD front-wheel bearings,
   Technician A says that most FWD cars use one tapered roller bearing, whereas RWD cars use two.
   Technician B says that most FWD front bearings are pressed onto the spindle end of the drive axle.
   Who is correct?
   A. A only
   B. B only
   C. Both A and B
   D. Neither A nor B

4. When discussing Rzeppa CV joints,
   Technician A says that the inner race of the joint is connected to the drive axle.
   Technician B says that the ball bearings rotate on a trunnion within the outer housing.
   Who is correct?
   A. A only
   B. B only
   C. Both A and B
   D. Neither A nor B

5. When explaining torque steer to a customer,
   Technician A says that joints on the longer axle shaft operate at less of an angle than the shorter axle shaft.
   Technician B says that cars with more powerful engines tend to have more torque steer issues.
   Who is correct?
   A. A only
   B. B only
   C. Both A and B
   D. Neither A nor B

6. When discussing the purposes of the protective boot on a CV joint,
   Technician A says that the boot prevents contamination of the joint's lubricant.
   Technician B says that the boot prevents the joint's lubricant from flying off when the shaft is rotating.
   Who is correct?
   A. A only
   B. B only
   C. Both A and B
   D. Neither A nor B

7. When discussing the differences between the various designs of CV joints,

    Technician A says that most joints use the tripod design and all are basically inter changeable.

    Technician B says that the fixed tripod-type joint uses needle bearings, not ball bearings like the ball type.

    Who is correct?

    A. A only
    B. B only
    C. Both A and B
    D. Neither A nor B

8. Which of the following statements about FWD drive axles is *not* true?

    A. The outer part of each drive axle is supported by a steering knuckle.
    B. The drive axle assembly comprises three pieces attached together in such a way as to allow the wheels to turn and move with the suspension.
    C. A short spindle shaft runs from the outer CV joint to mate with the wheel assembly.
    D. Half shafts operate at angles as high as 40 degrees for suspension travel and 20 degrees for turning.

9. While discussing the ways manufacturers try to reduce the effects of torque steer,

    Technician A says they sometimes use a combination of solid and hollow axles to help control torque steer.

    Technician B says some manufacturers divide the longer shaft into two pieces: one piece comes directly out of the transaxle and is supported by a bearing. The other piece is a half shaft made to the same length as the axle on the other side of the car.

    Who is correct?

    A. A only
    B. B only
    C. Both A and B
    D. Neither A nor B

10. While discussing FWD drive axles,

    Technician A says the axle shaft is connected to the differential by a fixed CV joint.

    Technician B says the axle shaft extends from the differential to an outer or plunge CV joint.

    Who is correct?

    A. A only
    B. B only
    C. Both A and B
    D. Neither A nor B

Chapter 6

DRIVE SHAFTS AND UNIVERSAL JOINTS

UPON COMPLETION AND REVIEW OF THIS CHAPTER, YOU SHOULD BE ABLE TO:

- Understand and describe the purpose and construction of common RWD drive shaft designs.
- Understand and describe the purpose and construction of common Universal joint designs.

- Explain the importance of drive shaft balance.
- Explain the natural speed variations inherent to a drive shaft.
- Describe the effects of canceling Universal joint angles.

INTRODUCTION

On front engined rear-wheel-drive cars, the rotary motion of the transmission's output shaft is carried through the drive shaft to the differential, which causes the rear drive wheels to turn (Figure 6-1).

The drive shaft (Figure 6-2) is normally made from seamless steel or aluminum tubing with Universal joint yokes welded to both ends of the shaft (Figure 6-3). To save weight, some

> The **drive shaft** is an assembly of one or two Universal joints connected to a shaft or tube used to transmit power from the transmission to the differential. It is also called the propeller shaft.

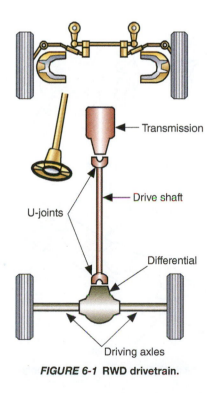

FIGURE 6-1 RWD drivetrain.

Labels in figure: Transmission, Drive shaft, U-joints, Differential, Driving axles

FIGURE 6-2 A two-piece drive shaft for a RWD light truck.

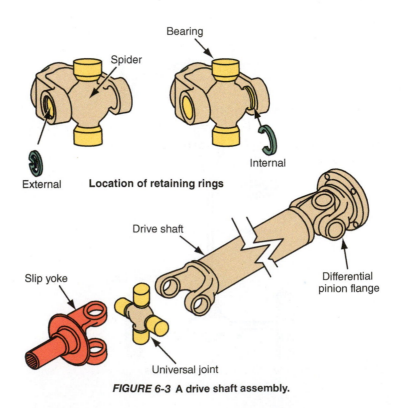

FIGURE 6-3 A drive shaft assembly.

FIGURE 6-4 A two-piece drive shaft, with a center bearing, is often used when there is a great distance between the transmission and the rear axle.

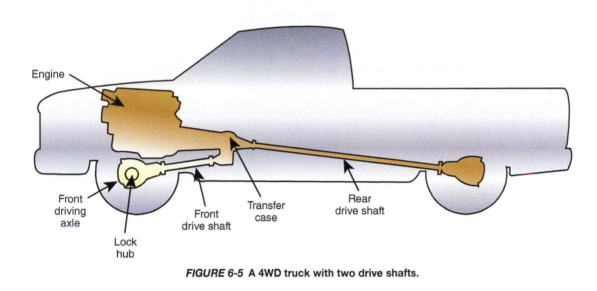

FIGURE 6-5 A 4WD truck with two drive shafts.

Engine

Front driving axle

Lock hub

Front drive shaft

Transfer case

Rear drive shaft

Although most late-model cars are front-wheel drive, which are not equipped with a drive shaft and Universal joints, the best-selling vehicles in the United States are pickup trucks, which do have a drive shaft and Universal joints. These, plus the many older cars on the road, give many opportunities to technicians trained in the diagnosis and repair of drive shafts and Universal joints.

The short drive shafts or axles used on front-wheel drive (FWD), four-wheel drive (4WD), and rear-wheel drive (RWD) with independent rear suspension are called half shafts.

A BIT OF HISTORY

In 1901, the Autocar was the first car in the United States to use a drive shaft.

Shop Manual
Chapter 6, page 256

In normal automotive terms, suspension spring compression is called jounce.

manufacturers use epoxy-and-carbon fiber shafts. Some drivelines have two drive shafts and three Universal joints and use a center support bearing mounted to a cross member that supports the two-piece assembly. A slip yoke near the center support bearing serves as the connecting link between the two halves (Figure 6-4).

Four-wheel-drive vehicles use two drive shafts, one to drive the front wheels and the other to drive the rear wheels (Figure 6-5). FWD cars, 4WD vehicles equipped with an independent front suspension, and RWD vehicles equipped with independent rear suspension use an additional pair of short drive shafts. These shafts are actually the car's drive axles, which transmit the torque from the differential to each drive wheel.

DRIVE SHAFT CONSTRUCTION

Two facts must be considered when designing a drive shaft: the engine and transmission are more or less rigidly attached to the car frame and the rear axle housing, with the wheels and differential attached to the frame by springs. As the rear wheels encounter irregularities in the road, the springs compress or expand. This changes the angle of the driveline between the transmission and the rear axle housing. It also changes the distance between the transmission and the differential.

In order for the drive shaft to respond to these constant changes, drive shafts are equipped with two or more Universal joints that permit variations in the angle of the shaft, and a slip joint that permits the effective length of the driveline to change.

A drive shaft is actually an extension of the transmission's output shaft, as its sole purpose is to transfer torque from the transmission to the drive axle assembly (Figure 6-6). It is usually made from seamless steel or aluminum tubing with a yoke welded or pressed onto each end, which provides a means of connecting two or more components together.

The drive shaft, like any other rigid tube, has a natural vibration frequency. This means that if one end of the tube were held tightly, the tube would vibrate at its own frequency when it is deflected and released or when it rotates. It reaches this natural frequency at its critical speed. The critical speed of a drive shaft depends on the diameter of the tube and the length of the drive shaft. Drive shaft diameters are as large as possible and shafts as short as possible to keep the critical speed frequency above the normal driving range.

Because the drive shaft rotates at high speeds and at varying angles, it must be balanced to reduce vibration. As the drive shaft's length and the operating angle and speed increase, the necessity for balance also increases. Half shafts found in FWD and IRS RWD vehicles are short and have gone through the speed reduction of a differential, and balance is not nearly as important as with a RWD drive shaft. Several methods are used to balance drive shafts. One of the most common techniques employed by manufacturers is to balance the drive shaft by welding balance weights to the outside diameter of the drive shaft (Figure 6-7).

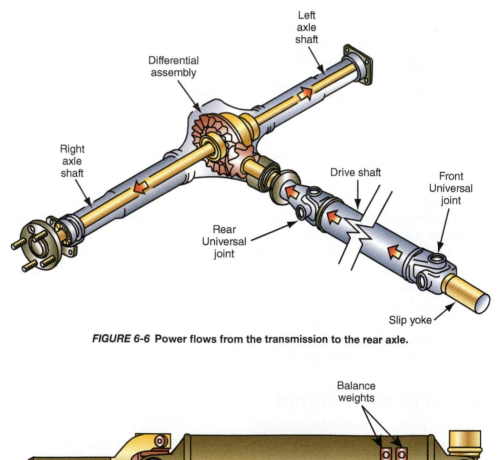

FIGURE 6-6 Power flows from the transmission to the rear axle.

FIGURE 6-7 Location of the balance weights on a drive shaft.

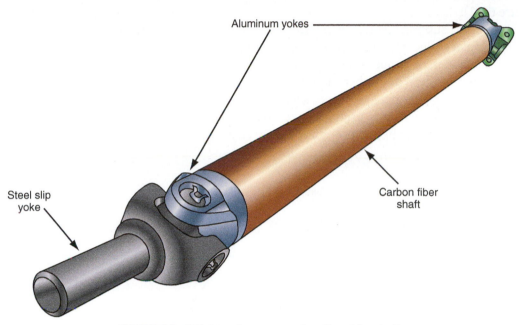

Aluminum yokes

Steel slip yoke

Carbon fiber shaft

FIGURE 6-8 A high-performance carbon fiber drive shaft.

To reduce the effects of vibrations and the resulting noises, manufacturers have used various methods to construct a drive shaft. An example of this is a drive shaft with cardboard liners inserted into the tube, which serve to decrease the shaft's vibration by damping the vibrations.

Drive shaft performance has also been improved by placing biscuits between the drive shaft and the cardboard liner. These biscuits are simply rubber inserts that reduce noise transfer within the drive shaft.

Another drive shaft design is the tube-in-tube, in which the input driving yoke has an input shaft that fits inside the hollow drive shaft. Rubber inserts are bonded to the outside diameter of the input shaft and to the inside diameter of the drive shaft. This design reduces the noise associated with drive shafts when they are stressed with directional rotation changes and also greatly reduces the vibration.

Drive shafts are often made with aluminum tubing or fiber composites. These composites give the shaft linear stiffness, while the positioning of the fibers provide for torsional strength. The advantages of a fiber composite drive shaft are weight reduction, torsional strength, fatigue resistance, easier and better balancing, and reduced interference from shock loading and torsional problems.

Linear stiffness means the shaft will resist deflection regardless of length.

TYPES OF DRIVE SHAFTS

A **Hotchkiss drive** system actually describes the entire drive shaft and rear axle assembly. This system allows the rear axle to move with the suspension while allowing torque to be transferred from the transmission to the rear axle. It is commonly called an open drive shaft system.

Three types of drive shafts have been used in automobiles. The first type, and the most commonly used, is the **Hotchkiss drive**. This type of drive shaft is readily recognized by its external shaft and U-joints (Figure 6-9). These shafts are either one- or two-piece assemblies consisting of a shaft with U-joints attached to each end. A Hotchkiss drive system can be used with either leaf or coil springs. When it is used with coil springs, additional braces, called control arms, must be used to control the movement of the rear drive axle (Figure 6-10).

A two-piece drive shaft is used on many long-wheelbase vehicles. It uses a third U-joint between the two shaft sections and a center bearing to support the middle of the shaft assembly (Figure 6-11).

The center support bearing maintains the alignment of the two pieces. In a center support bearing, a sealed ball or roller bearing allows the drive shaft to spin freely. The bearing is encased in rubber, or similar material, and the assembly is secured to a cross member of the frame or underbody. The rubber mount prevents noise and vibration from transferring into the passenger compartment of the vehicle.

The standard bearing is prelubricated and sealed and requires no further lubrication; however, some support bearings on heavy-duty vehicles have lubrication fittings.

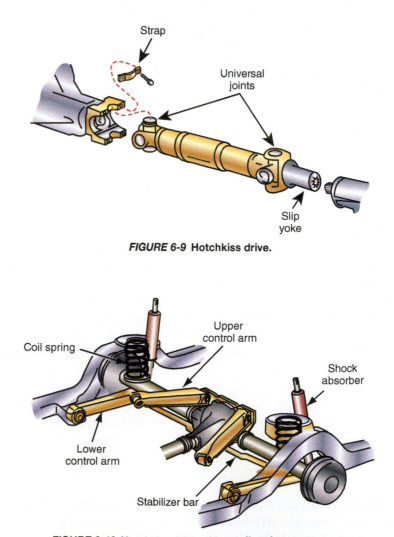

FIGURE 6-9 Hotchkiss drive.

FIGURE 6-10 Hotchkiss drive with a coil spring suspension.

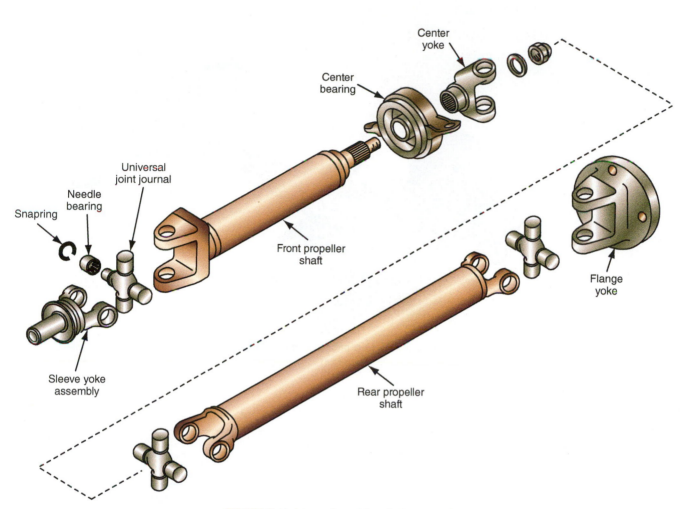

FIGURE 6-11 A two-piece drive shaft assembly.

AUTHOR'S NOTE: In their attempt to reduce vibrations set up by the drive shaft as it revolves in the center bearing, manufacturers are beginning to use magneto-rheological (MR) fluid in the mount. The MR fluid is used to change the stiffness of the mount while the vehicle is moving. In the presence of a magnetic field, MR fluids rapidly change their viscosity. The fluid becomes more viscous when the intensity of the magnetic field is increased; therefore, the mount becomes more rigid. To provide vibration isolation, a bracket with a very soft material, such as elastomer, is used in conjunction with the fluid's damping capability. This allows drive shaft vibrations to be dampened and isolated during all operating conditions, while keeping the drive shaft pieces securely aligned.

MR fluid is a synthetic oil with soft magnetic particles, such as iron, suspended in it.

The **torque tube** is a fixed tube over the drive shaft on some cars. It helps locate the rear axle and takes torque reaction loads from the drive axle so the drive shaft will not sense them.

A two-piece drive shaft is also referred to as a split drive shaft.

The second type of drive shaft is called a **torque tube**. Vehicles with independent rear suspension and a rear-mounted transaxle—such as late-model Corvettes and some Japanese RWD cars—use a torque tube. On these cars, the torque tube is rigidly connected at both ends. The Corvette setup has the clutch assembly behind the engine, and the torque tube attaches to the bell housing. The rear of the tube is bolted to a housing at the transaxle in the rear of the car. (Figure 6-12). The rotating inner drive shaft does

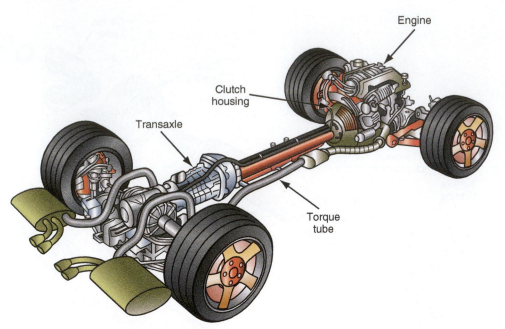

FIGURE 6-12 A front engine, rear transaxle setup with torque tube.

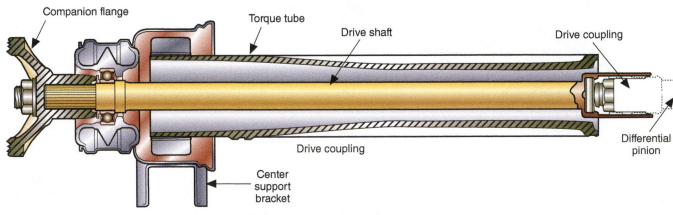

FIGURE 6-13 A typical torque tube assembly.

not need Universal joints because the transaxle location never changes relative to engine location (Figure 6-13).

The third and least commonly used type of drive shaft is the flexible type. This shaft is actually a flexible steel rope, much like an oversized speedometer cable that does not use Universal joints between the engine and the rear-mounted transaxle. The 1961–1963 Pontiac Tempest models used this type of drive shaft.

UNIVERSAL JOINTS

A drive shaft must smoothly transfer torque while rotating, changing length, and moving up and down. The different designs of drive shafts all attempt to ensure a vibration-free transfer of the engine's power from the transmission to the differential. This goal is complicated by the fact that the engine and transmission are bolted solidly to the frame of the car while the differential is mounted on springs. As the rear wheels go over bumps in the road or changes in the road's surface, the springs compress or expand. This changes the angle of the drive shaft between the transmission and the differential, as well as the distance between the two (Figure 6-14). To allow for these changes, the Hotchkiss-type drive shaft is fitted with at least two U-joints to permit variations in the angle of the drive and a slip joint that permits the

A torque tube system is commonly called an enclosed drive shaft system.

Shop Manual
Chapter 6, page 272

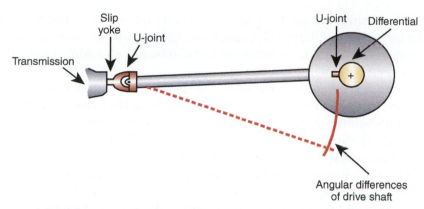

FIGURE 6-14 An illustration showing the changes in the length and angle of a drive shaft.

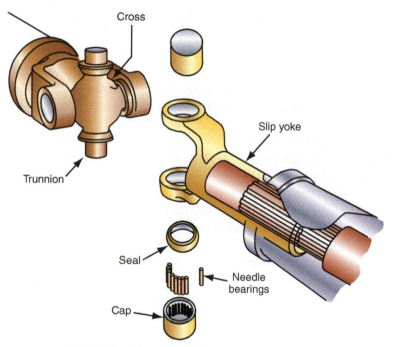

FIGURE 6-15 An exploded view of a Cardan U-joint.

effective length of the drive shaft to change. The Universal joint is basically a double-hinged joint consisting of two Y-shaped yokes, one on the driving or input shaft and the other on the driven or output shaft, plus a cross-shaped unit called the cross (Figure 6-15). A **yoke** is used to connect the U-joints together. The four arms of the cross are fitted with bearings in the ends of the two shaft yokes. The input shaft's yoke causes the cross to rotate, and the two other trunnions of the cross cause the output shaft to rotate. When the two shafts are at an angle to each other, the bearings allow the yokes to swing around on their trunnions with each revolution. This action allows two shafts, at a slight angle to each other, to rotate together.

Universal joints allow the drive shaft to transmit power to the rear axle through varying angles that are controlled by the travel of the rear suspension. Because power is transmitted on an angle, U-joints do not rotate at a constant velocity nor are they vibration free.

Speed Variations

Although simple in appearance, a U-joint is more intricate than it seems. Its natural action is to cause the shaft to which it is connected to speed up and slow down twice during each revolution, while operating at an angle. The amount that the speed changes varies according to the steepness of the Universal joint's angle.

The joint's cross is often called its spider.

The arms of the joint are also called trunnions.

The joint's **yoke** is a Y-shaped assembly into which two of the joint's arms fit.

A BIT OF HISTORY

The original Universal joint was developed in the sixteenth century by a French mathematician named Cardan. In the seventeenth century, Robert Hooke developed a cross-type Universal joint, based on the Cardan design. Then in 1902, Clarence Spicer modified Cardan and Hooke's inventions for the purpose of transmitting engine torque to an automobile's rear wheels. By joining two shafts with Y-shaped forks to a pivoting cruciform member, the problem of torque transfer through a connection that also needed to compensate for slight angular variations was eliminated. Both names, Spicer and Hooke, are at times used to describe a Cardan U-joint.

As a U-joint transmits torque through an angle, its output shaft speed increases and decreases twice on each revolution. These speed changes are not normally apparent, but may be felt as torsional vibration due to improper installation, steep and/or unequal operating angles, and high-speed driving.

If a U-joint's input shaft speed is constant, the speed of the output shaft accelerates and decelerates to complete a single revolution at the same time as the input shaft. In other words, the output shaft falls behind, then catches up with the input shaft during this revolution. The greater the angle of the output shaft, the more the velocity will change each shaft revolution.

U-joint **operating angle** is determined by the difference between the transmission **installation angle** and the drive shaft installation angle (Figure 6-16). When the U-joint is operating at an angle, the driven yoke speeds up and slows down twice during each drive shaft revolution. This acceleration and deceleration of the U-joint is known as speed variation.

These four changes in speed are not normally visible during rotation, but may be understood after examining the action of a U-joint. A Universal joint serves as a coupling between two shafts that are not in direct alignment. It would be logical to assume that the entire unit simply rotates. This is only true of the joint's input yoke.

The output yoke's rotational path looks like an **ellipse** because it can be viewed at an angle instead of straight on. This same effect can be obtained by rotating a coin with your fingers. The height of the coin stays the same even though the sides seem to get closer together.

This might seem to be merely a visual effect; however, it is more than that. The U-joint rigidly locks the circular action of the input yoke to the elliptical action of the output yoke. The result is similar to what would happen when changing a clock face from a circle to an ellipse.

Like the hands of a clock, the input yoke turns at a constant speed in its true circular path. The output yoke, operating at an angle to the other yoke, completes the path in the same amount of time (Figure 6-17). However, its speed varies, or is not constant, compared to the input.

Speed variation is more easily visualized when looking at the travel of the yokes by 90-degree quadrants (Figure 6-18). The input yoke rotates at a steady or constant speed through a complete 360-degree rotation. The output yoke quadrants alternate between shorter and longer distances of travel than the input yoke quadrants. When one point of the output yoke covers the shorter distance in the same amount of time, it must travel at a slower rate. Conversely, when traveling the longer distance in the same amount of time, it must move faster.

Because the average speed of the output yoke through the four 90-degree quadrants equals the constant speed of the input yoke during the same revolution, it is possible for the two mating yokes to travel at different speeds. The output yoke is falling behind and catching up constantly. The resulting acceleration and deceleration produces fluctuating torque and torsional vibrations and is characteristic of all Cardan U-joints. The steeper the U-joint angle, the greater the speed fluctuations. Conversely, smaller angles produce less change in speed. Most manufacturers maintain maximum operating angles of no more than three or four degrees. The difference between the front and rear angles should be no more than one degree.

The **operating angle** of a drive shaft is the amount the drive shaft deviates from the horizontal plane.

The **installation angle** of an object describes how far the object is tilted away from the horizontal plane.

An **ellipse** is merely a compressed form of a circle.

The speed variation of a Universal joint is sometimes called speed fluctuation.

A quadrant is another name for a quarter of something.

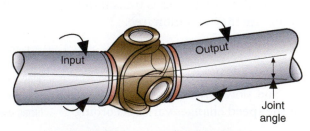

FIGURE 6-16 Universal joint angles.

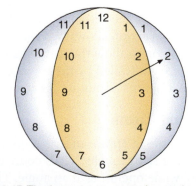

FIGURE 6-17 The face of a clock can be used to illustrate the elliptical action of the drive shaft's yokes.

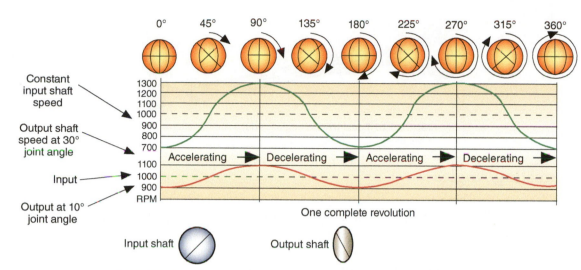

FIGURE 6-18 A chart showing the speed variations of a drive shaft's yoke.

In-phase describes the condition in which two events happen one after the other, regardless of speed.

SERVICE TIP: On a two-piece drive shaft, you may encounter problems if you are not careful. The center U-joint must be disassembled to replace the center support bearing. The center driving yoke is splined to the front drive shaft. If the yoke's position on the drive shaft is not indicated in some manner, the yoke could be installed in a position that is out of phase. Manufacturers use different methods of indexing the yoke to the shaft. Some use aligning arrows. Others machine a master spline that is wider than the others. When there are no indexing marks or master spline, the technician should always index the yoke to the drive shaft before disassembling the drive shaft assembly. This saves time and frustration during reassembly.

AUTHOR'S NOTE: Two-piece drive shafts present a problem when it comes to cancelling speed fluctuations at the front U-joint with equal speed changes at the rear U-joint. The front U-joint in a two-piece shaft is the "odd" joint, and must run at an angle small enough to prevent noticeable speed fluctuations but more than zero degrees. A U-joint must operate at a little angle or the rollers in the caps will not move, and the U-joint could seize.

Phasing of Universal Joints

The torsional vibrations set up by the changes in velocity are transferred down the drive shaft to the next U-joint. At this joint, similar acceleration and deceleration occurs. Because these speed changes take place at equal and reverse angles to the first joint, they cancel out each other whenever both occur at the same angle. To provide for this canceling effect, drive shafts should have at least two U-joints and their operating angles must be slight and equal to each other. Speed fluctuations can be canceled if the driven yoke has the same point of rotation, or same plane, as the driving yoke. When the yokes are in the same plane, the joints are said to be **in-phase** (Figure 6-19).

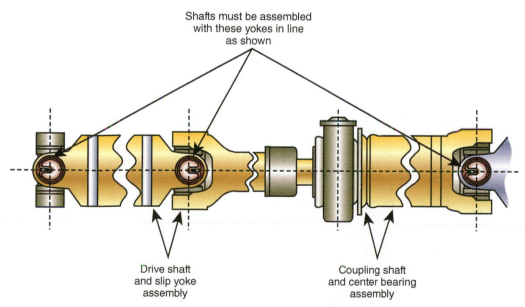

FIGURE 6-19 The U-joints should be in the same plane and in phase with each other.

Shop Manual
Chapter 6, page 272

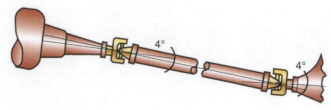

FIGURE 6-20 Equal U-joint angles reduce the vibrations of the shaft.

Canceling Angles

Canceling angles occur when the opposing operating angles of two Universal joints cancel the oscillations developed by the individual Universal joint.

Oscillations, resulting from speed variations, can be reduced by using **canceling angles** (Figure 6-20). The operating angle of the front U-joint is offset by the one at the rear of the drive shaft. When the front U-joint accelerates, causing a vibration, the rear U-joint decelerates, causing an equal but opposite vibration. These vibrations created by the two joints oppose each other and dampen the vibrations from one to the other. The use of canceling angles provides smooth drive shaft operation.

The correct operating angle of a U-joint must be maintained in order to prevent driveline vibration and damage. Shimming of leaf springs and the control arms on coil spring suspensions or adjusting the control arm eccentrics allow the operating angle of the drive shaft to be changed. Shimming at the transmission mount can also be done on some vehicles to change Universal joint angles.

Shop Manual
Chapter 6, page 275

Slip Joints

A **slip joint** is a variable-length connection that permits the drive shaft to change its effective length.

As road surfaces change, the rear axle assembly moves up and down with the rear suspension. Because the transmission is mounted to the frame, it does not move with the movement of the suspension and the relative distance between the transmission and rear drive axle changes. Drive shafts have a slip joint at one end of the drive shaft, which allow it to lengthen or shorten. The purpose of the slip joint is similar to the plunging CV joint used in FWD cars. The slip yoke is typically fitted to the front U-joint (Figure 6-21) but can also be positioned at the center of two-piece designs or at either end of the drive shaft. Some drive shafts are equipped with plunging-type CV joints in place of slip and Universal joints.

Some drive shafts are equipped with plunging-type CV joints in place of slip and Universal joints.

A slip joint assembly includes the transmission's output shaft, the slip joint itself, a yoke, a U-joint, and the drive shaft. The output shaft has external splines that match the internal splines of the slip joint. The meshing of the splines allows the two shafts to rotate together, but permits the ends of the shafts to slide along each other. This is why a slip yoke is also called a sliding yoke. This sliding motion allows for an effective change in the length of the drive shaft,

A slip yoke is also called a sliding yoke.

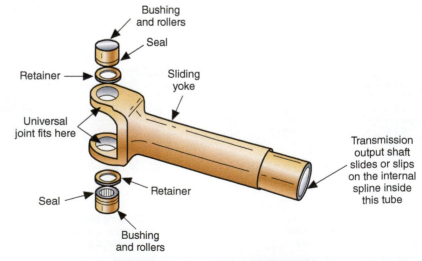

FIGURE 6-21 Typical slip joint assembly.

as the drive axles move toward or away from the car's frame. A U-joint connects the yoke of the slip joint to the drive shaft.

The slip yoke is supported in the transmission or transfer case rear housing by a bushing. Some circle-track cars with extreme amounts of rear suspension travel require long slip yokes to allow for large effective length changes of the drive shaft. These slip yokes are often supported by a bearing assembly in the extension housing instead of a bushing.

TYPES OF UNIVERSAL JOINTS

There are three common designs of Universal joints: single Universal joints retained by either an inside or outside snapring, coupled Universal joints (commonly called **double Cardan Universal joints**), and Universal joints held in the yoke by U-bolts or lock plates.

Single Universal Joints

The single **Cardan Universal joint**'s (Figure 6-22) primary purpose is to connect the two yokes that are attached directly to the drive shaft. The joint assembly forms a cross, with four machined trunnions or points equally spaced around the center of the axis. Needle bearings used to reduce friction and provide smoother operation are set into bearing cups. The trunnions of the cross fit into the cup assemblies, which fit snugly into the driving and driven Universal joint yokes. U-joint movement takes place between the trunnions, needle bearings, and bearing cups. There should be no movement between the bearing cup and its bore in the Universal joint yoke. The bearings are usually held in place by snaprings that drop into grooves in the yoke's bearing bores. The bearing caps allow free movement between the trunnion and yoke. The needle bearing caps also may be pressed into the yokes, bolted to the yokes, or held in place with U-bolts or metal straps.

Shop Manual
Chapter 6, page 262

Coupled Universal joints are commonly called **double Cardan Universal joints**. A double Cardan CV joint is actually two Cardan joints joined together by a yoke. By joining the two Cardan joints, rotational speed is not changed as the joint moves through its operating angles.

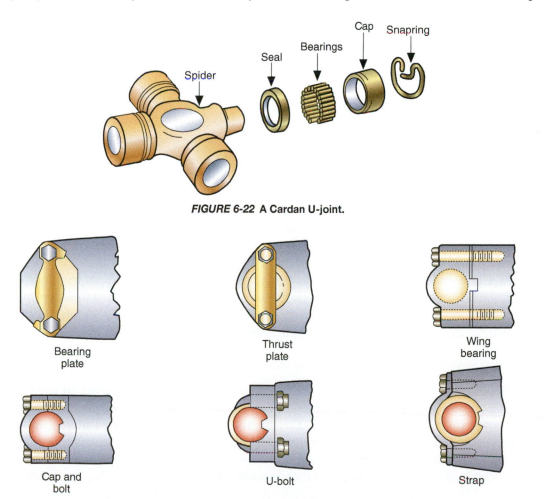

FIGURE 6-22 A Cardan U-joint.

SERVICE TIP:
There are many other methods used to retain the U-joint in its yoke, such as the use of a bearing plate, thrust plate, wing bearing, cap and bolt, U-bolt, and strap (Figure 6-23).

The **Cardan Universal joint** is also known as the cross or four-point joint. Its primary purpose is to connect the two yokes that are attached directly to the drive shaft.

Bearing plate

Thrust plate

Wing bearing

Cap and bolt

U-bolt

Strap

FIGURE 6-23 Various methods used to retain U-joints in their yokes.

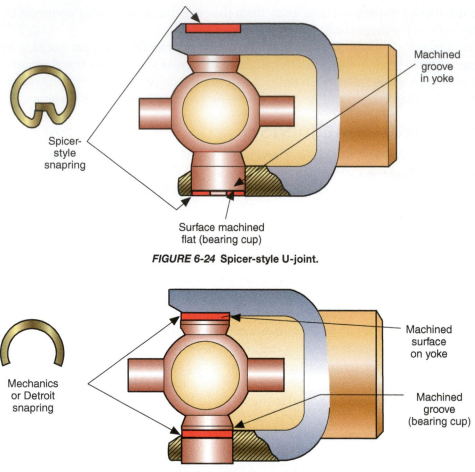

FIGURE 6-24 Spicer-style U-joint.

FIGURE 6-25 Mechanics-style U-joint.

There are other styles of single Universal joints. The method used to retain the bearing caps is the major difference between these designs. The Spicer style uses an outside snapring that fits into a groove machined in the outer end of the yoke (Figure 6-24). The bearing cups for this style are machined to accommodate the snapring.

The mechanics style uses an inside snapring or C-clip that fits into a groove machined in the bearing cup on the side closest to the grease seal (Figure 6-25). When installed, the clip rests against the machined portion of the yoke. The snaprings are retained by spring tension against the retaining ring grooves. Some joints have nylon injected in a groove in the bearing cap and yoke to retain it (Figure 6-26). The nylon may be heated to remove the joint, and an inside snap ring is used to retain the replacement joint.

The Cleveland style is an attempt to combine styles of Universal joints to obtain more applications from one joint. The bearing cups for this U-joint are machined to accommodate either Spicer- or mechanics-style snaprings. If a replacement U-joint comes with both style clips, use the clips that pertain to your application.

The mechanics-style U-joint is also called the Detroit/Saginaw style.

SERVICE TIP: In order to get the correct parts, you should know the type of U-joint and the yoke span originally installed in the car (Figure 6-28).

AUTHOR'S NOTE: Off-road performance vehicles really put U-joints to the test. U-joint operating angles far surpass OEM specifications as suspension components allow for large amounts of travel. The torque of powerful engines places added stress on the U-joints. Aftermarket manufacturers have designed a number of tough alternatives for the off-road enthusiast. One version uses bushings instead of needle bearings and alloy-retained trunnions instead of traditional caps (Figure 6-27).

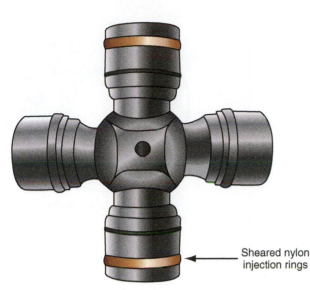

FIGURE 6-26 A U-joint retained by an injected nylon ring.

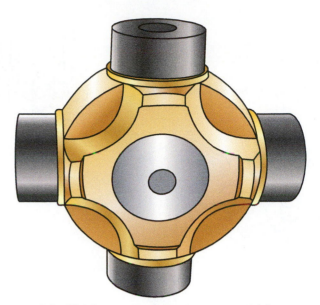

FIGURE 6-27 An aftermarket heavy-duty U-joint.

Double Cardan Universal Joint

A double Cardan U-joint is used with split drive shafts and consists of two individual Cardan U-joints closely connected by a centering socket yoke and a center yoke that functions like a ball and socket. The ball and socket splits the angle of two drive shafts between two U-joints (Figure 6-29). Because of the centering socket yoke, the total operating angle is divided equally between the two joints. Because the two joints operate at the same angle, the normal fluctuations that result from the use of a single U-joint are canceled out. The acceleration or deceleration of one joint is canceled by the equal and opposite action of the other.

Most often installed in front-engined rear-wheel-drive luxury cars and in the front drive shafts of some 4WD vehicles, the double Cardan Universal joint smoothly transmits torque regardless of the operating angle of the driving and driven members. It is therefore classified as a CV Universal joint (Figure 6-30). This joint is used when the U-joint operating angle is too large for a single joint to handle. On some vehicles, the double joint is used at both ends of the drive shaft. On other vehicles, it is used only on the drive input end of the drive shaft.

Shop Manual
Chapter 6, page 267

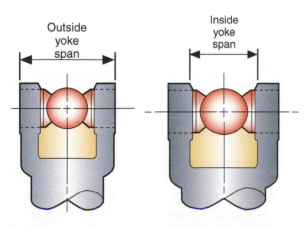

FIGURE 6-28 You should know both the outside (A) and inside (B) yoke span when ordering a new joint.

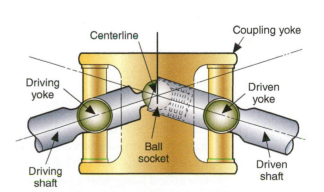

FIGURE 6-29 The ball and socket of a double Cardan joint splits the angle of the two shafts.

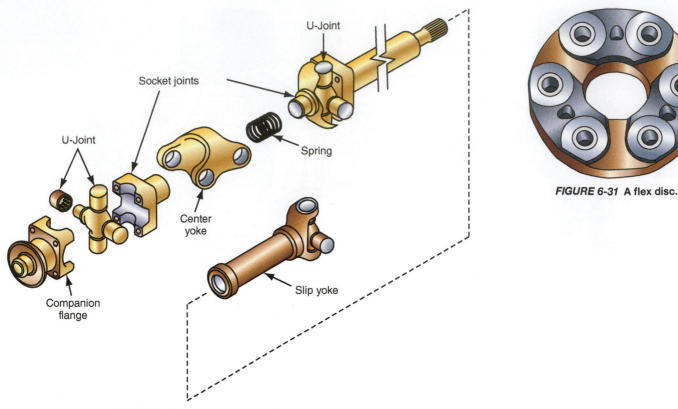

FIGURE 6-30 A double Cardan joint disassembled.

FIGURE 6-31 A flex disc.

Shop Manual
Chapter 6, page 269

Drive Shaft Flex Disc. A few RWD vehicles, mostly European, have a flexible coupler between the transmission's output shaft and the drive shaft to join the two together. This coupler is designed to isolate engine vibrations from the rest of the drivetrain. This coupler or disc serves the same purpose as a double Cardan joint, but relies on the flexibility of rubber to absorb the vibrations and flex with varying drive shaft angles (Figure 6-31). The rubber is reinforced to provide durability. However, they can only handle limited drive angles and will fail if they experience extreme operating angles. They are also found on a few 4WD vehicles.

Constant-Velocity Joints

For many years CV joints were primarily used on FWD drive axles; however, the half shafts of some RWD cars with IRS—such as Ford Thunderbird, Corvette, BMW, and Porsche, and on some front drive shafts on 4WD trucks—use them. The commonly used types of CV joints are the Rzeppa and the tripod. In an effort to reduce noise, vibration, and harshness concerns (NVH), some SUV and truck manufacturers use ball-style CV joints in place of U-joints on rear drive shafts due to their ability to smoothly transmit power (Figure 6-32).

AUTHOR'S NOTE: "Rock crawlers" and other extreme four-wheel-drive trucks have used constant-velocity joints in place of U-joints for many years. CV joints are able to handle the extreme drive shaft operating angles on vehicles with highly modified suspensions and the increased ride height needed to clear oversize tires and improve ground clearance.

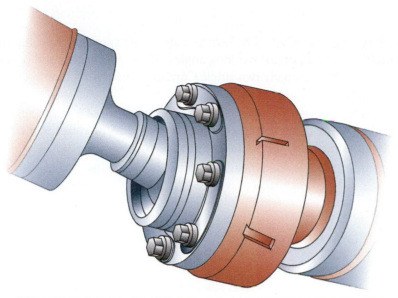

FIGURE 6-32 Rzeppa-style CV joints replace U-joints in some applications.

SUMMARY

- A drive shaft is normally made from seamless steel tubing with Universal joint yokes fastened to each end.
- U-joints are used to allow the angle and the effective length of the drive shaft to change.
- A drive shaft is actually no more than a flexible extension of the transmission's output shaft.
- Yokes on the drive shaft provide a means of connecting the shafts together.
- Balance weights are welded to the outside of a drive shaft to balance it and reduce its natural vibrations.
- Some drive shafts are internally lined with cardboard or cardboard with rubber inserts to help offset torsional vibration problems.
- A Hotchkiss drive system has an external drive shaft with at least two Universal joints.
- A torque tube consists of a tubular steel or small-diameter solid shaft enclosed in a larger steel tube and is rigidly connected to the rear axle housing.
- The operating angle of a Universal joint is determined by the installation angles of the transmission and rear axle assembly.
- U-joints vibrate if the connecting shafts are not on the same plane because one shaft will be accelerating and decelerating at different speeds than the other.
- Drive shaft vibrations can be reduced by using canceling angles. The operating angle of the front joint is offset by the angle of the rear joint.
- Speed fluctuations can be canceled if the driven yoke has the same point of rotation, or same plane, as the driving yoke. When the two yokes are in the same plane, the joints are said to be in-phase.
- There are three common designs of Universal joints: single U-joints retained by either an inside or outside snapring, coupled U-joints, and U-joints held in the yoke by U-bolts or lock plates.
- A single Cardan joint, the most common type of joint, uses a spider, four machined trunnions, needle bearings, and bearing caps to allow the transmission of power through slight shaft angle changes.

151

SUMMARY

TERMS TO KNOW

(continued)

Operating angle
Slip joint
Torque tube
Yoke

- A double Cardan joint is called a CV joint because shaft speeds do not fluctuate, regardless of the shaft's angle at normal working angles. These joints are used on two-piece shaft assemblies and are actually two single Cardan joints joined together by a centering socket yoke assembly.

- The methods used to retain a U-joint in its yoke are the use of a snapring, C-clip, bearing plate, thrust plate, wing bearing, cap and bolt, U-bolt, and/or strap. Some manufacturers use an injected nylon ring to retain the caps in the yoke.

- A center support bearing is used on all two-piece drive shafts.

REVIEW QUESTIONS

Short Answer Essays

1. State the purposes of a Universal joint.

2. What methods are used by some auto manufacturers to reduce drive shaft torsional vibration problems?

3. What determines the operating angle of a Universal joint?

4. Explain why and when U-joints vibrate.

5. What effect do in-phase Universal joints have on the operation of a drive shaft?

6. What is meant by canceling angles?

7. Describe the construction and operation of a single Cardan joint.

8. Describe the construction and operation of a double Cardan joint.

9. Why are some cars and trucks equipped with a two-piece drive shaft?

10. What methods are used to retain a U-joint in its yoke?

Fill-in-the-Blanks

1. A drive shaft is normally made from _____ _____ tubing with _____ _____ _____ fastened to each end.

2. A drive shaft is actually an extension of the transmission's _____ _____.

3. _____ on the drive shaft provide a means of connecting the shafts together.

4. _____ _____ are welded to the outside of a drive shaft to _____ it and reduce its natural _____.

5. A slip joint is normally fitted to the _____ _____ _____.

6. A Hotchkiss drive system has an external _____ _____ with at least _____ Universal joints.

7. A _____ _____ is made up of a small-diameter solid shaft enclosed in a larger steel tube and is rigidly connected to the transmission and the rear axle assembly.

8. There are three common designs of Universal joints: _____ retained by either an inside or outside snapring, _____, and _____ held in the yoke by U-bolts or lock plates.

9. U-joints may be retained in the drive shaft by _____, _____, or _____.

10. The basic styles of Cardan joints are _____ and _____.

11. The type of drive shaft with an input shaft bonded with rubber to the inside of a hollow drive shaft is a _____.

ASE-Style Review Questions

1. When discussing drive shaft design,
   Technician A says that the shorter the shaft is, the less likely it is to become out of balance.
   Technician B says that cardboard is often inserted into the shaft to dampen the vibrations of the shaft.
   Who is correct?
   A. A only
   B. B only
   C. Both A and B
   D. Neither A nor B

2. When discussing the Hotchkiss design of drive shaft,

   Technician A says that it consists of a small-diameter solid shaft enclosed in a larger tube.

   Technician B says that it has a Universal joint attached to each end of the shaft.

   Who is correct?

   A. A only   C. Both A and B

   B. B only   D. Neither A nor B

3. When discussing the phasing of Universal joints,

   Technician A says that this allows the speed changes at one U-joint to be canceled by the other U-joint.

   Technician B says that this means that both U-joints are positioned at opposite angles to each other.

   Who is correct?

   A. A only   C. Both A and B

   B. B only   D. Neither A nor B

4. When discussing the different types of Universal joints,

   Technician A says that the most commonly used type is the double Cardan.

   Technician B says that the double Cardan type consists of two U-joints connected by a centering socket yoke.

   Who is correct?

   A. A only   C. Both A and B

   B. B only   D. Neither A nor B

5. When discussing slip joints,

   Technician A says that they allow the drive shaft to change its effective length as the suspension goes through jounce and rebound.

   Technician B says that they are normally found at both ends of a Hotchkiss-style drive shaft.

   Who is correct?

   A. A only   C. Both A and B

   B. B only   D. Neither A nor B

6. When discussing the different types of drive shafts that have been used to reduce vibrations,

   Technician A says that the flexible rope design is the most commonly used.

   Technician B says that the tube-in-tube design is not fitted with Universal joints.

   Who is correct?

   A. A only   C. Both A and B

   B. B only   D. Neither A nor B

7. When discussing Universal joint speed fluctuations,

   Technician A says that Universal joint speed fluctuations can be canceled if the driven yoke has the same plane of rotation as the driving yoke.

   Technician B says that Universal joint speed fluctuations can be canceled by putting the joints in-phase.

   Who is correct?

   A. A only   C. Both A and B

   B. B only   D. Neither A nor B

8. When discussing slip joints,

   Technician A says that the yoke normally has external splines.

   Technician B says that the slip yoke is normally held in place on the transmission's output shaft by a snapring.

   Who is correct?

   A. A only   C. Both A and B

   B. B only   D. Neither A nor B

9. When discussing the cause of drive shaft vibration,

   Technician A says that excessive joint operating angles can cause an increase in drive shaft vibration.

   Technician B says two-piece drive shaft that is assembled out of phase can cause an increase in shaft vibration.

   Who is correct?

   A. A only   C. Both A and B

   B. B only   D. Neither A nor B

10. While discussing Universal joints,

   Technician A says the joint consists of two Y-shaped yokes plus a cross-shaped unit called the cross or spider.

   Technician B says the single Cardan Universal joint assembly has four machined trunnions equally spaced around the center of the axis. Needle bearings are set into bearing cups. The trunnions of the cross fit into the cup assemblies that fit into the Universal joint yokes.

   Who is correct?

   A. A only   C. Both A and B

   B. B only   D. Neither A nor B

Chapter 7

DIFFERENTIALS AND DRIVE AXLES

UPON COMPLETION AND REVIEW OF THIS CHAPTER, YOU SHOULD BE ABLE TO:

- Describe the purpose of a differential.

- Identify the major components of a differential and explain their purpose.

- Describe the various gears in a differential assembly and state their purpose.

- Describe the various methods used to mount and support the drive pinion shaft and gear.

- Explain the need for drive pinion bearing preload.

- Describe the difference between hunting, nonhunting, and partial nonhunting gear sets.

- Explain the purpose of the major bearings within a differential assembly.

- Describe the operation of a limited-slip differential.

- Describe the operation of a locking differential.

- Describe the construction and operation of a rear axle assembly.

- Identify and explain the operation of the two major designs of rear axle housings.

- Explain the operation of a FWD differential and its drive axles.

- Describe the different types of drive axles and the bearings used to support each of them.

INTRODUCTION

Shop Manual
Chapter 7, page 285

IRS stands for independent rear suspension.

Final drive is the final set of reduction gears the engine's power passes through on its way to the drive wheels.

The drive axle assembly of a rear-wheel-drive (RWD) vehicle is mounted at the rear of the car. Most of these assemblies use a single housing to mount the differential gears and axles (Figure 7-1). The entire housing is part of the suspension and helps to locate the rear wheels.

Another type of rear drive axle is used with **independent rear suspension (IRS)**. With IRS the differential is bolted to the chassis and does not move with the suspension. The axles are connected to the differential and drive wheel constant-velocity (CV) or Universal joints (Figure 7-2). Because the axles move with the suspension and the differential is bolted to the chassis, a common housing for these parts is impossible.

On RWD cars, the **final drive** is located in the rear axle housing. On most front-wheel-drive (FWD) cars, the final drive is located within the transaxle. Some current FWD cars mount the engine and transaxle longitudinally. These configurations use a differential that is similar to other FWD models. Some FWD cars have a longitudinally mounted engine fitted to a special transmission with a separate differential mounted to it.

A differential is needed between any two drive wheels, whether in a RWD, FWD, all-wheel-drive (AWD), or four-wheel-drive (4WD) vehicle. The two drive wheels must turn at different speeds when the vehicle is in a turn.

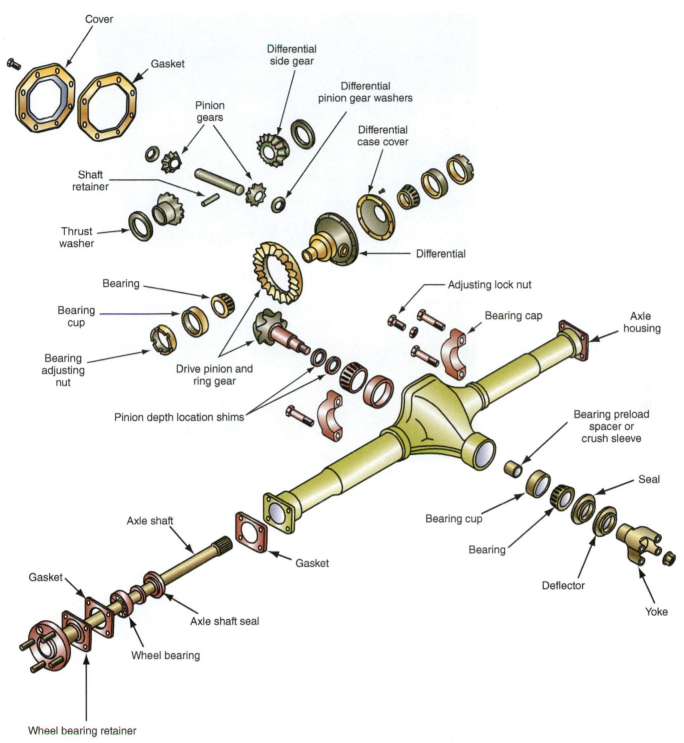

Cover

Gasket

Differential side gear

Differential pinion gear washers

Pinion gears

Differential case cover

Shaft retainer

Thrust washer

Differential

Bearing

Adjusting lock nut

Bearing cup

Bearing cap

Axle housing

Bearing adjusting nut

Drive pinion and ring gear

Pinion depth location shims

Bearing preload spacer or crush sleeve

Seal

Bearing cup

Bearing

Axle shaft

Gasket

Gasket

Deflector

Axle shaft seal

Yoke

Gasket

Wheel bearing

Wheel bearing retainer

FIGURE 7-1 **A typical RWD axle assembly.**

RWD final drives normally use a hypoid ring and pinion gear set that turns the power-flow 90 degrees from the drive shaft to the drive axles. A hypoid gear set allows the drive shaft to be positioned low in the vehicle because the final drive pinion gear centerline is below the ring gear centerline. Hypoid gear sets are also found in some longitudinal FWD and AWD transaxles. Spiral bevel gear sets are often found in quick-change racing differentials (Figure 7-3). Amboid differentials have the pinion gear centerline above the ring gear centerline. This arrangement is used in many large trucks, and is also used in some 4WD vehicles to help maintain workable drive shaft angles.

Normally, rear axles on RWD vehicles are called live axles because they transmit power.

FIGURE 7-2 An IRS assembly with suspension.

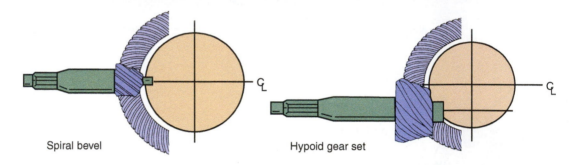

Spiral bevel   Hypoid gear set

FIGURE 7-3 In a hypoid gear set, the drive pinion meshes with the ring gear at a point below its centerline.

AUTHOR'S NOTE: Terminology used to describe the final drive unit of a vehicle varies widely, and can lead to some confusion. We just have to realize that all these names refer to the same thing. To many, the final drive unit is simply called "the diff." To others, it's the "rear end," although that wouldn't work for the final drive in the front! To some people from an earlier generation, the final drive unit is "the pumpkin," a descriptive term that fits a removable carrier differential. To them, the pumpkin is mounted in the "banjo housing." Other common terms for a removable carrier unit include "third member" and "center section." New or old, the good technician understands these various terms and can relate to just about any customer. In this chapter, the final drive unit is often referred to as the differential.

On FWD cars with transversely mounted engines, the power-flow axis is naturally parallel to that of the drive axles. Because of this, a simple set of helical gears in the transaxle serve as the final drive gears.

The differential is a geared mechanism located between the two driving axles. It rotates the driving axles at different speeds when the vehicle is turning a corner. It also allows both axles to turn at the same speed when the vehicle is moving straight. The drive axle assembly directs driveline torque to the vehicle's drive wheels. The gear ratio of the differential's ring and pinion gear is used to increase torque, which improves driveability. The differential serves to establish a state of balance between the forces or torques between the drive wheels and allows the drive wheels to turn at different speeds when the vehicle changes direction.

FUNCTION AND COMPONENTS

The differential allows for different speeds at the drive wheels when a vehicle goes around a corner or any time there is a change of direction. When a car turns a corner, the outside wheels must travel farther and faster than the inside wheels (Figure 7-4). If compensation is not made for this difference in speed and travel, the wheels would skid and slide, causing poor handling and excessive tire wear. Compensation for the variations in wheel speeds is made by the differential assembly. While allowing for these different speeds, the differential also must continue to transmit torque.

The differential of a RWD vehicle is normally housed with the drive axles in a large casting called the rear axle assembly. Power from the engine enters into the center of the rear axle assembly and is transmitted to the drive axles. The drive axles are supported by bearings and are attached to the wheels of the car. The power entering the rear axle assembly has its direction changed by the differential. This change of direction is accomplished through the hypoid gears used in the differential.

Power from the drive shaft is transmitted to the rear axle assembly through the pinion flange. This flange is the connecting yoke to the rear Universal joint. Power then enters the final drive on the **pinion gear** (Figure 7-5). The pinion teeth engage the ring gear, which is mounted upright at a 90-degree angle to the pinion. Therefore, as the drive shaft turns, so do the pinion and ring gears.

Not too long ago, a differential was something that was in the rear axle assembly. Now, with the popularity of FWD vehicles, the differential is part of the transaxle and is most often called the final drive.

When engines are placed longitudinally in the car, they are said to have "north/south" placement.

Helical gears are gears on which the teeth are at an angle to the gear's axis of rotation.

When engines are mounted transversely in the car, they are said to have "east/west" or sideways placement.

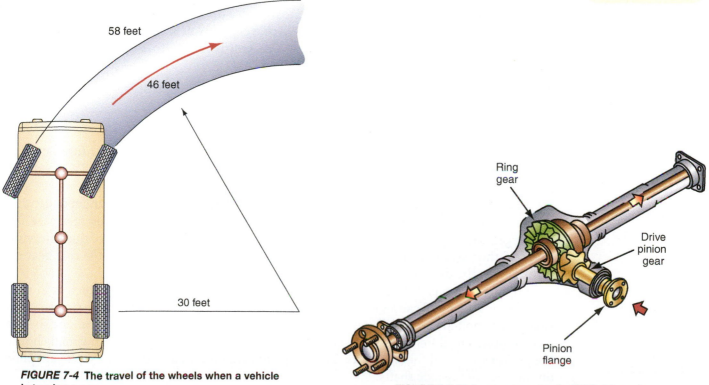

58 feet

46 feet

30 feet

FIGURE 7-4 The travel of the wheels when a vehicle is turning a corner.

Ring gear

Drive pinion gear

Pinion flange

FIGURE 7-5 Main components of a RWD drive axle.

In a **ring gear** and **pinion gear** set, the pinion is the smaller drive gear and the ring gear is the larger driven gear.

A BIT OF HISTORY

Early automobiles were driven by means of belts and ropes around pulleys mounted on the driving wheels and engine shaft or transmission shaft. As there was always some slippage of the belts, one wheel could rotate faster than the other when turning a corner. When belts proved unsatisfactory, automobile builders borrowed an idea from the bicycle and applied sprockets and chains. This was a positive driving arrangement, which made it necessary to provide differential gearing to permit one wheel to turn faster than the other.

The term *differential* means relating to or exhibiting a difference or differences.

Shop Manual

Chapter 7, page 286

When two beveled gears are meshed, the driving and driven shafts can rotate at a 90-degree angle.

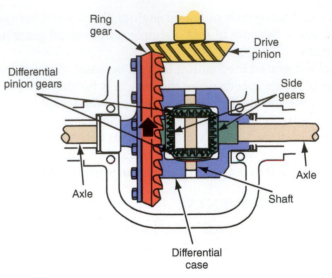

FIGURE 7-6 Components of a typical final drive assembly.

The **ring gear** is fastened to the differential case with hardened bolts or rivets. The differential case is made of cast iron and is supported by two tapered roller bearings in the rear axle housing. Holes machined through the center of the differential housing support the differential pinion shaft. The pinion shaft is retained in the housing case by a roll pin or a specially designed bolt. Two beveled differential pinion gears and thrust washers are mounted on the differential pinion shaft. In mesh with the differential pinion gears are two axle side gears splined internally to mesh with the external splines on the left and right axle shafts (Figure 7-6). Thrust washers are placed between the differential pinions, axle side gears, and differential case to prevent wear on the inner surfaces of the differential case.

> **AUTHOR'S NOTE:** Every manufacturer uses the term "case" to refer to the cast assembly that the ring gear bolts to. However, a great many technicians call this part the "carrier." Though technically incorrect, this practice is so common that the ASE A3 Manual Drive Train and Axles test refers to this component as the "case (carrier)."

DIFFERENTIAL OPERATION

The two drive wheels are mounted on axles that have a differential side gear fitted on their inner ends (Figure 7-7). To turn the power-flow 90 degrees, as is required for RWD vehicles, the side gears are bevel gears.

The **differential case** is mounted on bearings so that it is able to rotate independently of the drive axles. A pinion shaft, with small pinion gears, is fitted inside the differential case. The pinion gears mesh with the **side gears**. The ring gear is bolted to the flange of the differential case and the two rotate as a single unit. The drive pinion gear meshes with the ring gear and is rotated by the drive shaft (Figure 7-8).

Engine torque is delivered by the drive shaft to the drive pinion gear, which is in mesh with the ring gear and causes it to turn. Power flows from the pinion gear to the ring gear. The ring gear is bolted to the differential case, which drives the side gears, pinions, and axles as an assembly. The differential case extends from the side of the ring gear and normally houses the pinion gears and the side gears. The side gears are mounted so they can slip over splines on the ends of the axle shafts.

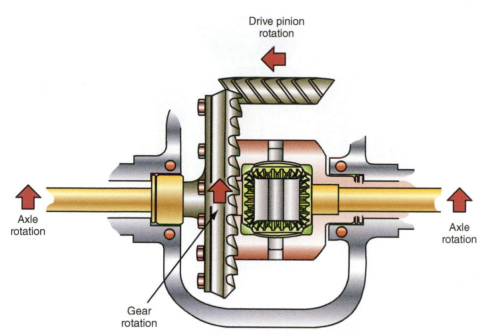

FIGURE 7-7 Power-flow through a RWD differential.

There is a gear reduction between the drive pinion gear and the ring gear, causing the ring gear to turn about one-third to one-fourth the speed of the drive pinion. The pinion gears are located between and meshed with the side gears (Figure 7-9), thereby forming a square inside the differential case. Differentials have two, three, or four pinion gears that are in mesh with the side gears (Figure 7-10). The differential pinion gears are free to rotate on their own centers and can travel in a circle as the differential case and pinion shaft rotate. The side gears are meshed with the pinion gears and are splined to the axles in a live axle assembly, to stub axles in an IRS setup and some FWD vehicles, and to CV shafts in most FWD vehicles.

> The **differential case** is the metal unit that encases the differential side gears and pinion gears and to which the ring gear is attached.

> **AUTHOR'S NOTE:** A differential with three or four pinions is stronger than a two-pinion setup as load is spread over the extra gears. Instead of a shaft, these gears ride on a "spider," so called because of its multiple legs. That's where we get the term "spider gears."

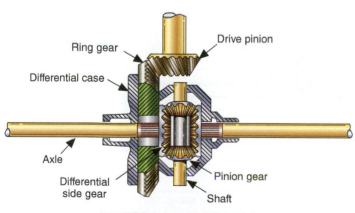

FIGURE 7-8 A basic differential.

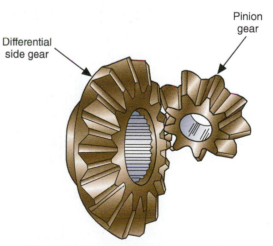

FIGURE 7-9 Pinion gears in mesh with the side gears.

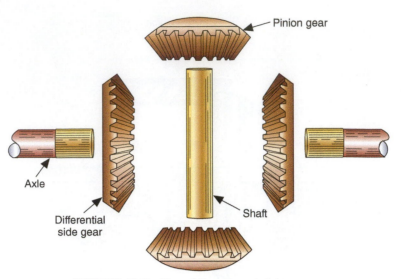

FIGURE 7-10 Position of the side and pinion gears.

The two side gears are placed on the side of the differential case, which is why they are called **side gears**.

The gear ratio in a differential is known as the axle ratio.

The small pinion gears are mounted on a pinion shaft that passes through the gears and the case. The pinion gears are in mesh with the axle side gears, which are splined to the axle shafts. In operation, the rotating differential case causes the pinion shaft and pinion gears to rotate end over end with the case (Figure 7-11). Because the pinion gears are in mesh with the side gears, the side gears and axle shafts are also forced to rotate.

When a car is moving straight ahead, both drive wheels are able to rotate at the same speed. Engine power comes in on the pinion gear and rotates the ring gear. The differential case is rotated with the ring gear. The pinion shaft and pinion gears are carried around by the case and all of the gears rotate as a single unit. Each side gear rotates at the same speed and in the same plane as does the case and they transfer their motion to the axles. The axles are thus rotated, and the car moves. Each wheel rotates at the same speed because each axle receives the same rotation.

As the vehicle goes around a corner, the inside wheel travels a shorter distance than the outside wheel. The inside wheel must therefore rotate more slowly than the outside wheel. In this situation, the differential pinion gears will "walk" forward on the slower turning or inside side gear (Figure 7-12). As the pinion gears walk around the slower side gear, they drive the other side gear at a greater speed. An equal percentage of speed is removed from one axle and given to the other (Figure 7-13); however, the torque applied to each wheel is equal.

When one of the driving wheels has little or no traction, the torque required to turn the wheel without traction is very low. The wheel with good traction in effect is holding the axle

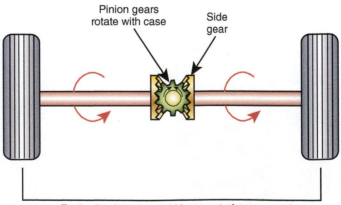

Each wheel rotates at 100 percent of case speed

FIGURE 7-11 The position of pinion gears inside the case causes the side gears to rotate.

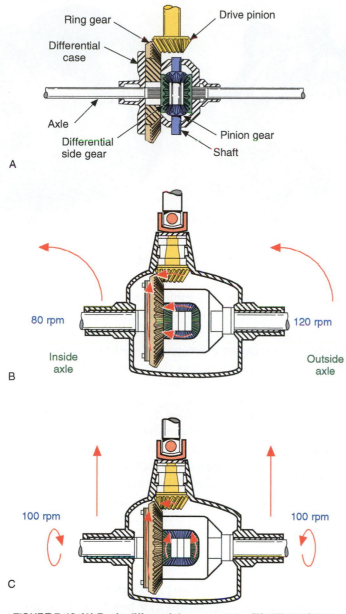

FIGURE 7-12 (A) Basic differential components; (B) differential action while the vehicle is turning left; and (C) differential action while the vehicle is moving straight.

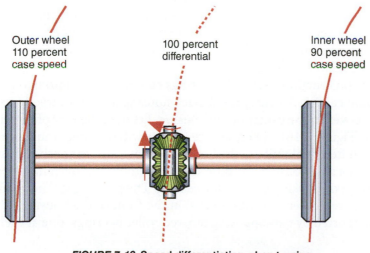

FIGURE 7-13 Speed differentiation when turning.

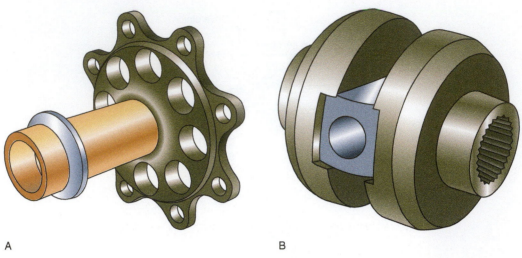

A   B

FIGURE 7-14 (A) Spool. (B) Mini-spool.

Shop Manual
Chapter 7, page 292

gear on that side stationary. This causes the pinions to walk around the stationary side gear and drive the other wheel at twice the normal speed but without any vehicle movement. With one wheel stationary, the other wheel turns at twice the speed shown on the speedometer. Excessive spinning of one wheel can cause severe damage to the differential. The small pinion gears can actually become welded to the pinion shaft or differential case.

> **AUTHOR'S NOTE:** Believe it or not, some drivers do not want any differential action to occur in the final drive! Drag racers and some dirt track drivers want to ensure equal torque is being applied to each drive wheel, and often install a "spool" that directly connects the axles. A true spool replaces the case, whereas a mini-spool fits in the existing case and connects the axles (Figure 7-14).

Integral carriers are commonly referred to as unitized or Salisbury-type differentials.

AXLE HOUSINGS

Live rear axles use a one-piece housing with two tubes extending from each side. These tubes enclose the axles and provide attachments for the axle bearings. The housing also shields the parts from dirt and retains the differential lubricant.

The rear axle housing is sometimes called a banjo because of the bulge in the center of the housing. The bulge contains the final drive gears and differential gears.

In IRS (Figure 7-15) or FWD systems, the housing is in three parts. The center part houses the final drive and differential gears. The outer parts support the axles by providing attachments for the axle bearings. These parts also serve as suspension components and attachment points for the steering gear or brakes. In FWD applications, the differential and final drive are either enclosed in the same housing as the transmission or in a separate housing bolted directly to the transmission housing.

Based on their construction, rear axle housings can be divided into two groups: integral carrier or removable carrier. An **integral carrier** housing attaches directly to the rear suspension. A service cover, in the center of the housing, fits over the rear of the differential and rear axle assembly (Figure 7-16). When service is required, the cover must be removed. The components of the differential unit are then removed from the rear of the housing.

In the rear, the outer sections of the housing may be called the uprights and in the front they are usually called the steering knuckle.

In an integral-type axle housing, the differential assembly and the pinion bearing retainer are supported by the axle housing in the same casting. The pinion gear and shaft is supported by two opposing tapered roller bearings located in the front of the housing. The differential assembly is also supported by two opposing tapered roller bearings, one at each side.

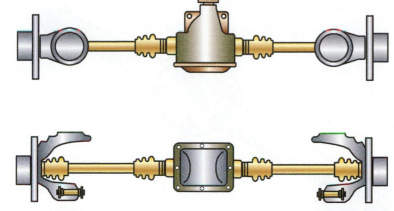

FIGURE 7-15 A drive axle assembly on a RWD vehicle with IRS.

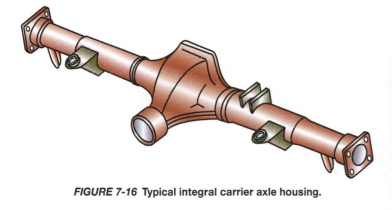

FIGURE 7-16 Typical integral carrier axle housing.

FIGURE 7-17 Typical removable carrier axle housing.

The differential assembly of a **removable carrier** assembly can be removed from the front of the axle housing as a unit (Figure 7-17). The differential is serviced on a bench and then installed into the axle housing. The differential assembly is mounted on two opposing tapered roller bearings retained in the housing by removable caps. The pinion gear, pinion shaft, and the pinion bearings are typically assembled in a pinion retainer, which is bolted to the carrier housing (Figure 7-18).

A typical housing has a cast-iron center section with axle shaft tubes pressed and welded into either side. The rear axle housing encloses the complete rear-wheel driving axle assembly. In addition to housing the parts, the axle housing also serves as a place to mount the vehicle's rear suspension and braking system. With IRS, the differential housing is mounted to the vehicle's chassis and does not move with the suspension.

Differential Fluids

On most nonremovable carrier-type rear axle assemblies, there is a differential inspection cover mounted to the rear of the housing. Normally, there is a fill plug near the center of the inspection cover or on the side of the axle housing. The opening for the plug is used to check the fluid level and to add fluid to the housing. When the housing is filled with oil, its level is at the bottom of the hole.

Hypoid gears require hypoid gear lubricant of the extreme pressure type and high viscosity. Limited slip differentials require special limited slip lubricant that provides the required

> **Removable carriers** are often referred to as the third member, dropout carrier, or pumpkin.

> In appearance the two designs of rear axle housing look similar except that the opening for the differential unit on a removable type is at the front, and the rear of the housing is solid.

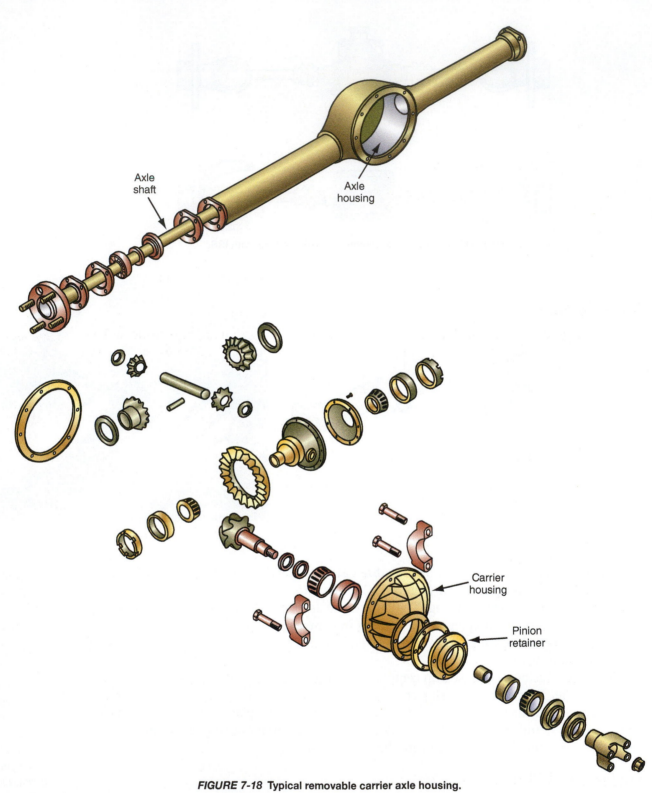

Axle shaft

Axle housing

Carrier housing

Pinion retainer

FIGURE 7-18 Typical removable carrier axle housing.

Shop Manual
Chapter 7, page 291

coefficient of friction for the clutch discs or cones as well as proper lubrication. Transaxles and some RWD differentials may require lower-viscosity oil, such as automatic transmission fluid. In addition, some transaxles may require separate lubricants for the transmission and differential.

The oil is circulated by the ring gear and thrown over all the parts (Figure 7-19). Special troughs or gullies are used to return the oil to the ring and pinion area. The housing is sealed with gaskets and oil seals to keep the fluid in and dirt and moisture out.

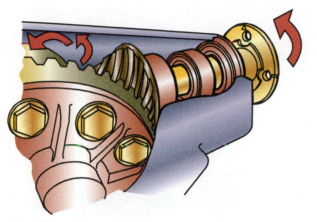

FIGURE 7-19 The flow of oil in a hypoid gear set as it spins.

HYPOID GEARS

Hypoid gear sets are commonly used in RWD passenger car and light truck applications. The pinion gear in a hypoid gear set is mounted well below the centerline of the ring gear. Hypoid gears are quiet running.

The teeth of a hypoid gear are curved to follow the form of a spiral, causing a wiping action when meshing. As the gears rotate, the teeth slide against each other. Because of this sliding action, the ring and pinion gears can be machined to allow for near perfect mating, which results in smoother action and a quiet-running gear set. Because this sliding action produces extremely high pressures between the gear teeth, only a hypoid-type lubricant should be used with hypoid gear sets.

The spiral-shaped teeth result in different tooth contacts as the pinion and ring gear rotate. The drive side of the teeth is curved in a convex shape, and the coast side of the teeth is concave (Figure 7-20). The inner end of the teeth on the ring gear is known as the toe and the outer end of the teeth is the heel (Figure 7-21).

When engine torque is being applied to the drive pinion gear, the pinion teeth exert pressure on the drive side of the ring gear teeth. During coast or engine braking, the concave side of the ring gear teeth exerts pressure on the drive pinion gear.

Upon heavy acceleration, the drive pinion attempts to climb up the ring gear and raises the front of the differential. The suspension's leaf springs or the torque arm on coil spring suspensions absorb much of the torque to limit the movement of the axle housing (Figure 7-22).

This phenomenon actually always occurs; it is just more noticeable during heavy acceleration. It is caused by an "equal and opposite" reaction of the wheels pushing one way while

> The drive side of the teeth is the side that has engine power working on it when moving forward, whereas the coast side is the side of the teeth that has contact during deceleration.

> When there is no torque applied either in drive or in coast, the condition is known as float.

> The climbing action of the drive pinion is sometimes called windup.

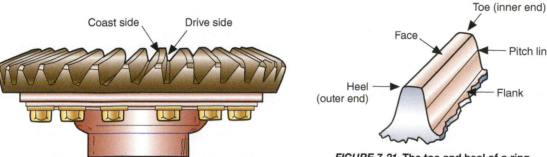

FIGURE 7-20 The drive and coast side of ring gear teeth.

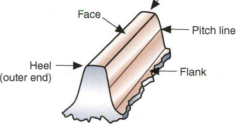

FIGURE 7-21 The toe and heel of a ring gear's tooth.

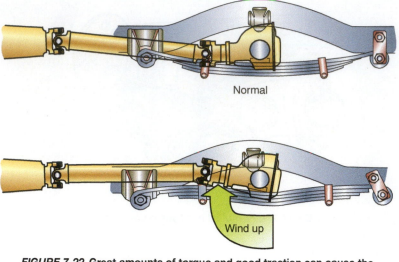

Normal

Wind up

FIGURE 7-22 Great amounts of torque and good traction can cause the rear axle assembly to "wind up."

trying to push the axle housing in the other direction. This is the same action that lifts the front of a dragster or motorcycle during acceleration.

AUTHOR'S NOTE: The drive shaft in a RWD vehicle spins clockwise, and hard acceleration will cause the right rear side of the axle to lift while the left is planted by axle windup. Since the right rear tire now has less traction, the tire may be forced to spin. We call this the "one-wheel peel."

Gear Ratios

Gear ratios express the number of turns the drive gear makes compared to one turn of the driven gear it mates with. The ring gear is driven by the pinion gear, therefore causing torque multiplication. The ring gear is always larger than the pinion. This combination causes the ring gear to turn more slowly but with greater torque. Remember that common terminology uses the term "low" gears to refer to high numerical gear ratios. These gears provide more torque multiplication at the expense of top speed. "High" gears refer to low numerical ratios, resulting in lower engine speed at highway speeds and improved fuel economy.

Many different final drive ratios are used. A final drive ratio of 2.73:1 is commonly used, especially on cars equipped with automatic transmissions. A 2.73:1 final drive ratio means the drive pinion must turn 2.73 times to rotate the ring gear one time. On cars equipped with manual transmissions, more torque multiplication is often needed, therefore a 3.23:1 final drive ratio is often used. To allow a car to accelerate more quickly or to move heavy loads, a final drive ratio of over 4:1 can be used. Also, small-engine cars with overdrive fourth and fifth gears often use over a 4:1 final drive ratio, which allows them to accelerate reasonably well in spite of the engine's low power output. The overdrive in fourth and fifth gear effectively reduces the final drive ratio when the car is moving in those gears. Trucks also use a final drive ratio of 4:1 or 5:1 to provide more torque to enable them to pull or move heavy loads.

It is important to remember that the actual final drive or overall gear ratio is equal to the ratio of the ring and pinion gear multiplied by the ratio of the speed gear the car is operating in.

For example, if a car has a final drive ratio of 3:1, the total final drive ratio for each transmission speed is as follows:

	Transmission Ratio	×	Final Drive Ratio	=	Total Final Drive Ratio
First gear	3:1		3:1		9:1
Second gear	2.5:1		3:1		7.5:1
Third gear	1.5:1				4.5:1
Fourth gear	1:1		3:1		3:1
Fifth gear	0.75:1		3:1		2.25:1

Notice that, in this example, the only time the total final drive ratio is the same as the ratio of the ring and pinion gear is when the transmission is in fourth gear, which has a speed ratio of 1:1.

Many factors are considered when a manufacturer selects a final drive ratio for a vehicle. Some of these factors are vehicle weight, engine rpm range, designed vehicle speed, frontal area of the body, fuel economy requirements, engine power output, and transmission type and gear ratios. Cars with final drive ratios around 2.5:1 will take longer to accelerate but will typically give a higher top speed. At the other end of the scale, a 4.11:1 ratio will give faster acceleration with a lower top speed. An emphasis on fuel economy means most cars are equipped with low numerical (high) gears to allow for lower engine speeds at normal driving speeds.

Determining Final Drive Ratio

To replace a ring and pinion gear set with one of the correct ratio, the ratio of the original set must be known. There are several ways to determine the final drive ratio of a ring and pinion gear set. If a shop manual is available, you can decipher the code found on the assembly or on a tag attached to it (Figure 7-24). Normally a table is given that lists the various codes and the ratios each represents.

Most axles are shipped with an identification tag bolted to them. These tags contain all of the information needed to identify the axle for diagnosis and service. The tags are located under the housing-to-carrier stud nut or are attached by a cover-to-carrier bolt. Manufacturers also often stamp identification numbers into the axle housing. These codes are normally located on the front side of an axle tube (Figure 7-25). Always refer to your shop manual to locate and decipher the codes.

FIGURE 7-23 A vehicle's VIN can be used to identify the ratio and type of axle the vehicle was originally equipped with.

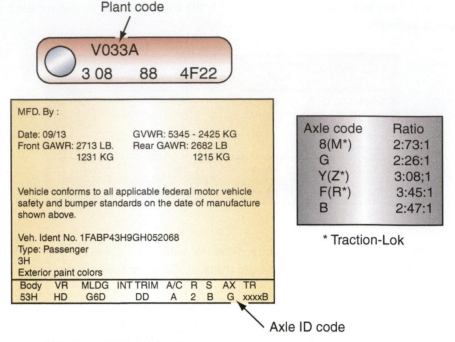

Plant code

Axle ID code

FIGURE 7-24 Deciphering differential codes from information given on the differential tag.

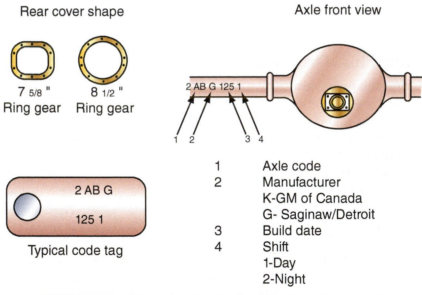

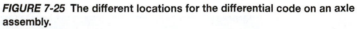

FIGURE 7-25 The different locations for the differential code on an axle assembly.

Another way to determine the final drive ratio is to compare the number of revolutions of the drive wheels with those of the drive shaft. While turning both wheels simultaneously, note how many times the drive shaft turns to complete one revolution of the drive wheels. This count basically represents the ratio of the gears.

The gear ratio also can be determined when the differential is disassembled. Count the number of teeth on both the drive pinion and the ring gear. Divide the ring gear teeth number by the pinion drive number to calculate the final drive ratio. Many manufacturers stamp the actual gear counts on the outside diameter of the ring gear.

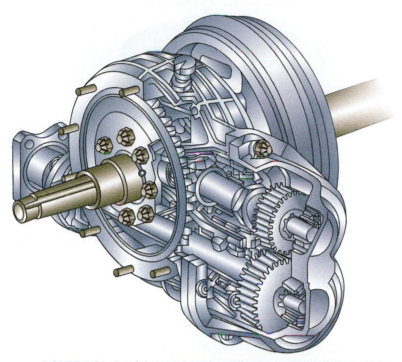

FIGURE 7-26 A quick-change racing differential with removable drive and driven spur gears.

AUTHOR'S NOTE: Sometimes having a fixed gear ratio can be a bad thing. Many types of racing require the cars to operate on different tracks, and finding the exact final drive ratio to match the car with the track is a real science. Ratio changes must be made quickly, and a "quick-change" diff is just the ticket for best performance. The quick-change has a fixed ring and pinion ratio, but there is a pair of gears in the assembly that can be switched out in minutes simply by pulling a cover. The drive gear is splined to the input from the drive shaft, and the driven gear is splined to the pinion gear. This arrangement allows rapid adjustments to the final drive ratio to be made depending on conditions (Figure 7-26).

Hunting and Nonhunting Gears

Ring and pinion gear sets are usually classified as hunting, nonhunting, or partial nonhunting gears. Each type of gear set has its own requirements for a satisfactory gear tooth contact pattern. These classifications are based on the number of teeth on the pinion and ring gears. Note that all ring and pinion gear sets are just that: they are manufactured as a matched pair.

A **nonhunting gear set** is one in which any one pinion tooth comes into contact with only some of the ring gear teeth. One revolution of the ring gear is required to achieve all possible gear tooth contact combinations. As an example, if the ratio of the ring gear teeth to the pinion gear teeth is 39 to 13 (or 3.00:1), the pinion gear turns three times before the ring gear completes one turn. One full rotation of the pinion gear will cause its 13 teeth to mesh with one-third of the ring gear's teeth. On the next revolution of the pinion gear, its teeth will mesh with the second third of the ring gear's teeth and the third revolution will mesh with the last third of the ring gear. Each tooth of the pinion gear will return to the same three teeth on the ring gear each time the pinion rotates.

A **partial nonhunting gear set** is one in which any one pinion tooth comes into contact with only some of the ring gear teeth, but more than one revolution of the ring gear is required

Shop Manual
Chapter 7, page 300

The alignment of nonhunting and partial nonhunting gears is often referred to as timing the gears.

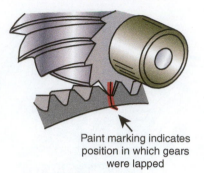

Paint marking indicates
position in which gears
were lapped

FIGURE 7-27 Index marks on a
ring and pinion gear set.

to achieve all possible gear tooth contact combinations. If the ratio of the ring gear teeth to the pinion gear teeth is 35 to 10 (or 3.5:1), any given tooth of the pinion will meet seven different teeth (seven complete revolutions of the pinion gear) of the ring gear before it returns to the space where it started.

When **hunting gear sets** are rotating, any pinion gear tooth will contact all the ring gear teeth. If the ring gear has 37 teeth and the pinion gear has 9, the gear set has a ratio of 37 to 9 (or 3.89:1). Any given tooth in the pinion gear meets all of the teeth in the ring gear before it meets the first tooth again. Hunting gear sets are by far the most common type used by manufacturers.

During assembly the nonhunting and partial nonhunting gears must be assembled with the index marks properly aligned (Figure 7-27). When these gear sets were manufactured, they were probably **lapped** to ensure proper meshing and because specified teeth on the pinion will always meet specific teeth on the ring gear, a noisy gear will result if they are not properly aligned. Hunting gears do not need to be aligned because any tooth on the pinion may mesh with any tooth on the ring gear.

DIFFERENTIAL BEARINGS

At least four bearings are found in all differentials. Two fit over the drive pinion shaft to support it and the other two support the differential case and are usually mounted just outboard of the side gears (Figure 7-28). The drive pinion and case bearings are typically tapered roller bearings.

Different forces are generated in the differential due to the action of the pinion gear. As the pinion gear turns, it tries to climb up the ring gear and pull the ring gear down. Also, as the pinion gear rotates, it tends to move away from the ring gear and pushes the ring gear equally as hard in the opposite direction. Because of these forces, the differential must be securely mounted in the carrier housing. The bearings on each end of the differential case support the case and absorb the thrust of the forces. The pinion gear and shaft are mounted on bearings to allow the shaft to rotate freely without allowing it to move in response to the torque applied to it. All of these bearings are installed with a preload to prevent the pinion gear and ring gear from moving out of position.

Pinion Mountings

As torque is applied to a pinion gear, the pinion gear rotates; three separate forces are produced by its rotation and the torque applied to it. These forces make it necessary to securely mount the pinion gear.

The **drive pinion flange** is splined to the rear axle's drive pinion gear. The drive pinion gear is placed horizontally in the axle housing and is positioned by one of two types of mounting: straddle or overhung. The straddle-mounted pinion gear is used in some removable carrier-type axle housings. The **straddle-mounted pinion** has two opposing tapered roller

Lapping is the process of using a grinding paste to produce a fine finish on the teeth of the two gears that will be in full contact with each other.

The rear Universal joint on the drive shaft attaches to the drive pinion flange or the companion flange.

Shop Manual

Chapter 7, page 318

The **drive pinion flange** is a rim used to connect the rear of the drive shaft to the rear axle drive pinion.

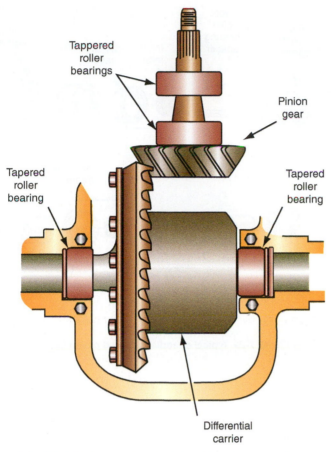

Tappered roller bearings

Pinion gear

Tapered roller bearing

Tapered roller bearing

Differential carrier

FIGURE 7-28 The position of pinion drive and side bearings in a typical differential assembly.

bearings positioned close together with a short spacer between their inner races and ahead of the pinion gear. A third bearing, usually a straight roller bearing, is used to support the rear of the pinion gear (Figure 7-29).

The **overhung-mounted pinion** also uses two opposing tapered roller bearings but does not use a third bearing. The two roller bearings must be farther apart than the opposing bearings of a straddle-mounted pinion because a third bearing is not used to support the pinion gear (Figure 7-30). This type of pinion gear mounting can be found on either the removable carrier or integral-type driving axle.

Some pinion shafts are mounted in a bearing retainer that is removable from the carrier housing. This type of pinion assembly utilizes a pilot bearing to support the rear end of the pinion and is equipped with two opposing tapered roller bearings.

Drive Pinion Bearing Preload

A spacer is placed between the opposing tapered bearings to control the distance between them (Figure 7-31). This spacer also controls the amount of preload or loading pressure applied to the bearings. Preload prevents the pinion gear from moving back and forth in the bearing retainer.

Some differentials use a solid noncollapsible spacer with selective thickness shims to adjust pinion bearing preload.

When the pinion shaft nut is tightened to specifications, pressure is exerted by the pinion drive flange against the inner race of the front pinion bearing. This applies pressure against the spacer and the rear bearing, or in some designs against a machined shelf on the drive pinion. The purpose of the collapsible spacer is to maintain a spring-like force on the inner race of

Preload is a fixed amount of pressure constantly applied to a component. Preload on bearings eliminates looseness. Preload on limited-slip differential clutches also provides torque transfer to the driven wheel with the least traction.

SERVICE TIP:
Collapsible spacers should never be reused. After they have been compressed once, they are not capable of maintaining preload when they are compressed again. The spacers should always be replaced when servicing the differential.

End play is the amount of axial or end-to-end movement in a shaft due to clearance in the bearings.

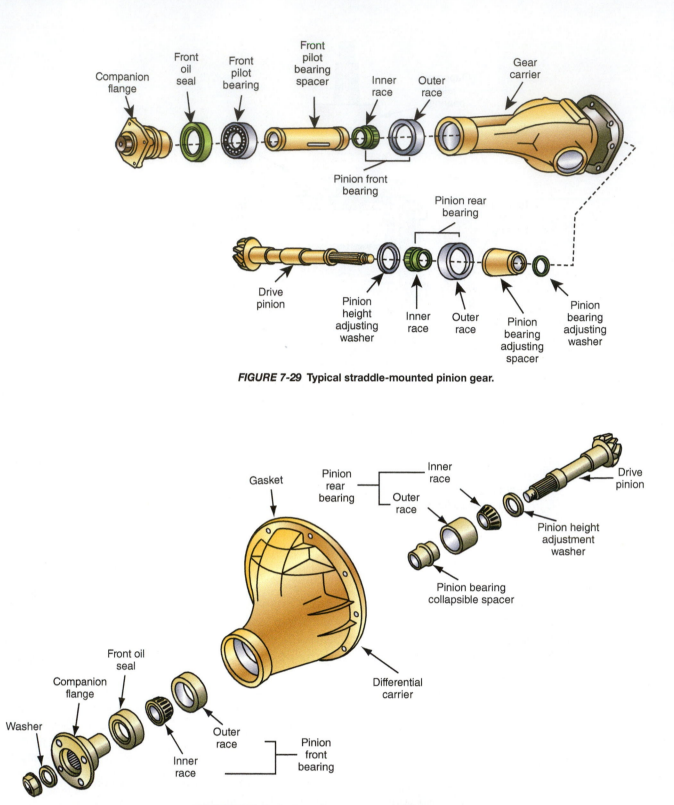

FIGURE 7-29 Typical straddle-mounted pinion gear.

FIGURE 7-30 Typical overhung-mounted pinion gear.

the front bearing to prevent it from spinning on the drive pinion shaft. Preload on the two pinion bearings ensures that there will be no pinion shaft end play. Any pinion shaft end play will result in rapid gear and bearing failure and noise.

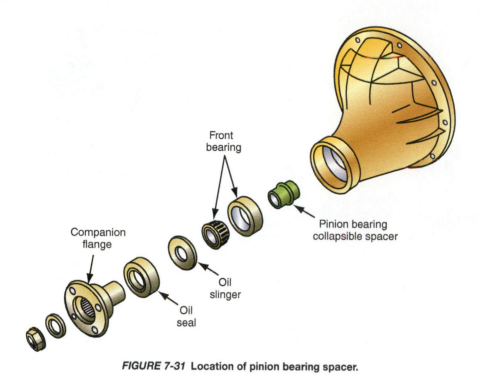

FIGURE 7-31 Location of pinion bearing spacer.

Differential Case

The differential case is supported in the carrier by two tapered roller side bearings. This assembly can be adjusted from side to side to provide the proper backlash between the ring gear and pinion and the required side bearing preload. This adjustment is achieved by threaded bearing adjusters (Figure 7-32) on some units and the placement of selective shims and spacers (Figure 7-33) on others.

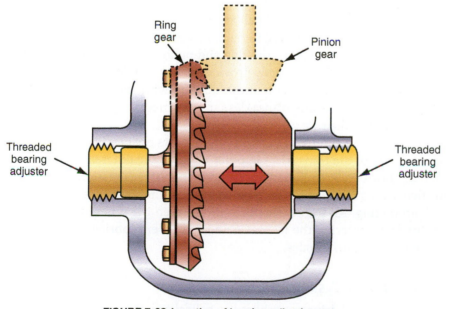

FIGURE 7-32 Location of bearing adjusting nuts.

Shop Manual
Chapter 7, page 319

Backlash is the clearance or play between two gears in mesh. It is the amount one of the gears can be moved without moving the other (Figure 7-34).

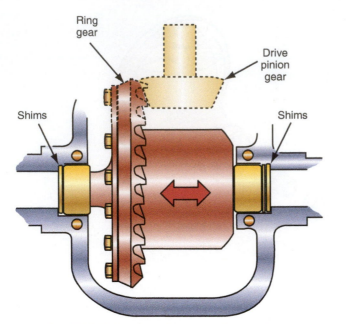

FIGURE 7-33 Location of bearing selective shims.

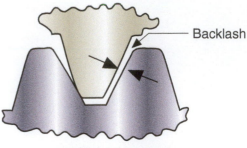

FIGURE 7-34 Gear backlash.

TRANSAXLE FINAL DRIVE GEARS AND DIFFERENTIAL

Shop Manual
Chapter 7, page 327

Transaxle final drive gears provide the means for transmitting transmission output torque to the differential section of the transaxle.

The differential section of the transaxle has the same components as the differential gears in a RWD axle and basically operates in the same way. The power-flow in transversely mounted powertrains is in line with the wheels and therefore the differential unit does not need to turn the power 90 degrees.

Another name for collapsible spacer is crush sleeve.

The drive pinion and ring gears and the differential assembly are normally located within the transaxle housing of FWD vehicles. There are three common configurations used as the final drives on FWD vehicles: helical, planetary, and hypoid. The helical and planetary final drive arrangements are usually found in transversely mounted powertrains. Hypoid final drive gear assemblies are used with longitudinal powertrain arrangements.

The ring gear in a transaxle is sometimes referred to as the differential drive gear.

The drive pinion gear is connected to the transmission's output shaft and the ring gear is attached to the differential case. Like the ring and pinion gear sets in a RWD axle, the drive pinion and ring gear of a FWD assembly provide for a multiplication of torque.

The teeth of the ring gear usually mesh directly with the transmission's output shaft (Figure 7-35). However on some transaxles, an intermediate shaft is used to connect the transmission's output to the ring gear.

On some models, the differential and final drive gears operate in the same lubricant as the transmission section of the transaxle. On other designs, the differential section is separately enclosed and is lubricated by a different lubricant than the transmission section. These designs require positive sealing between the differential unit and the transmission to keep the different lubricants from mixing. All transaxles use seals between the differential and the drive axles to prevent dirt from entering the transaxle and to prevent lubricant from leaking past the attachment point of the drive axles.

Helical Final Drive Assembly

Helical (Figure 7-36) final drive assemblies use helical gear sets that require the centerline of the pinion gear to be at the centerline of the ring gear. The pinion gear is cast as part of the main shaft and is supported by tapered roller bearings. The pinion gear is meshed with the

174

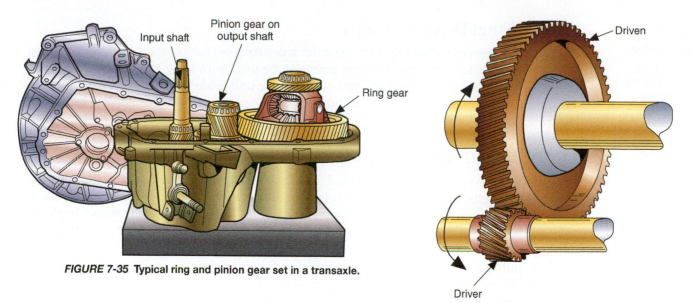

FIGURE 7-35 Typical ring and pinion gear set in a transaxle.

FIGURE 7-36 Helical gear set.

ring gear to provide the required torque multiplication. Because the ring is mounted on the differential case, the case rotates in response to the pinion gear.

Planetary Final Drive Assembly

The ring gear of a planetary final drive assembly has lugs around its outside diameter. These lugs fit into grooves machined inside the transaxle housing. These lugs and grooves hold the ring gear stationary. The transmission's output shaft is splined to the planetary gear set's sun gear. The planetary pinions are in mesh with both the sun gear and ring gear and form a simple planetary gear set (Figure 7-37). The planetary carrier is constructed so that it also serves as the differential case.

In operation, the transmission's output drives the sun gear, which, in turn, drives the planetary pinions. The planetary pinions walk around the inside of the stationary ring gear. The rotating planetary pinions drive the planetary carrier and differential housing. This combination provides maximum torque multiplication from a simple planetary gear set. The planetary final drive design is very common in automatic transaxles.

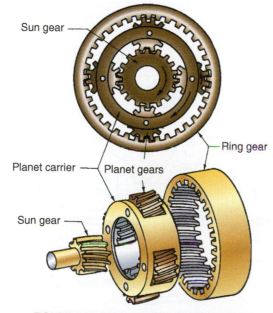

FIGURE 7-37 Planetary final drive gear set.

Helical gears are gears with teeth that are cut at an angle to the gear's axis of rotation.

A BIT OF HISTORY

In 1932, Ferdinand Porsche designed a Grand Prix race car for the Auto Union Company. The high engine output caused one of the rear wheels to spin at any speed up to 100 mph. In 1935, Porsche commissioned ZF Friedrichshafen to design a limited-slip differential that would reduce the wheel spin and allow the car to perform better.

An open differential is a standard-type differential, often called the "peg leg type" by racers.

Many names are used for limited-slip differentials, including Posi-Traction, Traction-Lok, and Posi-units.

Many names are used for limited-slip differentials, some of which are Twin-Grip, Gov-Lock, E-Diff, Posi-Traction, Equa-Lock, Trak-Lok, Traction-Lok, Powr-Lok, Tru-Lok, Sure Grip, Safe-T-Track, No-Spin, Hydratrak, Anti-Spin, and Posi-units.

Shop Manual
Chapter 7, page 334

A **clutch pack** consists of a complete set of alternating clutch plates and discs.

Hypoid Final Drive Assembly

Hypoid gears have the advantage of being quiet and strong because of their thick tooth design. And due to their strength, hypoid-type gears can be used with large engines that are longitudinally mounted in vehicles. This type of final drive unit is identical to those used in RWD vehicles. Because they require a special lubricant, they are contained in a separate housing in the transaxle.

LIMITED-SLIP DIFFERENTIALS

With an open or conventional differential, as long as the drive wheels are gripping the road, there is some resistance on the drivetrain and the vehicle moves forward. When one tire is on a slippery surface, it loses traction, rotational resistance decreases, and the wheel begins to spin. With the decreased resistance, the amount of torque delivered to each wheel changes to the point where the tire that has no traction is the one that receives all of the torque.

This puts a lot of stress on the differential as the differential case drives the differential pinion gears around the stationary side gear. While differential pinion gears are designed to rotate around the shaft during cornering, they are not designed to spin at rapid speeds. Heat is rapidly produced and the film of lubrication on the gears breaks down, resulting in metal-to-metal contact. This damages the inner bore of the gears and the shaft or spider. If the spinning wheel suddenly has traction, the shock of the sudden traction can destroy the gears and axle shaft, as well as other parts in the drivetrain. To overcome these problems, some vehicles have a **limited-slip differential (LSD)**. LSDs provide more torque to the wheel that has traction when the other tire loses it.

Limited-slip differentials merely limit the amount of differential action between the side gears. LSDs are primarily used on performance-oriented cars for increased traction while cornering and on off-road vehicles where the drive wheels are constantly losing traction. They can be found in front, rear, and/or center differentials.

The main types of LSDs are torque sensitive (geared or clutch-based), speed sensitive (viscous/pump and clutch pack), or mechanically locked.

Clutch-Based Units

Many LSDs use friction material to transfer the torque applied to a slipping wheel to the one with traction. They typically respond to torque differences between the side gears of the differential. They use a **clutch pack** that consists of a combination of steel plates and friction plates (Figure 7-38). There is a pack for each side gear (Figure 7-39).

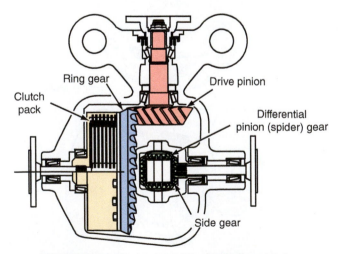

FIGURE 7-38 A late-model LSD with friction clutches.

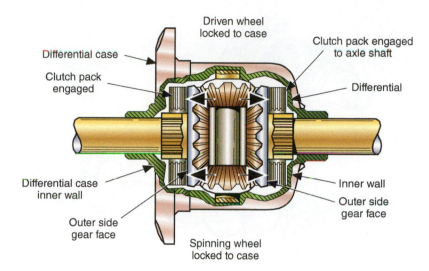

Energized clutches cause locked differential

FIGURE 7-39 Action of the clutches in a limited-slip differential.

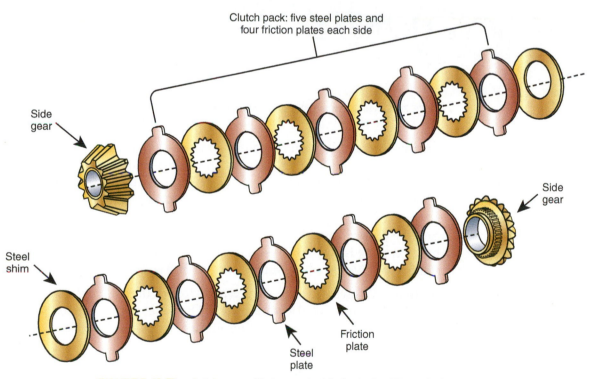

FIGURE 7-40 The clutch assembly in a typical limited-slip differential assembly.

The friction discs have an abrasive coating on both sides. They fit over the external splines on the side gears' hub. The plates are stacked on the hub and are housed in the differential case (Figure 7-40). Pressure is kept on the clutch packs by either a preload S-shaped spring (Figure 7-41), a disc or Bellville spring, or coil springs.

As long as the friction discs maintain their grip on the steel plates, the differential side gears are locked to the differential case. Under normal driving conditions, the clutch slips as the torque generated by differential action during cornering easily overcomes the capacity of the clutch assembly.

This allows for normal differential action when the vehicle is turning.

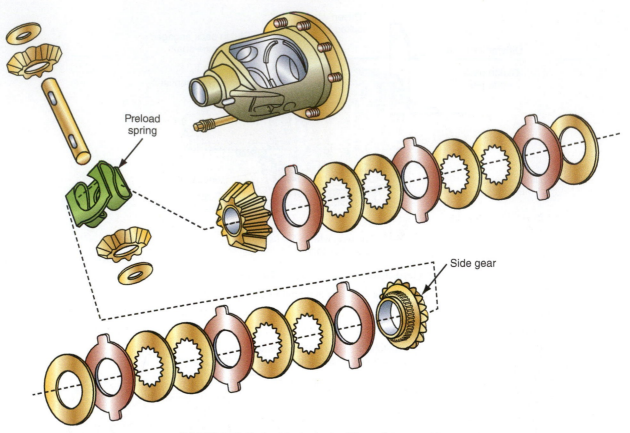

FIGURE 7-41 Typical limited-slip differential assembly.

These are load-sensitive units; they require load to apply and the greater the load, the tighter they apply. The teeth of the differential's pinion and side gears are cut at an angle. As the pinion gears transfer torque to the side gears, the shape of the teeth causes them to attempt to push away from each other. This separating force causes the side gears to compress the clutches.

Unfortunately, when the desired gain in traction is the highest, such as when driving on extreme surfaces such as ice, there is very little resistance to the turning of the axles and side gears, which results in little separating force. The result is one axle and wheel receiving the torque.

Cone Clutches. A common LSD uses two cone-shaped parts to lock the side gears to the differential case (Figure 7-42); this unit is called a **cone clutch**. The cones are splined to the side gear hubs. The exterior surface of the cones is a friction surface that grabs the inside surface

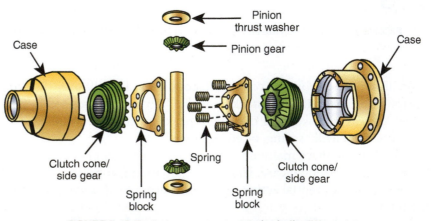

FIGURE 7-42 Typical cone-clutched limited-slip differential.

178

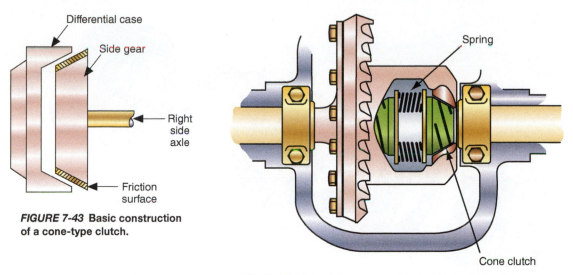

FIGURE 7-43 Basic construction of a cone-type clutch.

Differential case

Side gear

Right side axle

Friction surface

Spring

Cone clutch

FIGURE 7-44 An Auburn cone-clutched limited-slip differential.

of the differential case (Figure 7-43). The frictional surfaces have spiral grooves cut in them to allow lubricant to flow through the cones. Four to six coil springs mounted in thrust plates between the side gears maintain a preload on the cones.

When the cones are forced against the case, the cones and axles rotate with the case. When the vehicle is moving straight, spring pressure and the separating force created by the pinion gears pushes each clutch cone against the internal cone in the differential case. During cornering, normal differential action overcomes the pressure of the springs, and allows the inner axle to slow down while the outer axle speeds up.

Some cone-clutched LSDs have beveled ends on the differential's pinion shafts and matching "ramps" cut into the shaft openings of the differential case. When torque is applied to the "ramp-type" differential, the ramps tend to force the side gears apart and apply pressure to the cone assembly on the axle with the best traction. The cone clutch simultaneously grip the side gears and the inside of the differential case.

The Auburn (Figure 7-44) differential is offered as an option by automobile manufacturers and uses interlocking cones to provide holding power. These units are made with low spring pressures for street use or with higher spring pressures for racing applications only.

The cone clutches in an Auburn Gear limited-slip differential are coupled to beveled side gears. As torque is transmitted through the differential side gears to the axle shafts, the side gear separating forces and spring preload firmly seat the cones into the differential case. The cone design, along with the applied force, determines the torque transfer of the differential. When torque levels decrease, as in a corner, the gear-separating forces also decrease, allowing the axle shafts to rotate independently. When a vehicle is moving straight ahead, the axle shafts are linked to the differential case through the clutch, and each wheel gets equal torque.

Normally, each axle gets an equal amount of torque through the differential. However, when one wheel slips, some of that wheel's torque is lost through the pinion gears spinning on the pinion shaft. The clutch on the other wheel remains applied and some of the torque from the slipping side is applied to the wheel with traction. The actual amount of torque applied to that wheel is determined by the frictional capabilities of its clutch assembly. Power is delivered to that wheel only until torque overcomes the frictional characteristics of the clutch assembly, at which time it begins to slip. This action limits the maximum amount of torque that can be applied to the wheel with traction.

Viscous Clutch. Some vehicles use a viscous clutch in their limited-slip differentials (Figure 7-45). A viscous LSD has a **viscous coupling** with alternately positioned perforated steel and friction plates connected to the two drive axles. The application of the plates relies on the resistance generated by a high-viscosity silicon fluid. When there is no rotational difference between the

Sure-Grip (limited-slip) differentials are often called **ramp-type differentials**. BMW also uses a differential similar in design to the Sure-Grip.

Locked differentials are often called "lockers." When speed differential increases, the viscous torque also increases.

Late-model, high-performance cars from GM may be equipped at the factory with an Auburn limited-slip differential.

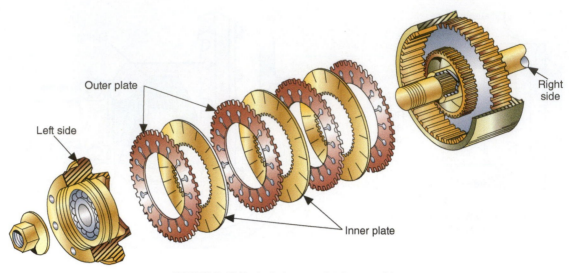

FIGURE 7-45 Typical viscous clutch assembly.

left- and right-side wheels, power is distributed evenly to both axles. When one wheel has less traction than the other, there is a difference in rotational speeds between the axles. This speed differential causes the silicon fluid to shear, causing the fluid to heat rapidly and become a near-solid. This action in effect "glues" the clutches together and connects the two axles.

Gerodisc (Gerotor) Differentials. Gerodisc differentials are speed-sensitive slip-limiting differential units. They contain a clutch pack and a hydraulic pump. The pump is gerotor-type, whose pressure output depends on rotational speed. An axle shaft drives the pump; therefore the output from the pump depends on the speed of that axle. The pump's output is fed to a sealed piston against the clutch pack and the amount of pressure determines how tightly the clutches will be squeezed together (Figure 7-46). When the clutch pack is fully engaged, the two drive axles are locked together. When one wheel spins faster than the other, the pump's speed and output pressure increases. The fluid then works on the clutch pack at the wheel with the most traction. The amount of torque transfer is determined by the amount of pressure applied to the clutch pack. When an axle is not slipping, the pump does not deliver pressure and the differential functions as an open unit.

Electronic. There are two basic types of electronic limited-slip systems. Brake-based systems use wheel speed sensors, antilock brakes, and microcomputers to electronically monitor slipping. If a wheel on an axle is rotating faster than the other, the computer will assume that

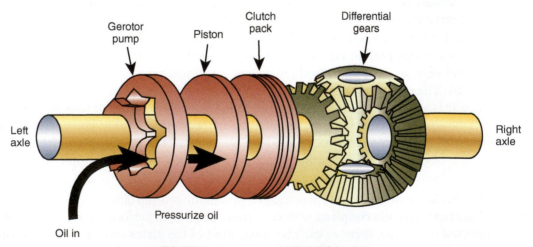

FIGURE 7-46 A gerodisc differential setup.

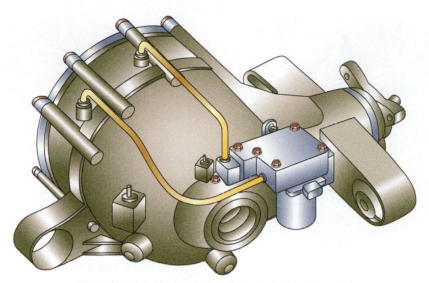

FIGURE 7-47 An electrohydraulic limited-slip differential.

it is slipping and will briefly apply that wheel's brake. This slows the spinning wheel and allows the other wheel to gain speed and keep traction.

This type of system has little effect on the steering and control of the vehicle, and it also generates less stress on the drivetrain compared with that generated by a mechanical locking device. It can also be tuned for specific applications on and off road and at different speeds. Brake-based electronic LSDs are often used in high-torque FWD applications to help reduce torque steer.

A more recent development has been LSD systems that rely on electrohydraulic application of limited-slip clutches inside the differential housing (Figure 7-47). These units use a computer-controlled pump that is used to apply pressure to the limited-slip clutches. There are many advantages to these systems. The rate of clutch apply can often be adjusted by the driver to meet different driving conditions. The limited-slip action is instantaneous when needed, and the differential functions as an open unit unless the system is activated. These units have found favor in front, rear, and center differentials.

Gear-Based Units

Geared, torque-sensitive mechanical limited-slip differentials use worm gears that react to an increase or decrease in torque at one of two shafts. The most commonly used are Torsen, Quaife Automatic Torque Biasing Differential (ATB), and the Eaton Detroit Truetrac. Their action is entirely dependent on the torque, not speed, difference between the output shafts.

Geared LSDs can be used to reduce torque steer on FWD vehicles, to maximize traction and make oversteer easier to manage (as in drifting) on RWD vehicles, and serve as a center differential in 4WD vehicles. Road racers often prefer a geared limited-slip differential because it does not lock the two output shafts and cause them to spin at the same rate. Rather, they bias the torque, up to 80 percent, to the tire that has the most grip.

Helical-geared LSDs respond very quickly to changes in traction. However, both output shafts must be loaded to maintain proper torque distribution. If an axle or output shaft spins freely, zero torque is transmitted to the other shaft and the torque-sensitive differential performs like an open differential.

Torsen Units. A commonly used LSD with a parallel-axis helical gear set is a torque-bias or torque-sensing unit called a Torsen (Figure 7-48). Torsen is a contraction of _Torque-Sensing_. These units are used in front, rear, and/or center differential assemblies.

A Torsen differential moves the torque that is available at the wheel that is starting to spin or has lost traction and sends it to the wheel that has better traction and is turning slower. This action begins as soon as there is some resistance between the gears in the differential.

Torsen-style differentials may be found as standard equipment on some models from Audi, BMW, Mercedes, the Honda S2000, and Ford Raptor pickups.

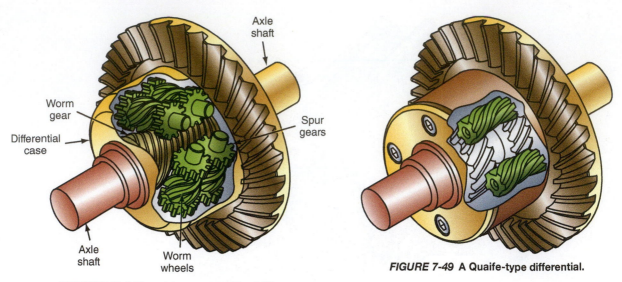

FIGURE 7-48 A Torsen torque-sensitive LSD.

FIGURE 7-49 A Quaife-type differential.

QUAIFE Automatic Torque Biasing Differential. The ATB differential is made by Quaife Engineering. The QUAIFE ATB differential (Figure 7-49) is an automatic gear-operated torque-biasing differential. This is an all-mechanical geared unit that does not use clutch packs or preloading to transfer torque from one axle to the other, and torque transfer occurs automatically when one wheel loses traction. Torque transfer also occurs gradually, without steps. It is particularly effective in reducing the effects of FWD torque steer. Some 4WD vehicles also use this LSD.

Detroit Truetrac. The Detroit Truetrac, manufactured by Tractech, uses parallel axis planetary helical gears to provide a quiet, automatic splitting of drive torque. This is a gear-to-gear LSD that behaves like an open differential during normal operating conditions and automatically transfers torque to the wheel with better traction when the other wheel begins to slip.

Truetrac unit has helical gears and associated pinions mounted in the recesses of the differential case. When one wheel begins to lose traction, the pinions separate slightly from the side gear and wedge in the recessed area, locking the axle.

Locked Differentials

A **locked differential** is a type of differential with the side and pinion gears locked together.

Another type of differential is the **locked differential**. This provides very limited differential action, if any. It is designed to provide both drive axles with nearly the same amount of torque, regardless of traction. Needless to say, this differential is designed only for off-road use and for racing applications.

Some trucks and off-road equipment use differentials that can be locked and unlocked by pressing a button. This results in a unit that can behave as an open or locked differential. Compressed air, a mechanical cable, an electric motor, or hydraulic fluid controls the locking mechanism. Normally, the controls activate a pump that applies pressure on the clutches and locks them to the side gears.

Detroit Locker. A commonly found, or at least much talked about, locked differential is the "Detroit Locker." This unit is a ratcheting-type differential. It is very strong and will almost always provide equal torque and speed to each of the drive wheels. It does not allow for much differential action; therefore, cornering is hampered. However, good drivers know when to lift off the throttle right before turning to minimize scrubbing and power their way through turns. This action allows time for the locker to unlock and provide some differential action during the turn. Detroit Lockers are primarily used in vehicles built for oval racing, such as NASCAR, and off-road ventures.

FIGURE 7-50 An Eaton mechanical automatic locking differential.

During normal conditions, both axles are locked to the differential case. If one wheel is forced to rotate faster, such as the outer wheel during a turn, the ratcheting mechanism unlocks that axle and allows it to rotate freely. This means that during a turn, the inside tire is providing the drive torque. The lack of torque at the outside tire causes the unit to lock again. This action tends to straighten the path of the vehicle until the unit senses that tire is traveling faster. Then it unlocks again. Needless to say, the constant and unpredictable switching from an open-type differential to a locked one makes it difficult to control the vehicle.

Eaton Automatic Locking Differential. A very common mechanical automatic locking differential has been used for years by Chevrolet and GMC trucks for many years. The most recent version is called the MLocker (Figure 7-50). This design uses a flyweight mechanism and self-energizing clutches. When a wheel speed difference of more than 100 rpm occurs, a speed-sensitive latching bracket forces a ramped cam plate to apply clutches that bring each side gear to case speed. The device automatically unlocks when the need for improved traction is gone.

Spool. To eliminate all differential action, cars built for drag racing and drifting use a spool. A spool is basically a ring gear mounted to an empty differential case. The ring gear is driven by a pinion gear. Both the right and left axles are splined to the case, providing for a solid connection between them. With a spool, even the slightest of turns causes the tires to scrub.

DRIVE AXLE SHAFTS AND BEARINGS

Located within the hollow horizontal tubes of the axle housing are the axle shafts (Figure 7-51). The purpose of an axle shaft is to transmit the driving force from the differential side gears to the drive wheels. Axle shafts are heavy steel bars splined at the inner end to mesh with the axle side gear in the differential. The driving wheel is bolted to the wheel flange at the outer end of the axle shaft. The drive wheels rotate to move the vehicle forward or reverse.

The drive axles in a transaxle usually have two CV joints to allow independent front-wheel movement and steering of the drive wheels. These CV joints also allow for lengthening and shortening of the drive axles as the wheels move up and down.

Shop Manual
Chapter 7, page 338

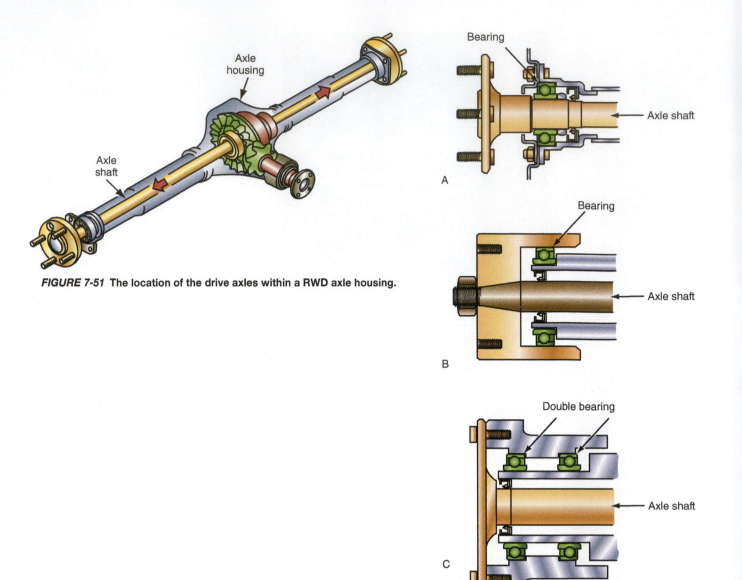

FIGURE 7-51 The location of the drive axles within a RWD axle housing.

FIGURE 7-52 Types of rear axle shafts:
(A) semifloating; (B) three-quarter floating;
and (C) full-floating.

Dead axles are found on trailers and are the type of axle found in the rear of FWD vehicles.

The **live axle** transmits power from the differential to the wheels.

Thrust loads are loads placed on a part that is parallel to the center of the axis of rotation.

The purpose of the axle shaft is to transfer driving torque from the differential assembly to the vehicle's driving wheels. There are two types of axles: the dead axle that supports a load and the **live axle** that supports and drives the vehicle.

There are basically three designs by which axles are supported in a live axle: full-floating, three-quarter floating, and semifloating. These refer to where the axle bearing is placed in relation to the axle and the housing. The bearing of a **full-floating rear axle** is located on the outside of the housing (Figure 7-52). This places all of the vehicle's weight on the axle housing with no weight on the axle.

Three-quarter floating axles and **semifloating rear axles** are supported by bearings located in the housing and thereby carry almost all of the weight of the vehicle (Figure 7-53). Most passenger cars are equipped with three-quarter or semifloating axles. Full-floating axles are commonly found on heavy-duty trucks.

The axle shaft bearing supports the vehicle's weight and reduces rotational friction. With semifloating axles, radial and thrust loads are always present on the axle shaft bearing when the vehicle is moving. Radial bearing loads act at 90 degrees to the axle's center of axis. Radial

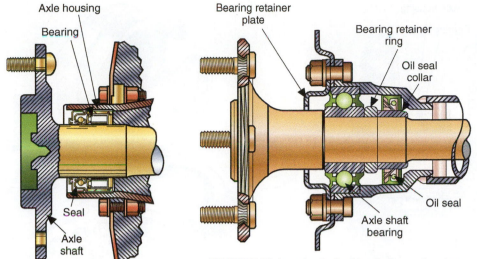

FIGURE 7-53 Locating the axle bearings in the axle housing places some of the vehicle's weight on the bearing.

FIGURE 7-54 An axle shaft with a ball-type bearing.

loading is always present whether or not the vehicle is moving. Thrust loading acts on the axle bearing parallel with the center of axis. It is present on the driving wheels, axle shafts, and axle bearings when the vehicle turns corners or curves.

Three designs of axle shaft bearings are used on semifloating axles: ball-type, straight-roller, and tapered-roller bearings. The load on a bearing that is of primary concern is the axle's end thrust. When a vehicle moves around a corner, centrifugal force acts on the vehicle's body, causing it to lean to the outside of the curve. As the body leans outward, a thrust load is placed on the axle shaft and axle bearing. Each type of axle shaft handles axle shaft end thrust differently.

The end-to-end movement of the axle is controlled by a C-type retainer on the inner end of the axle shaft or by a bearing retainer and retainer plate at the outer end of the axle shaft.

Ball-Type Axle Bearings

An axle with ball-type axle bearings has the axle shaft and bearing held in place inside the axle housing by a stamped metal bearing retainer plate (Figure 7-54). The plate is bolted to the axle housing and is held in place on the axle shaft by a retaining ring, which is pressed onto the axle shaft. The retaining ring is not re-useable, and must be replaced whenever the axle bearing is replaced.

The operation of the ball-type bearing is designed to absorb radial load as well as the axle shaft end thrust. Because both bearing loads are taken at the bearing, there is no axle shaft end thrust absorption or adjustment designed into the rear axle housing.

To seal in the lubricant, an oil seal collar and oil seal is used. The oil seal collar is a machined sleeve or finished portion of the axle on which the lips of the seal ride. The oil seal retains the gear lubricant inside the axle housing. The axle seal prevents the lubricant from leaking into the brakes.

Straight Roller Axle Bearings

The straight roller bearing uses the axle shafts as its inner race (Figure 7-55). The outer bearing race and straight rollers are pressed into the axle tubes of the rear axle housing. The inner end of the axle shaft at the differential has a groove machined around its outside diameter where the **C-type retainer** fits.

The bearing is lubricated by hypoid lubricant from the differential area of the axle housing. The grease seal, located outside the axle shaft bearing, prevents the lubricant from leaking out of the housing.

Radial loads are loads applied at 90 degrees to an axis of rotation.

Body lean is also called body roll.

Shop Manual
Chapter 7, page 339

A **C-type retainer** controls the end-to-end or side-to-side movement of the axle. The axle is referred to as a "C-clip axle" in this design.

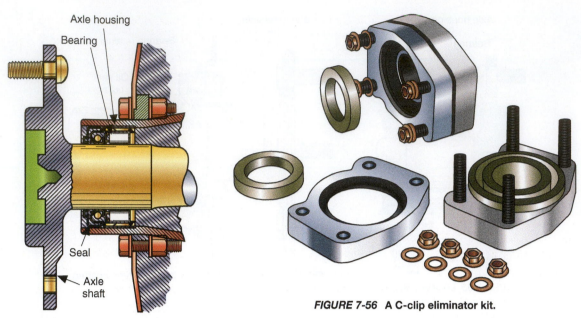

FIGURE 7-55 An axle shaft with a
straight roller-type bearing.

FIGURE 7-56 **A C-clip eliminator kit.**

When the vehicle takes a turn, the body and axle housing move outward and the axle shaft moves inward on the bearings. The inner end of the axle shaft contacts the differential pinion shaft. The axle shaft end thrust exerted against the differential pinion shaft tries to move the differential housing and differential side bearing assembly against the integral housing axle tube. The axle housing absorbs the axle shaft end thrust. There is no end thrust adjustment designed into the rear axle housing.

AUTHOR'S NOTE: The C-clip axle arrangement is not considered strong enough for extreme horsepower applications. Aftermarket manufacturers solve this problem with C-clip eliminator kits that modify the vehicle with a bolt-on retainer plate and a pressed-on axle bearing with a retaining ring (Figure 7-56).

Tapered Roller Axle Bearings

The tapered roller bearing and axle shaft assembly are held inside the axle housing by a flange, which is bolted to the axle housing (Figure 7-57). The inside of the flange may be threaded to receive an adjuster or it is machined to accept adjustment shims.

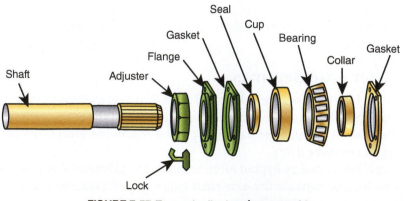

FIGURE 7-57 Tapered roller bearing assembly.

The axle shaft is designed to float, based on the slight in-and-out movements of the axles. As the axle shaft moves inward, it contacts a thrust block that separates both axle shafts at the center of the differential. The inward-moving axle shaft contacts the thrust block, which passes the thrust force to the opposite axle shaft. There the axle shaft end thrust becomes an outward-moving force, which causes the opposite tapered roller axle bearing to seat further in its bearing cup. Axle end thrust adjustments can be made by a threaded adjuster or thin metal shims placed between the brake assembly plate and the axle housing.

The tapered roller bearing is lubricated before installation in the axle housing. A seal and two gaskets keep the hypoid lubricant and foreign matter out of the bearing operating area. A collar holds the rear axle bearing in place on the axle shaft.

IRS Axle Shafts

The drive axles on most newer IRS systems use two U- or CV-joints per axle to connect the axle to the differential and the wheels. They are also equipped with linkages and control arms to limit camber changes. The axles of an IRS system are much like those of a FWD system. The outer portion of the axle is supported by an upright or locating member that is also part of the suspension.

SUMMARY

- The drive axle of a RWD vehicle is mounted at the rear of the car. It is a single housing for the differential gears and axles. It also is part of the suspension and helps to locate the rear wheels.
- The final drive is the final set of reduction gears the engine's power passes through on its way to the drive wheels.
- A differential is needed between any two drive wheels, whether in a RWD, FWD, or 4WD vehicle, because the two drive wheels must turn at different speeds when the vehicle is in a turn.
- RWD final drives use a hypoid ring and pinion gear set, which turns the power-flow 90 degrees from the drive shaft to the drive axles. A hypoid gear set also allows the drive shaft to be positioned low in the vehicle.
- The differential rotates the driving axles at different speeds when the vehicle is turning and at the same speed when the vehicle is traveling in a straight line.
- The differential is normally housed with the drive axles in the rear axle assembly. Power from the engine enters into the rear axle and is transmitted to the drive axles, which are attached to the wheels of the car.
- The differential allows for different speeds between the two drive wheels.
- The pinion gear meshes with the ring gear, which is fastened to the differential case. The pinion shaft and gears are retained in the differential case and mesh with side gears splined to the drive axles.
- When both driving wheels are rotating at the same speed, the differential pinions do not rotate on the differential pinion shaft; the differential assembly rotates as one and the driving wheels, axles, and axle side gears rotate at the same speed.
- When the vehicle turns, the drive wheels rotate at different speeds because the differential case forces the pinion gears to walk around the slow-turning axle side gear. This action causes the outside axle side gear to reach a higher speed than the inside wheel. The amount of differential action taking place depends on how sharp the corner or curve is. Differential action provides control on corners and prolongs drive tire life.
- Live rear axles use a one-piece housing with two tubes extending from each side. These tubes enclose the axles and provide attachments for the axle bearings. The housing shields all parts from dirt and retains the differential lubricant.

SUMMARY

- Rear axle housings can be divided into two groups: integral carrier or removable carrier.
- An integral carrier housing has a service cover that fits over the rear of the differential and rear axle assembly.
- The differential assembly of a removable carrier assembly can be removed from the front of the axle housing as a unit and is serviced on a bench and then installed into the axle housing.
- The types of gears currently used as final drive gears are the helical, spiral bevel, and hypoid gears.
- With hypoid gears, the drive side of the teeth is curved in a convex shape, whereas the coast side of the teeth is concave. The inner end of the teeth on a hypoid ring gear is known as the toe and the outer end of the teeth as the heel.
- The gear ratio of the pinion and ring gear is often referred to as the final drive ratio.
- Gear ratios express the number of turns the drive gear makes compared to one turn of the driven gear it mates with.
- Ring and pinion gear sets are usually classified as hunting, nonhunting, or partial nonhunting gears.
- A nonhunting gear set is one in which any one pinion tooth comes into contact with only some of the ring gear teeth and is identified by a x.00 gear ratio.
- A partial nonhunting gear set is one in which any one pinion tooth comes into contact with only some of the ring gear teeth, but more than one revolution of the ring gear is required to achieve all possible gear tooth contact combinations. These gears are identified by a x.50 ratio.
- When hunting gear sets are rotating, any pinion gear tooth will contact all the ring gear teeth.
- During assembly the nonhunting and partial nonhunting gears must be assembled with the timing marks properly aligned. Hunting gears do not need to be aligned because any tooth on the pinion may mesh with any tooth on the ring gear.
- At least four bearings are found in all differentials. Two fit over the drive pinion shaft to support it while the other two support the differential case.
- The drive pinion gear is placed horizontally in the axle housing and is positioned by one of two types of mounting, either straddle or overhung.
- The straddle-mounted pinion gear is used in most removable carrier-type axle housings and uses two opposing tapered roller bearings positioned close together with a short spacer between their inner races and a third bearing to support the rear of the pinion gear.
- The overhung-mounted pinion uses two opposing tapered roller bearings but not a third bearing.
- A spacer is placed between the opposing pinion shaft bearings to control the amount of preload applied to the bearings. Preload prevents the pinion gear from moving back and forth in the bearing retainer.
- The differential section of the transaxle has the same components as the differential gears in a RWD axle and basically operates in the same way, except that the power-flow in transversely mounted powertrains does not need to turn 90 degrees.
- The final drive gears and differential assembly are normally located within the transaxle housing of FWD vehicles.
- There are three common configurations used as the final drives on FWD vehicles: helical, planetary, and hypoid. The helical and planetary final drive arrangements are usually found in transversely mounted powertrains. Hypoid final drive gear assemblies are normally used with longitudinal powertrain arrangements.

TERMS TO KNOW

C-type retainer
Clutch pack
Cone clutch
Differential case
Drive pinion flange
Final drive
Full-floating rear axle
Hunting gear set
Integral carrier
Independent rear suspension (IRS)
Lapping
Limited-slip differential (LSD)
Live axle
Locked differential
Nonhunting gear set
Overhung-mounted pinion

- A limited-slip unit provides more driving force to the wheel with traction when one wheel begins to spin by restricting the differential action.
- Most limited-slip differentials use either a clutch pack, a cone clutch, or a viscous clutch assembly.
- There are a variety of limited-slip differentials used by various manufacturers. These are torque or speed sensing.
- A locked differential provides very limited differential action and is designed to provide both drive axles with nearly the same amount of power regardless of traction.
- Gerodisc differentials are speed-sensitive slip-limiting differential units that have a clutch pack and a hydraulic pump.
- The purpose of the axle shaft is to transfer driving torque from the differential assembly to the vehicle's driving wheels.
- There are basically three ways drive axles are supported by bearings in a live axle: full-floating, three-quarter floating, and semifloating.
- There are three designs of axle shaft bearings used on semifloating axles: ball-type, straight roller, and tapered roller bearings.

TERMS TO KNOW

(continued)

Partial nonhunting gear set

Pinion gear

Ramp-type differential

Removable carrier

Ring gear

Semifloating rear axle

Side gears

Straddle-mounted pinion

Three-quarter floating axle

Viscous coupling

REVIEW QUESTIONS

Short Answer Essays

1. Define the term *final drive*.

2. Explain why a differential prolongs tire life.

3. List the reasons why hypoid gears are used in nearly all RWD final drives.

4. List the three major functions of a typical RWD differential.

5. Describe the main components of a differential and state their locations.

6. Explain the major differences between an integral carrier housing and a removable carrier housing.

7. Explain the differences between hunting, nonhunting, and partial nonhunting gears.

8. What do limited-slip differential units with a viscous coupling rely on to send power to the axle with the most traction?

9. What applies the clutch pack in a gerodisc-type limited-slip differential assembly?

10. List the different ways drive axles are supported in an axle housing and explain the major characteristics of each one.

Fill-in-the-Blanks

1. The differential's _____ gear meshes with the _____ gear, which is fastened to the differential case, which houses the _____ shafts and gears, which are in mesh with the _____ gears, which are splined to the drive axles.

2. Rear axle housings can be divided into two groups: _____ carrier and _____ carrier.

3. The types of gears currently used as differential gears are the _____, _____ _____, and _____.

4. The drive side of a hypoid gear tooth is curved in a _____ shape, and the coast side of the tooth is _____. The inner end of the teeth on a hypoid ring gear is known as the _____ and the outer end of the teeth as the _____.

5. Gear ratios express the number of turns the _____ gear makes compared to one turn of the _____ gear it mates with.

6. Ring and pinion gear sets are usually classified as _____, _____, or _____ gears.

7. A straddle-mounted pinion gear is usually mounted on _____ bearings, whereas an overhung-mounted pinion is mounted on _____.

8. The three common configurations used as the final drives on FWD vehicles are _____, _____, and _____.

9. Most limited-slip differentials use either a _____, a _____, or a _____ to limit the action of the differential.

10. _____ differentials are speed-sensitive slip-limiting differential units that contain a clutch pack and a hydraulic pump.

ASE-Style Review Questions

1. When discussing differential action,
   Technician A says that when a car is making a turn, the outside wheel must turn faster than the inside wheel.
   Technician B says that a locked differential may cause the car to slide around a turn.
   Who is correct?
   A. A only
   B. B only
   C. Both A and B
   D. Neither A nor B

2. When discussing the torque multiplication factor of the differential,
   Technician A says that all of the gears in a differential affect torque multiplication.
   Technician B says that there is a gear reduction as the power flows from the pinion to the ring gear.
   Who is correct?
   A. A only
   B. B only
   C. Both A and B
   D. Neither A nor B

3. When discussion rear axle operation,
   Technician A says that when a car is moving straight ahead, all differential gears rotate as a unit.
   Technician B says that when a car is turning a corner, the inside differential side gear is locked to the differential case, causing the outside side gear to rotate faster.
   Who is correct?
   A. A only
   B. B only
   C. Both A and B
   D. Neither A nor B

4. When discussing the mounting of the drive pinion shaft,
   Technician A says that drive pinion shafts may be mounted in a long bushing.
   Technician B says that drive pinion shafts may be held by two tapered roller bearings.
   Who is correct?
   A. A only
   B. B only
   C. Both A and B
   D. Neither A nor B

5. When discussing gear ratios,
   Technician A says that they express the number of turns the driven gear makes compared to one turn of the drive gear.
   Technician B says that the gear ratio of a differential unit expresses the number of teeth on the ring gear compared to the number of teeth on the pinion gear.
   Who is correct?
   A. A only
   B. B only
   C. Both A and B
   D. Neither A nor B

6. When discussing numerical final drive gear ratios,
   Technician A says that lower gear ratios allow for better acceleration.
   Technician B says that higher gear ratios allow for improved fuel economy but lower top speeds.
   Who is correct?
   A. A only
   B. B only
   C. Both A and B
   D. Neither A nor B

7. When discussing different types of ring and pinion gear sets,
   Technician A says that with a hunting gear set, each tooth of the pinion will always return to the same few teeth on the ring gear each time the pinion rotates.
   Technician B says that when a nonhunting gear set rotates, any pinion gear tooth is likely to contact each and every tooth on the ring gear.
   Who is correct?
   A. A only
   B. B only
   C. Both A and B
   D. Neither A nor B

8. When discussing electrohydraulic limited-slip differentials,
   Technician A says that these differentials improve handling on slippery surfaces.
   Technician B says that these differentials limit the amount of differential action between the side gears.
   Who is correct?
   A. A only
   B. B only
   C. Both A and B
   D. Neither A nor B

9. When discussing the different designs of rear axles,
   Technician A says that the bearings for full-floating shafts are located within the axle tubes of the rear axle housing.
   Technician B says that C-type retainer axles are held in place by a pressed retainer ring against the axle bearing.
   Who is correct?
   A. A only
   B. B only
   C. Both A and B
   D. Neither A nor B

10. When discussing LSD lubricants,
    Technician A says that limited-slip differentials require a special lubricant.
    Technician B says that all RWD differential units require a special hypoid-compatible lubricant.
    Who is correct?
    A. A only
    B. B only
    C. Both A and B
    D. Neither A nor B

Chapter 8

FOUR-WHEEL-DRIVE SYSTEMS

A BIT OF HISTORY

The first known gasoline-powered four-wheel-drive automobile was the Spyker, built in the Netherlands in 1902.

UPON COMPLETION AND REVIEW OF THIS CHAPTER, YOU SHOULD BE ABLE TO:

- Explain the advantages and disadvantages of four-wheel drive.

- Use the correct terminology when discussing four-wheel-drive systems.

- Describe the different designs of four-wheel-drive systems and their applications.

- Compare and contrast the components of part- and full-time four-wheel-drive systems.

- Describe the operation of various transfer case designs and their controls.

- Identify the differences in operation and construction between different front-axle actuating systems.

- Identify the suspension requirements of vehicles equipped with four-wheel drive.

INTRODUCTION

The primary focus of this chapter is on the transfer cases and related systems used for four-wheel-drive (4WD) on light trucks (Figure 8-1), automobiles, and off-road vehicles. Although the principles of operation are similar for all 4WD units, the components, location, and controls of the various systems differ according to manufacturer and application.

Four-wheel-drive (4WD) and all-wheel-drive (AWD) systems can dramatically increase a vehicle's traction and handling ability in rain, snow, and off-road driving. Consider that the

With the popularity of SUVs and pickup trucks, the need for technicians that can diagnose and service four-wheel-drive systems has drastically increased. Although all-wheel-drive passenger cars are available, most prospective buyers for all-wheel- and four-wheel-drive vehicles are opting for truck-based SUVs and pickups (Figure 8-2).

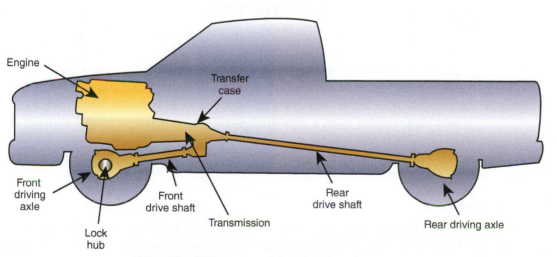

FIGURE 8-1 Typical arrangement of 4WD components.

FIGURE 8-2 A 4WD sport utility vehicle.

The designation of 4×4 is used with vehicles to indicate the number of wheels on the ground and the number of wheels that can be driven. For example, the typical car has four wheels, two of which can deliver power; therefore they could be designated as 4×2 vehicles. A 4WD vehicle is designated as a 4×4 because it has four wheels and power can be applied to all four wheels.

SERVICE TIP:
Whenever diagnosing, servicing, or repairing a 4WD performance car, light truck, or off-road vehicle, refer to the appropriate service manual for specific information and service procedures for that vehicle.

Shop Manual
Chapter 8, page 362

The **center differential** is commonly referred to as the interaxle differential. It allows the front and rear wheels to turn at different speeds and with different amounts of torque.

vehicle's only contact with the road is the small areas of the tires. Driving and handling is vastly improved when the workload is spread out among four wheels rather than one or two. Other factors such as the side forces created by cornering and wind gusts also have less effect on vehicles with four driving wheels.

The increased traction also makes it possible to apply greater amounts of energy through the drive system. Vehicles with 4WD and AWD can maintain control while transmitting levels of power that would cause two wheels to spin either on takeoff or while rounding a curve.

Both 4WD and AWD systems add cost and weight to a vehicle. A typical AWD system adds approximately 150 pounds to a passenger car. The additional weight in larger 4WD trucks can be as much as 400 pounds or more. The systems also add to the initial cost of the vehicle. They also require special services and maintenance, and use more fuel, making the cost of operating a 4WD or an AWD vehicle higher than operating a 2WD vehicle. For many, these disadvantages are heavily outweighed by the traction and performance these systems offer.

Where the engine's torque is sent depends on the components of the 4WD system. If the vehicle is equipped with an open differential, power is applied to two drive wheels (one in the front and one in the rear). If the vehicle has a locking differential, it is possible for torque to be delivered to all four wheels. However, it is also possible for AWD vehicles to drive only one wheel. Understanding how each of the components works will allow you to determine if the system is working properly.

FOUR-WHEEL-DRIVE DESIGN VARIATIONS

Four-wheel drive is most useful when a vehicle is traveling off the road or on any slippery surface. 4WD vehicles designed for off-road use are normally RWD vehicles equipped with a transfer case (Figure 8-3), a front drive shaft, and a front differential and drive axles.

Some high-performance cars are equipped with 4WD to improve the handling characteristics of the car. Many of these cars are FWD models converted to 4WD. FWD cars are modified by adding a transfer case, a rear drive shaft, and a rear axle with a differential (Figure 8-4). Although this is the typical modification, many cars are also equipped with a **center differential**. This differential unit allows the rear and front wheels to turn at different speeds and with different amounts of torque.

Terminology

Like other terms used by the automotive industry, the term all-wheel drive (AWD) can mean many different things. It is often difficult to clearly define the difference between 4WD and AWD. Both have the ability to send torque to all four wheels. For clarity, the primary difference

FIGURE 8-3 A typical transfer case.

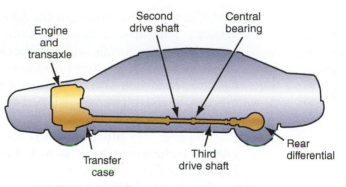

FIGURE 8-4 An AWD system based on a FWD platform.

between the two is the transfer case. A 4WD vehicle offers two speed ratios in four-wheel drive: high and low (Figure 8-5). An AWD vehicle does not have a "low" operating range because it is designed for light off-road use. The most common understanding of AWD is that it is a system that constantly provides power to all four wheels. To the contrary, 4WD usually is understood as a system that needs to be engaged by the driver in order to drive all four wheels. Plus these systems have a low range and are suited for heavy off-road use.

Other terms are also used to help define the type of the drive system. But for many, this only confuses things. This is especially true when manufacturers are not very technically accurate about how they market their vehicles. Of course, as they develop new technologies for 4WD or AWD, they want to make sure the public knows that a change or advance has been made. Selling automobiles is their business, so who can blame them? To help clear this up, let us look at what the authorities say.

The Society of Automotive Engineers (SAE) has the role of recommending specifications and practices for the automotive industry. Most are totally accepted by the manufacturers and some are included in government safety or emission control mandates. The SAE defines all-wheel drive as a system that sends power to four wheels. Based on that definition, all of the 4WD systems discussed in this text would be called all-wheel-drive systems.

The SAE does attempt to clear this up by offering other terms to label the various systems. These terms are what will be used in text, unless the manufacturer calls it

Shop Manual
Chapter 8, page 374

A BIT OF HISTORY

The first 4WD car produced in the United States was the F.W.D. in 1908.

FIGURE 8-5 A 4WD-mode selector switch.

something else. According to the SAE, **part-time 4WD** systems are manually controlled systems. They operate as a 2WD until the driver does something to activate or engage the other axle. Then torque is divided between the two axles and all four wheels can be driven. These systems have a fixed torque split between the front and rear. They do not corner well at speed when 4WD is engaged because tire slip is the only way these systems accommodate for the differences in travel distance between the front and rear tires as the vehicle goes through a turn.

The SAE says **full-time 4WD** means the system is permanently AWD. Basically these systems make torque available to all four wheels regardless of conditions and the driver cannot select 2WD.

On-demand 4WD systems are those systems that are automatically controlled. Typically these systems drive one axle until some wheel slip is detected. Some systems are even designed to send torque to the opposite axle when slip is anticipated. To overcome slip at one axle, some torque is sent to the other axle. The amount of torque transmitted to the other axle depends on the amount of slip and the system.

Some 4WD systems give the driver the option of selecting full-time 4WD or 2WD. In many cases, the full-time mode is really on-demand 4WD. When the driver selects 4WD Auto, nearly all of the power is sent to one drive axle. As soon as that axle experiences some slip, some torque is transferred to the other axle. For example, with Ford's Control-Trac system, when the 4WD Auto mode is selected, 96 percent of the engine's torque is sent to the rear wheels until slippage or wheel spin is detected. In response to the slippage, the system energizes an electromagnetic clutch in the transfer case. The action of the clutch transmits torque toward the front axle. The amount of torque sent to the front axle depends on the amount the rear wheels are slipping. Up to 96 percent of the torque can be sent to the front axle.

Various Designs

While most full-sized 4WD trucks and utility vehicles are design variations of basic RWD vehicles, many passenger cars and smaller SUVs equipped with AWD are based on FWD designs. These modified FWD systems consist of a transaxle and differential to drive the front wheels plus some type of mechanism for connecting the transaxle to a rear driveline. In many cases, this mechanism is a simple clutch or differential. Normally, FWD cars are modified by adding a transfer case, a rear drive shaft, and a rear axle with a differential (Figure 8-6).

Some systems have an interaxle or center differential (Figure 8-7) in the transfer case. This differential unit allows the rear and front wheels to turn with different amounts of torque. Also, other AWD systems are fitted with a viscous coupling inside or in place of the center differential.

Limited-Slip Differentials

With full-time 4WD, an open center differential allows the front wheels to travel farther or turn faster than the rear wheels without slipping. However, if the wheels on either axle lose traction and begin to spin, the differential continues to deliver maximum torque to the axle with the minimum traction. As a result, insufficient torque to move the vehicle may be provided to the wheels that still have traction.

To overcome this problem, many full-time systems use a limited-slip center differential (Figure 8-8). These units may use a viscous clutch, cone clutch, or a multiple-plate clutch assembly. By using a limited-slip differential, the amount of torque delivered to the axle with the least traction can be limited and allows more torque to the axle with the most traction.

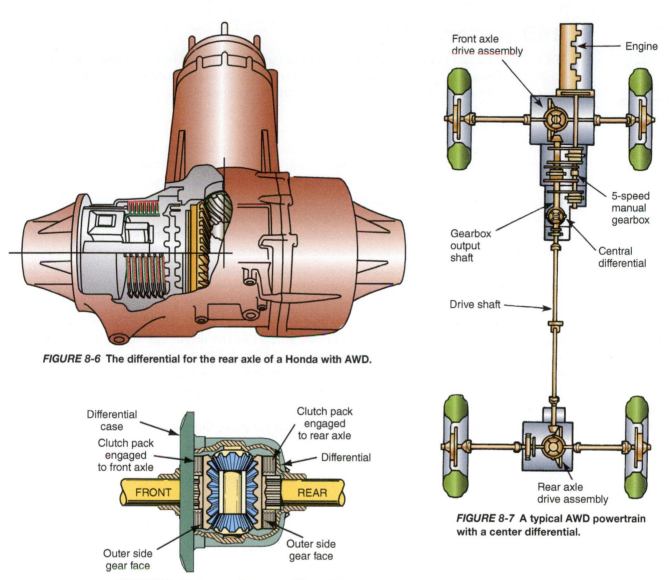

FIGURE 8-6 The differential for the rear axle of a Honda with AWD.

FIGURE 8-7 A typical AWD powertrain with a center differential.

FIGURE 8-8 A cutaway of a center differential with a clutch pack.

When differential action is not desired, the center differential (on some models) can be locked to provide torque to all four wheels. Because a locked center differential will not allow a speed difference between the front and rear axle, the lock position is intended for and must only be used on soft surfaces, such as dirt, mud, and snow. Typically, the differential is locked by an electromagnetic clutch. The clutch may be driver-controlled through a switch or may be computer-controlled in response to wheel speed sensors (normally the same speed sensors as used for ABS).

> **AUTHOR'S NOTE:** Customers need to be made aware that wear on 4WD systems is much greater than on 2WD systems. This is especially true if the driver keeps the vehicle in 4WD while it is traveling on dry pavement. Some manufacturers will not warranty parts of the 4WD system if there is evidence of abuse or misuse.

4WD Systems

4WD systems can have several names, including 4×4, and can be classified as part time or full time. These describe when power is sent to the four wheels. Both of these have a transfer case normally mounted to the side or rear of the transmission.

Part-Time Systems

Part-time systems are often found on 4WD pickups and older SUVs. These are basically RWD vehicles fitted with a two-speed transfer case and axle connects/disconnects. The transfer case does two things in response to the driver's commands: it engages/disengages 4WD and adjusts the overall gear ratio for the final drive. Normally, the system operates in 2WD until the driver selects 4WD.

When 4WD is engaged, the front and rear-drive axles in a part-time system rotate with the same torque, and there is no allowance for axle speed differences. When the vehicle is turning a corner on a slippery surface, this doesn't present a problem as the tires can easily skid or spin across the slick surface to accommodate the speed differences. However, on dry surfaces, the required speed differentials cause the tires to scrub against the pavement.

An electric switch or shift lever (Figure 8-9) controls the transfer case so that power is directed to the axles selected by the driver. Normally, the driver can select between 2WD-High, which engages only the rear axle and is used for all dry-road driving; 4WD-High, which engages both axles and is used at any speed on slippery surfaces; and 4WD-Low, which engages both axles and lowers the ratios of the entire driveline and should only be used at low speeds and on very demanding terrain.

The high range could be called the normal range as the torque from the transmission is not altered by the transfer case. This gear selection is used to provide 4WD traction on roads covered with ice or packed snow. The low range provides extra torque at very low output speed. This allows drivers to slowly and smoothly climb very steep hills or pull heavy loads.

Other transfer cases are a single speed and only allow the driver to select between 2WD and 4WD modes. Some systems require the vehicle to be at a stop before shifting into or out of 4WD; others allow the change at any speed. The latter systems are typically called "shift on the fly" systems.

The transmission is connected directly to the transfer case. A transfer case is similar to a standard transmission. It uses shift forks to select the operating mode, plus shafts, gears, shims, bearings, and other components found in transmissions. The shift fork that moves to engage 4WD is typically called the **mode fork**. The shift fork that is used to engage low range is called the **range fork**. The outer case is made of cast iron, magnesium, or aluminum. It is filled with lubricant that reduces friction on all moving parts. Seals hold the lubricant in the

"Shift on the fly" 4WD is simply a system that can be shifted from two- to four-wheel drive while the vehicle is moving.

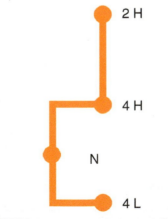

FIGURE 8-9 Typical shift lever positions.

case and shims that maintain the proper clearances between the internal components and the case. A transfer case also has a gear set or chain drive to transmit power to one or both of the drive axles.

The transfer case is connected to the front and rear axle assemblies by two separate drive shafts. One sends torque from the transfer case to the front axle and the other sends it to the rear axle (Figure 8-10). When the system is operating in 4WD, the front drive shaft is locked to the rear drive shaft. Therefore each axle receives half of the torque coming from the transmission and the axles rotate at the same speed.

The drive shafts from the transfer case connect to differentials at the front and rear axles. Universal joints or constant-velocity (CV) joints are used to connect the drive shafts to the

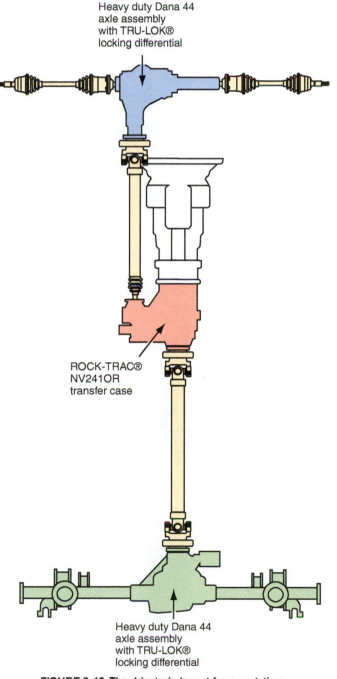

Heavy duty Dana 44
axle assembly
with TRU-LOK®
locking differential

ROCK-TRAC®
NV241OR
transfer case

Heavy duty Dana 44
axle assembly
with TRU-LOK®
locking differential

FIGURE 8-10 The drivetrain layout for a part-time heavy-duty 4WD vehicle.

differentials and transfer case. The rear-drive axle of a 4WD vehicle is identical to those used in 2WD vehicles. The front-drive axle is also like a conventional rear axle, except that it is modified to allow the front wheels to steer (Figure 8-11). Universal joints or CV joints are used to connect the front axles to the wheel hubs on heavy-duty trucks. Light-duty vehicles and 4WD passenger cars generally use half shafts and CV joints in their front-drive axle assembly (Figure 8-12). The differential and final drive units in the axle assemblies are similar to those found in a RWD vehicle. The front differential is generally an open differential because of the great differences in wheel rotational speed when steering around corners. Some off-the-road vehicles have lockable front differentials for extreme conditions.

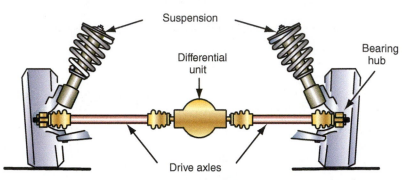

FIGURE 8-11 Typical front-drive axle assembly for a 4WD.

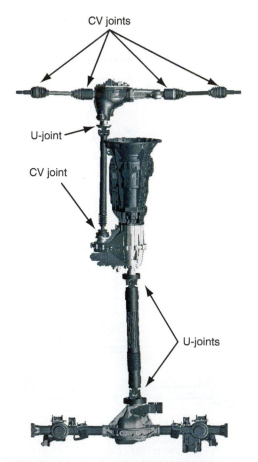

FIGURE 8-12 The location on the CV joints and U-joints on a typical 4WD vehicle.

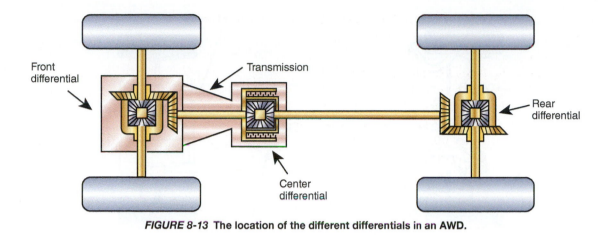

FIGURE 8-13 The location of the different differentials in an AWD.

Locking Hubs. On some 4WD vehicles, the front axles are engaged by locking wheel hubs. The hubs are designed to stop the rotation of the front axles and differential when 4WD is not selected. This increases fuel economy and decreases the wear on those parts. Locking hubs can be either manual or automatic. Newer systems have automatic hubs that engage when the driver switches into 4WD. Older vehicles had manual hubs that required the driver to turn a knob on the front wheels. Most locking hubs have a sliding collar that locks the front-drive axles to the hub.

Driveline Windup. It must be kept in mind that although vehicles with two driving axles have the same gear ratios at both driving axles, the distance the tires must travel are different when the vehicle is making a turn. The outside front wheel will travel farther than the outside rear wheel and the result is called **driveline windup**. This is why new full-time transfer case designs use clutch packs or viscous couplings to permit full-time use on dry pavement. On wet or slippery roads, the front and rear wheels slide enough to prevent damage to the driveline components. However, this may not be the case on dry surfaces.

Full-Time 4WD. In a full-time 4WD system, all four wheels have power delivered to them, all of the time. They do well off-road and can be used on all surfaces, including dry pavement. In addition to a transfer case, these systems have a center or interaxle differential or viscous clutch, which allows for a difference in speeds between the front and rear axles (Figure 8-13). These units decrease the amount of wheel slip when the vehicle is turning a corner. The center differential is located between the front and rear drive shafts or may be an integral part of the transfer case. Like all 4WD systems, the transfer case has a low- and high-speed range. However, the driver does not have the option of shifting into 2WD and there are no locking hubs at the front wheels.

FWD-Based Systems. Some FWD vehicles fitted for 4WD use a compact transfer case bolted to the transaxle. A drive shaft then carries the power to the rear differential. The driver can switch from 2WD to 4WD by pressing a switch that activates a solenoid vacuum valve or electric motor.

All-Wheel-Drive Systems

All-wheel-drive systems do not give the driver the option of 2WD or 4WD (Figure 8-14). AWD gives the driver maximum control in adverse operating conditions by biasing the driving torque to the axle with traction. The advantage of all-wheel drive can be compared to walking on snowshoes. Snowshoes prevent the user from sinking into the snow by spreading body

> **Driveline windup** is a reaction that takes place as a result of the transfer of torque to the rear wheels.

> **2WD** is an acronym commonly used for two-wheel-drive vehicles.

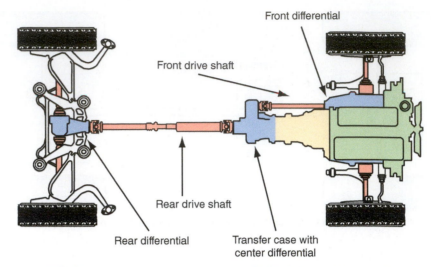

FIGURE 8-14 An AWD system based on a RWD platform with a transfer case and three differentials.

weight over a large surface. AWD vehicles spread the driving force over four wheels when needed, rather than two wheels.

When a vehicle travels, the driving wheels transmit a tractive force to the road's surface. The ability of each tire to transmit tractive force is a result of vehicle weight pressing the tire onto the road's surface and the coefficient of friction between the tire and road. If the tire and the road's surface are dry, the coefficient of friction is high and four driving wheels are not needed. If the road's surface is wet and slippery, the coefficient of friction between the tire and road is low. This could result in loss of driver control. Unlike a two-wheel-drive vehicle, an AWD vehicle spreads the tractive effort to all four driving wheels. In addition to spreading the driving torque, the AWD vehicle biases the driving torque to the axle that has the traction only when it is needed.

AWD systems automatically react to normal or slippery road conditions and are not designed for off-road use; rather, they are designed to increase vehicle performance in poor traction situations.

AWD systems rely on a single-speed transfer case added to the transaxle and/or an inter-axle differential to split power between the front and rear axles. Many AWD systems are fitted with a viscous coupling in place of a center differential. The viscous coupling contains two sets of alternating perforated steel clutch plates in a sealed unit with a special silicone-based medium (Figure 8-15). One set of clutches is connected to the front-drive wheels and the other to the rear. When one axle slips, the speed difference causes the clutches to shear and heat the viscous fluid, causing it to become a near-solid and connect the clutches and therefore the front and rear axles. Normal speed differences caused by cornering do not lock the clutches. The viscous coupling provides power to all four wheels but prevents driveline windup (Figure 8-16).

Some AWD systems are called On-Demand 4WD systems. In these systems, torque is transmitted to primarily one axle. Torque is automatically transferred to the other axle when traction is poor on the primary drive axle. Because this system responds to present road conditions, it may be operated on all road surfaces.

Typically, these automatic AWD systems are electronically controlled. To split the power between axles, a multiple-disc clutch is used. This clutch serves as an interaxle differential and permits a speed difference between the front and rear-drive axles. Sensors monitor front- and rear-axle speeds, engine speed, and load on the engine and driveline. An electronic control unit (often called the **transmission control unit [TCU]**) receives information from the sensors and controls a solenoid that operates on a **duty cycle** to control the fluid flow that engages

FIGURE 8-15 A disassembled viscous coupling.

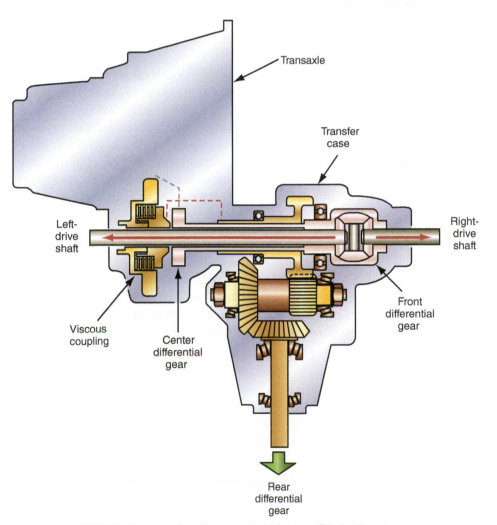

Transaxle

Transfer case

Left-drive shaft

Right-drive shaft

Front differential gear

Viscous coupling

Center differential gear

Rear differential gear

FIGURE 8-16 Power-flow through a transfer case fitted with a viscous coupling and an interaxle differential.

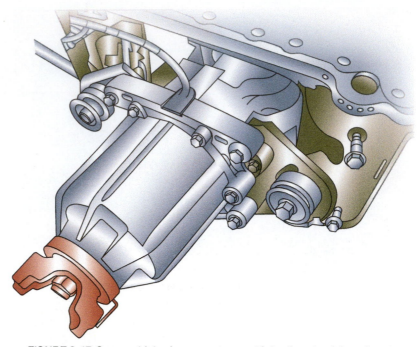

FIGURE 8-17 Some vehicles incorporate a multiple-disc clutch into the rear differential assembly. This clutch assembly eliminates the need for a center differential, and operates mechanically, not electronically.

the transfer clutch. The **duty solenoid** pulses, cycling on and off very rapidly, which develops a controlled slip condition. As a result, the transfer clutch operates like an interaxle differential and allows for a power split from 95 percent FWD and 5 percent RWD to 50 percent FWD and 50 percent RWD. This power split takes place so rapidly that the driver is unaware of the traction problem.

Some AWD vehicles do not have a center differential; rather, there is a multiple-disc clutch at the input of the rear differential (Figure 8-17). In these systems, all torque is normally sent to the front wheels. If one of the front wheels begins to slip, the clutch can send torque to the rear wheels.

> **AUTHOR'S NOTE:** Matching tire size and proper tire inflation are critical to the proper operation of AWD vehicles. The rotational speed differences caused by different tire size, uneven wear, or incorrect inflation can quickly cause a viscous clutch to overheat and be destroyed. The technician must be aware of this possibility when recommending tires or performing vehicle service.

TRANSFER CASES

Shop Manual
Chapter 8, page 374

The transfer case delivers power to both the front- and rear-drive axle assemblies and is constructed much like a conventional transmission (Figure 8-18). It uses shift forks to select the operating mode, plus splines, gears, shims, bearings, and other components commonly found in transmissions. The outer case is made of magnesium, aluminum, or cast iron and is filled with lubricant. Seals are used to hold the lubricant in and keep dirt out. Shims are used to maintain proper clearances or preloads between the parts of the transfer case.

The purpose of a transfer case is to transfer torque from the output of the transmission to the vehicle's front and rear axles. The transfer case is normally connected to each axle by two drive shafts—one between the transfer case and the front axle and the other between the

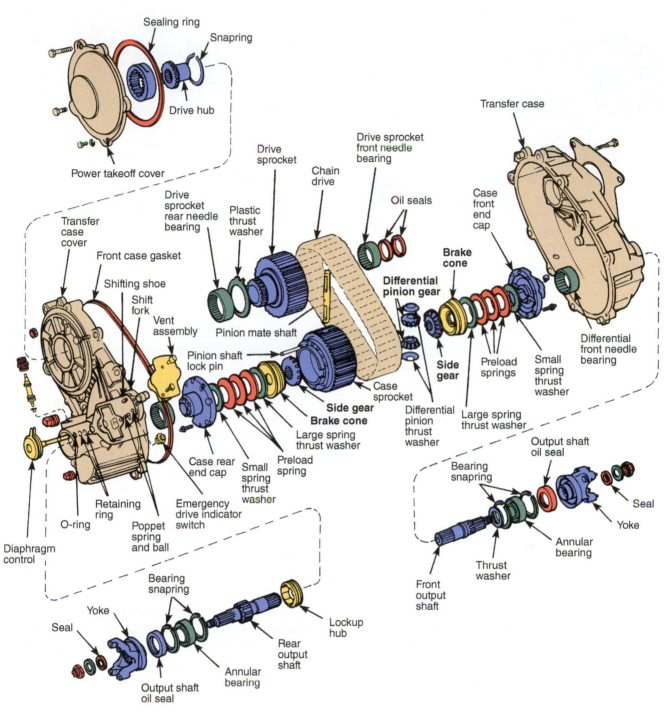

FIGURE 8-18 4WD transfer case with integral differential and cone brakes for limited slip.

transfer case to the rear axle. Each axle has its own differential unit, which then turns the wheels. Torque is then multiplied by the gear reduction of the differential units and sent to the wheels.

Modes of Operation

A transfer case is an auxiliary transmission mounted to the side, or in the back of the main transmission (Figure 8-19). There are basically three classifications of transfer cases: part time, which provides the following ranges—Neutral, 2WD-High, 4WD-High, and 4WD-Low; full time, which provides 2WD-High, 4WD-High; and 4WD-Low; and part time/full time, which provides 2WD-High, full-time 4WD-High, and part-time 4WD-Low.

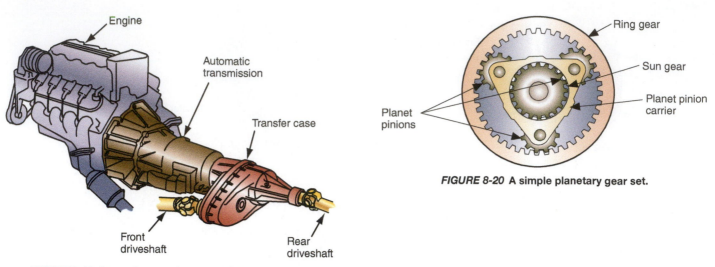

FIGURE 8-19 A transfer case integrated into the rear of a transmission.

FIGURE 8-20 A simple planetary gear set.

The change from the high-speed range to the low-speed range is made by moving the shift lever of the transfer case, which moves a gear on the transfer case's main drive shaft from the high-speed drive gear to the low-speed drive gear. High speed provides direct drive, whereas low speed usually produces a gear ratio of about 2:1.

Low speed is a torque multiplication gear that allows for increased power to the wheels and is an addition to the gear reductions of the transmission and final drive gears. When the transfer case is in high speed, torque is only multiplied by the transmission and final drive gears. In addition to these two speeds, transfer cases may also have a neutral position. When the transfer case is in neutral, regardless of which speed gear the transmission is in, no power is applied to the wheels. Some 4WD vehicles are equipped with a single-speed transfer case and do not have a low range.

Most transfer cases use a planetary gear set (Figure 8-20) to provide for low range. Although the shifting mechanisms vary with the different models of transfer cases, the power-flow through nearly all transfer cases is the same. The shift mechanism moves the planetary gear set carrier and the sun gear along a splined section of the input shaft. This movement engages and disengages various gears, which results in the different gear selections. A description of the power-flow through a typical two-speed transfer case follows.

When in neutral, the transfer case is driven directly by the main output shaft of the transmission. Power is not transmitted to the driving axles when the transfer case is in neutral, regardless of transmission gear position (Figure 8-21). With the transfer case in neutral, the sun gear turns the planetary gears, which drive the ring gear (often called the annulus gear).

The sun gear is the central gear in a planetary gear set around which the rest of the gears rotate.

The ring gear of a planetary gear set is often called the **annulus gear**.

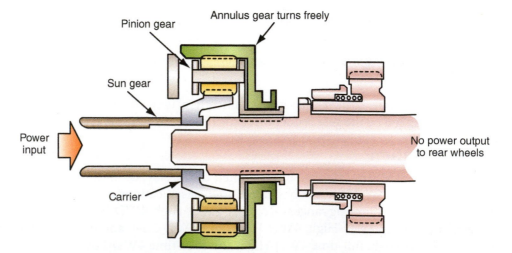

FIGURE 8-21 Power-flow through a transfer case while it is in neutral.

This gear set rotates with the input shaft. The planetary gear case, which is splined to the rear output shaft, remains stationary because the gear set is positioned away from and not engaged with the case. Therefore, power is not transmitted to the rear output shaft.

When in 2WD-High, the planetary gear set moves rearward and the mode shift fork holds the sliding clutch in the 2WD position. As the planetary gear set moves rearward, it locks into the planetary gear case. This prevents the rotation of the planetary gears on their axes, causing the planetary gears, planetary carrier, and the ring gear to rotate as a single unit. As the gear set rotates, it drives the rear output shaft at the same speed as the input shaft (Figure 8-22). This provides for direct drive.

When in 4WD-High, the planetary gear set is in the same position as in 2WD-High, but the 4WD mode shift fork is moved forward. This movement releases the sliding clutch, which is splined to the driving chain **sprocket** carrier gear (Figure 8-23). The clutch shift spring pushes the sliding clutch into engagement with the rear output shaft. This causes the chain to drive the front output shaft at the same speed as the rear output shaft, thus sending power to both axles and providing 4WD.

A **sprocket** is a projecting tooth on a wheel or cylinder that engages with the links of a chain to make it move forward.

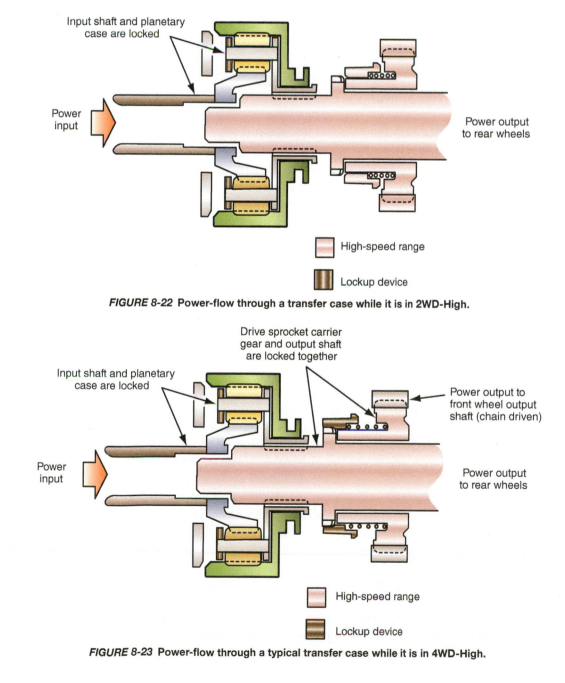

FIGURE 8-22 Power-flow through a transfer case while it is in 2WD-High.

FIGURE 8-23 Power-flow through a typical transfer case while it is in 4WD-High.

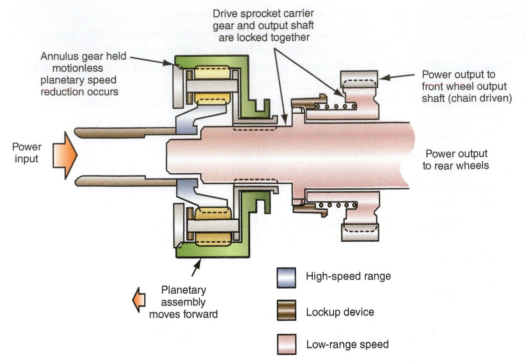

Drive sprocket carrier
gear and output shaft
are locked together

Annulus gear held
motionless
planetary speed
reduction occurs

Power output to
front wheel output
shaft (chain driven)

Power
input

Power output
to rear wheels

Planetary
assembly
moves forward

☐ High-speed range

☐ Lockup device

☐ Low-range speed

FIGURE 8-24 Power-flow through a transfer case while it is in 4WD-Low.

When in 4WD-Low, the 4WD mode shift collar is positioned to provide for 4WD. However, the range shift fork moves the sun gear and planetary gear set assembly rearward, causing the ring gear to engage with a locking ring that is part of the bearing retainer assembly (Figure 8-24) that holds the ring gear stationary and allows the planetary gears to "walk around" the inside of the ring gear. This causes the planetary pinion gears to drive the planetary carrier around more slowly than the input shaft speed. Because the planetary carrier hub is splined directly to the output shaft, the output shaft rotates at the slower speed. This speed reduction results in increased torque at the drive wheels.

Spur-Helical Gear Drives. Some transfer cases use a spur or helical gear set to provide for the speed ranges and for engagement and disengagement of 4WD. The front axle is engaged by shifting a sliding or clutching gear inside the transfer case into engagement with the driven gear on the drive shaft for the front wheels (Figure 8-25). The sliding gears and clutches are driven through splines on the shafts.

> **AUTHOR'S NOTE:** Planetary gear sets can be difficult to figure out. The best way to get a good feel for how all of the potential gear combinations are possible is to play with a planetary unit. Hold and turn various parts of the gear set and watch what happens. Then duplicate the above positions to verify them.

Shop Manual
Chapter 8, page 375

A sliding gear on the main shaft locks either the low-speed gear or the high-speed gear to the main shaft. In many transfer cases, this shift cannot be made unless the transmission is in neutral. Because the transfer case main shaft and the sliding gear, which is splined to it, will be turning at a different speed than the low- or high-speed gear, gear clash will occur.

To engage and disengage the front axle, another sliding gear or clutch is used to lock the front-axle drive gear to the front-axle drive shaft. On many transfer cases, the front axle can be engaged and disengaged when the vehicle is moving. This is done by releasing the accelerator pedal to remove the torque load through the gears and moving the shift lever. However, both

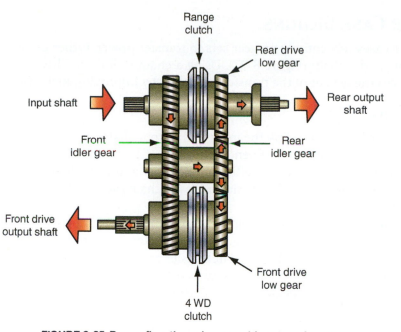

FIGURE 8-25 Power-flow through a gear-driven transfer case while it is in 4WD-Low.

the front and the rear wheels must be turning at the same speed. If the rear wheels have lost traction and are spinning, or if the brakes are applied and either the front or the rear wheels are locked and sliding, gear clashing will occur when engagement of the front axle is attempted. On some 4WD systems, the vehicle must be stopped before the change can be made. Also, the shift into 4WD-Low range normally requires that the vehicle be stopped first.

Some transfer cases are single speed and only allow for a change between 2WD and 4WD. An integral main drive gear in the transfer case, which is driven by the output shaft of the transmission, provides power to the rear driveline at all times. This main drive gear is in mesh with an idler gear that, in turn, meshes with a front-axle drive gear to rotate the front driveline when it is in 4WD. When the vehicle is in 2WD, the idler gear is moved out of mesh with the front-axle drive gear.

Some "shift-on-the-fly" systems use a magnetic clutch in the transfer case to bring the front drive shaft, differential, and drive axles to the same speed as the transmission. When the speeds are synchronized, an electric motor in the transfer case (Figure 8-26) completes the shift. The system will not shift until the speeds are synchronized.

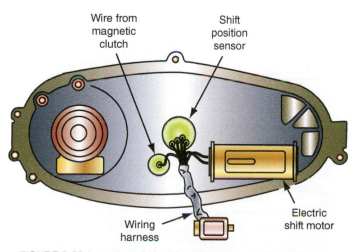

FIGURE 8-26 Location of electric shift motor on a transfer case.

Shop Manual
Chapter 8, page 376

TRANSFER CASE DESIGNS

Some transfer cases rely entirely on gear sets to transfer power. Other designs use a combination of gears and a chain (Figure 8-27). Using a chain to link the drive axles, instead of a gear set, reduces the weight of the transfer case, thereby improving fuel economy.

Drive Chains

Drive chains are often used to link the input and output shafts in a transfer case. The chain only serves as a link and does not influence gear ratios. Chains are commonly used with planetary gear sets because they are very efficient and quiet and allow for flexible positioning of the transfer case's components. Two basic designs of chains are used: the round-pin style and the pin-and-rocker joint style.

In the **round-pin** design, a single pin is inserted into mating holes at each end of the link plate. Load is distributed over a large area, allowing greater chain tensions and higher dynamic loads. The round-pin joint is widely used in transfer cases on part-time 4WD vehicles.

The **pin-and-rocker joint** design uses two convex joints that roll against one another as the chain moves. This type of chain is very efficient at continuous high speeds and is used on full-time 4WD systems.

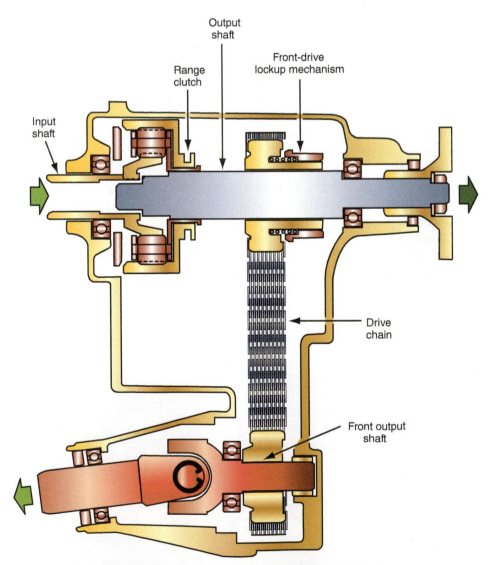

FIGURE 8-27 Basic setup for a chain-driven transfer case.

Planetary Gear Drives

A typical part-time transfer case uses an aluminum case, a chain drive, and a planetary gear set for reduced weight and increased efficiency. Further efficiency is gained because the internal components of the transfer case do not rotate while the vehicle is operating in 2WD.

A simple planetary gear set is made up of three gears. In the center of the gear set is the sun gear. All other gears in the set revolve around the sun gear. Meshing with the sun gear are three or four planetary pinion gears. The pinions are held together by the planetary pinion carrier or the planetary carrier. The carrier holds the gears in place while allowing them to rotate around their shafts. On the outside of the planetary pinions is the ring gear. The ring gear has teeth around its inside circumference that mesh with the teeth of the planetary pinions. Each planetary pinion gear is mounted to the planetary carrier by a pin or shaft. The carrier assembly and the sun gear are mounted on their own shafts.

When the transfer case is in neutral, the rotation of the input shaft spins the planetary pinion gears and the ring gear around them. With both the pinion gears and the ring gear spinning freely, no power is transmitted through the planetary gear set.

Low- and high-range speeds are available from the planetary gear set. Speed reduction provides the low range and direct drive provides the high range. Speed reduction is provided when the ring gear is held stationary (Figure 8-28). When the transfer case's shift lever is moved to low range, the planetary gear assembly slides forward on its shaft and engages with a locking plate. The locking plate is bolted to the case and engages with the teeth of the ring gear to hold it stationary.

The driving gear of the gear set is the sun gear, which is connected to the output of the transmission. When the ring gear is held stationary and the sun gear is driving, the planetary pinions rotate on their pins. As the pinions rotate, they must walk around the ring gear because they are in mesh with it. This action causes the pinion carrier to rotate in the same direction as the sun gear.

However, the planetary carrier turns more slowly than the sun gear because of the size difference between the pinions and the sun gear. As the pinions move around the inside of the ring gear, the shaft attached to the planetary carrier is driven in the same direction as the sun gear, but at a lower speed. This action causes speed reduction and torque multiplication, thereby providing the transfer case with low range.

The transfer case can be selected into the high range. In this operating mode, the ring gear and planetary carrier are locked together. As a result, the pinion gears cannot rotate on their pins and the entire assembly turns as a solid unit. Therefore, the output shaft rotates at the same speed as the input shaft, providing for direct drive.

Electronically Controlled Planetary Gear Sets. Some vehicles are equipped with an electronically controlled 4WD system that couples the front and rear axles through a planetary gear set that allows each axle to run at its own speed (Figure 8-29). The planetary gear set is

In any set of two or more gears, the smallest gear is often called the pinion gear.

Shop Manual
Chapter 8, page 381

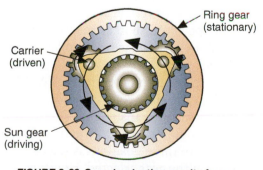

FIGURE 8-28 Speed reduction results from holding the ring gear and driving the sun gear.

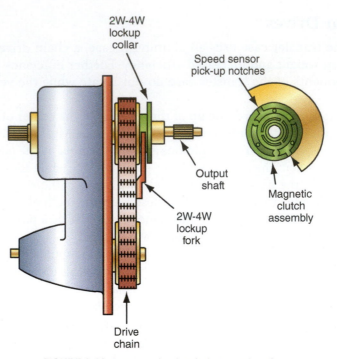

2W-4W lockup collar

Speed sensor pick-up notches

Output shaft

2W-4W lockup fork

Drive chain

Magnetic clutch assembly

FIGURE 8-29 A magnetic clutch that synchronizes the shaft speeds before allowing a shift.

Shop Manual
Chapter 8, page 364

The **electro-magnetic clutch** locks the planetary gear when wheel sensors detect a difference between the speed of the front and rear wheels. It is also used on some models to lock up the center differential when power is needed at all four wheels.

not used for speed reduction; rather, it is used as a differential. The transmission output shaft drives the planetary carrier. The ring gear is connected to the driveline for the rear wheels and the sun gear is connected by a chain to the front driveline.

This type of system uses a speed sensor at each side of the planetary gear, in which power is split one-third to the front axle and two-thirds to the rear axle under normal driving conditions. When the sensors detect a difference between the speed of the front and rear wheels—a sign that one of the wheels has lost traction—an electronic controller sends current to an **electromagnetic clutch** that locks the planetary gear.

Half of the clutch's plates are splined to the sun gear and the other half are splined to the ring gear. By locking the sun gear to the ring gear, the axles are locked together and torque is split equally to the two driving axles. After a few seconds, the AWD electronic controller disengages the clutch and rechecks the speed of the axles. If slippage is still present, the controller reengages the clutch.

An electromagnetic clutch is also used on some models to lock up the center differential when power is needed at all four wheels. The clutch may be activated by a switch on the dash or it may be controlled electronically by a computer in response to wheel speed sensor signals.

A few vehicle models use a magnetic clutch to match the speed of the front axle with that of the rear axle. This speed matching takes very little time and allows a shift to be made from 2WD-High to 4WD-High at any speed. As soon as the front- and rear-output shafts of the transfer case reach the same speed, a spring-loaded shift collar automatically engages the main shaft hub to the driveline sprocket.

Many newer RWD-based vehicles have an automatic 4WD feature that switches from 2WD to 4WD when the transfer case shift control module receives wheel rotating slip information from the wheel sensors. The transfer case shift control module then engages the transfer case motor/encoder to go into 4WD. The transfer case has the typical three gear selection positions. When any of these positions is selected, the transfer case motor is locked and the transfer case stays at the gear selected. However, when the control switch is in the AUTO (or A4WD) position, the transfer case is in the adaptive mode and will cycle to 4WD when necessary.

Some pickups and SUVs are fitted with an AWD transfer case. In this system, AWD is always activated. A viscous clutch is used as a torque distribution device. Normal torque distribution is 35 percent to the front axle and 65 percent to the rear.

A takeoff from electronically operated AWD systems for trucks is Chrysler's Quadra-Drive. This system uses a gerotor pump to react to variations between front- and rear-axle speeds. When the front and rear drive shafts are rotating at the same speed, the gerotor pump produces no pressure and everything is normal. If a rear wheel loses traction and begins to spin, the speed difference between the front and rear axles causes the inner and outer sections of the gerotor pump to spin at different speeds, building hydraulic pressure. This pressure gradually locks the clutch pack inside the transfer case, transferring power to the front axle. This system also uses gerotor pumps with clutch packs in each axle. This provides limited-slip operation at both axles. All of the torque splitting takes place without driver intervention. The Quadra-Drive II system allows the driver to electronically lock the transfer case.

LOCKING HUBS

Some 4WD systems on trucks and utility vehicles use front-wheel locking hubs. These hubs connect the front wheels to the front-drive axles when they are in the locked position. Manual locking hubs require that a lever or knob (Figure 8-30) be turned by hand to the 2WD or 4WD position. Automatic locking hubs can be locked by shifting into 4WD and moving forward slowly. Some are unlocked by slowly backing up the vehicle. On other 4WD systems, a front-axle lock is used in place of individual locking hubs.

To reduce wear on the driveline of the front axle the hubs are unlocked when the vehicle is in 2WD, which allows the wheels to rotate independently of the drive axles. Locking hubs are also called freewheeling hubs because their purpose is to disengage the front axle and allow it to free wheel while the vehicle is operating in 2WD. If the hubs remain in the 4WD or locked position when the vehicle is in 2WD, the driver will notice a drop in fuel economy and excessive noise from the front axle.

A manual **locking hub** is a type of clutch that engages or disengages the outer ends of the front-axle shafts from the wheel hub. A handle located in the center of manual hubs is turned to lock or unlock the hubs. This handle applies or releases spring tension on the hub's clutch. When the hub is in the locked position, the ring of the clutch is set onto the splines of the axle shaft (Figure 8-31). When the hub is in the unlocked position, spring pressure forces the clutch ring away from the axle shaft, thereby disconnecting the wheel hub and the axle.

Automatic Locks

Automatic hubs can be mechanically engaged by using a vacuum supplied when the driver puts the vehicle into 4WD or by using a centrifugal design that engages the hub when the front axles are driven. Ford uses Integrated Wheel End (IWE) vacuum hub assemblies that

Shop Manual
Chapter 8, page 364

Locking hubs are also called free-wheeling hubs because their purpose is to disengage the front axle and allow it to free wheel when the vehicle is operating in 2WD.

Free running position

Lock position

FIGURE 8-30 Knob positions for manual locking hubs.

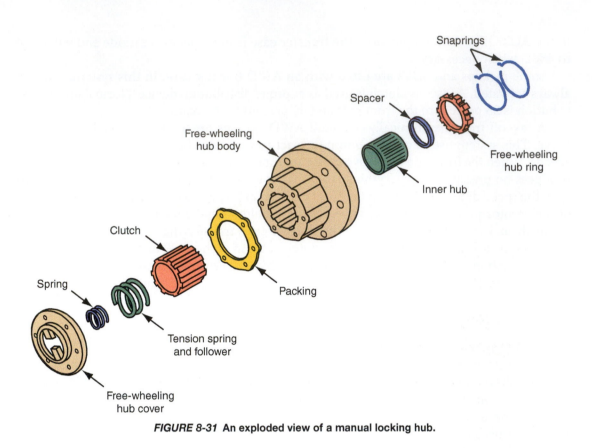

FIGURE 8-31 An exploded view of a manual locking hub.

are controlled by a vacuum solenoid. A vacuum supply from the solenoid keeps the hubs disengaged. When the driver commands 4WD, the solenoid vents the vacuum and the hubs engage. Although automatic hubs (Figure 8-32) are more convenient for the driver, they do have a disadvantage. Some self-locking hubs are designed to unlock when the vehicle is moved in reverse. Therefore, if the vehicle is stuck and needs to back out of a trouble spot, only RWD will be available to move it. Other automatic hubs unlock immediately when 4WD is disengaged without the need to back up. On these systems, the hubs are automatically locked, regardless of the direction in which the vehicle is moving. Aftermarket suppliers offer manual locking hub replacement kits for many vehicles.

Locking hubs are not needed with full-time 4WD. The wheels and hubs are always engaged with the axle shafts. The interaxle differential or transfer case prevents damage and undue wear to the parts of the powertrain.

Axle Disconnects. Many 4WD vehicles use a vacuum motor or mechanical linkage to move a splined sleeve to connect or disconnect the front-drive axle (Figure 8-33). With this system, locking hubs are not needed. When 2WD is selected, one axle is disconnected from the front differential. As a result, all engine torque moves to the side of the differential with the axle disconnected. This is due to normal differential action. When the vehicle is shifted into 4WD, the shift collar connects the two sections of the axle shaft. When the sleeve is fully engaged, a switch on the actuator causes a light to go on to inform the driver that the vehicle is in 4WD.

Other axle disconnects are operated electrically. An electric motor can be used to connect and disconnect the axle (Figure 8-34). This system allows for a smooth transition from 2WD to 4WD. General Motors uses a system whereby selecting 4WD on the selector switch energizes a heating element in the axle disconnect that heats a gas, causing the plunger to operate the shift mechanism.

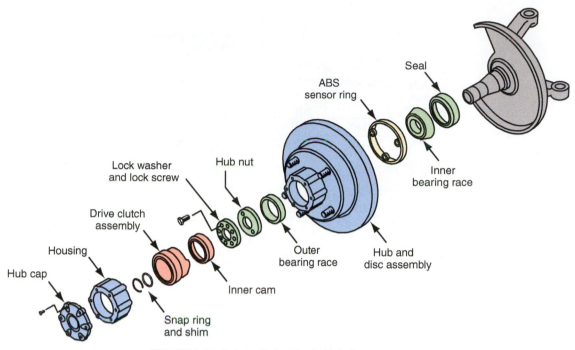

FIGURE 8-32 Automatic locking front hub components.

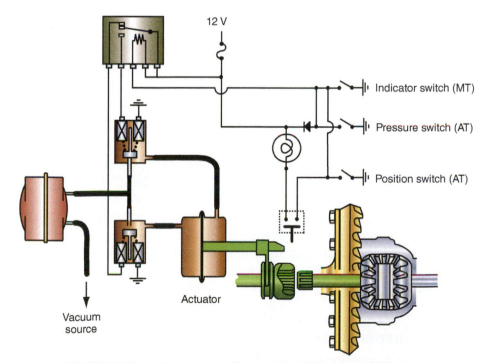

FIGURE 8-33 Toyota's Automatic Disconnecting Differential (ADD) system.

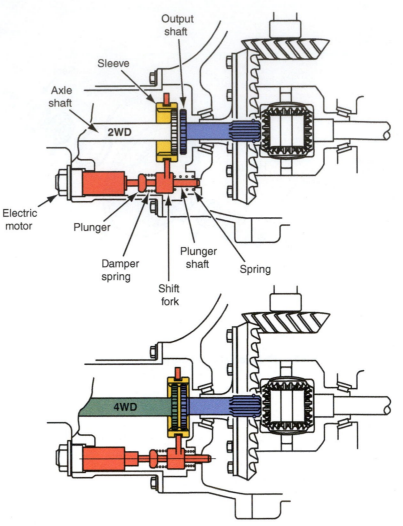

Output
shaft

Sleeve

Axle
shaft

2WD

Electric
motor

Plunger

Damper
spring

Shift
fork

Plunger
shaft

Spring

4WD

FIGURE 8-34 An electric motor disconnects the axles in this system.

FOUR-WHEEL-DRIVE SUSPENSIONS

Shop Manual
Chapter 8, page 391

The suspension components necessary for 4WD vehicles are basically the same as for FWD and RWD vehicles. When the vehicle is based on a FWD setup, the addition of the rear axle requires a rear suspension similar to those found on RWD models (Figure 8-35).

When the 4WD vehicle is based on a RWD platform, provisions for the front-drive axle must be made. Basically there are two types of front-drive axles used on 4WD vehicles: the solid axle and the independent front suspension axle assembly. The solid type is basically a rear axle turned so that the pinion gear faces the center of the vehicle. The front-drive shaft is equipped with a slip yoke at the transfer case end to accommodate the movement of the axle. The ends of the axle housing are fitted with steering knuckles that allow for steering (Figure 8-36).

Independent Suspensions

Shop Manual
Chapter 8, page 368

An independent front suspension (IFS) axle (Figure 8-37) has the differential housing mounted solidly to the vehicle's frame. Short drive shafts or half shafts connect the differential to the wheels. These half shafts are suspended by springs and are able to respond independently to road surfaces. The half shafts are normally equipped with CV joints and U-joints to accommodate this movement. The ends of the shafts fit into the steering knuckles. IFS suspension improves the vehicle's handling characteristics (Figure 8-38).

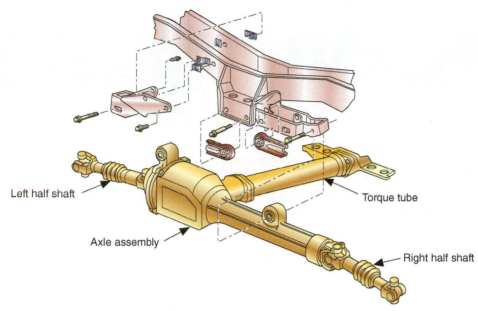

FIGURE 8-35 AWD rear-drive axle assembly.

Left half shaft

Torque tube

Axle assembly

Right half shaft

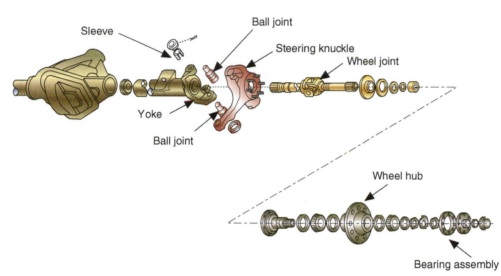

FIGURE 8-36 Exploded view of a front-drive axle's steering knuckle and hub.

Sleeve

Ball joint

Steering knuckle

Wheel joint

Yoke

Ball joint

Wheel hub

Bearing assembly

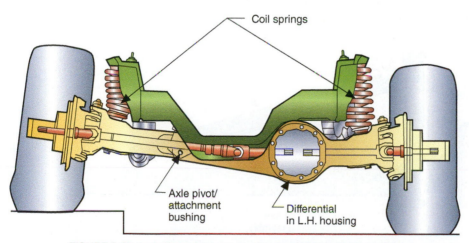

FIGURE 8-37 An independent suspension front-drive axle assembly.

Coil springs

Axle pivot/
attachment
bushing

Differential
in L.H. housing

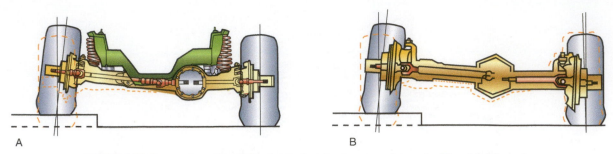

FIGURE 8-38 Comparison of an (A) independent front suspension and a (B) solid drive axle assembly in response to a bump.

To provide independent front suspension, some vehicles have one half-shaft and one solid axle for the front-drive axle. The half shaft is able to move independently of the solid axle, thereby giving the vehicle the ride characteristics desired from an IFS suspension.

The Twin-Traction Beam IFS on 4WD vehicles used by Ford Motor Company from 1980 until 1997 utilizes two steel axle carriers and a third U-joint in the right axle shaft next to the differential. The carrier and joints allow each front wheel to move up and down independently. Instead of a one-piece tube to serve as the axle housing, the tube is split into two shorter sections and joined together at a pivot point. The axle shaft can flex near the pivot because of the centrally located third U-joint. The most common IFS has a short axle housing with stub axles located below and/or between the frame rails. These have a double A-frame front suspension and front-wheel-drive axle shafts with CV joints connecting the drive axle to the wheels.

With IFS, each wheel is able to react to the road conditions individually. Bumps are absorbed by each front wheel instead of some of the shock being transferred through the solid axle to the other wheel. Normally, these vehicles are equipped with coil springs, although some heavy-duty models use leaf springs.

The wheel bearings in front- and rear-axle assemblies of 4WD vehicles are typically located so that they do not support the weight of the vehicle and are held in the hub assembly in the same way as other full-floating axle bearings.

Suspension Modifications

When a vehicle is available in either 2WD or 4WD, the suspension will differ with the drivetrain. As an example of this, let us study the suspension of a 2007 Dodge Caravan. The regular model is a 2WD with FWD. This model has an AWD option. The front suspensions of these two models are identical. The rear suspension, however, is different, mainly to support the weight and resist the torque of the rear-drive axle. The rear axle used on FWD applications is mounted to the rear leaf spring using isolator bushings at the axle mounting brackets (Figure 8-39). FWD applications also have a track bar connected to the rear axle and the frame.

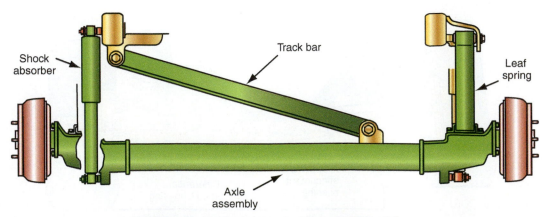

FIGURE 8-39 A typical rear suspension of a FWD minivan.

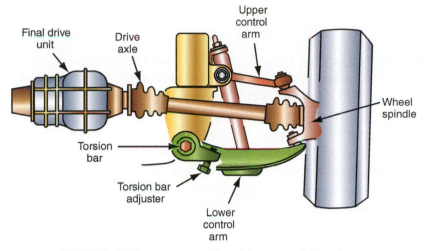

FIGURE 8-40 The rear suspension of the same minivan shown in Figure 8-39 when it is equipped with AWD.

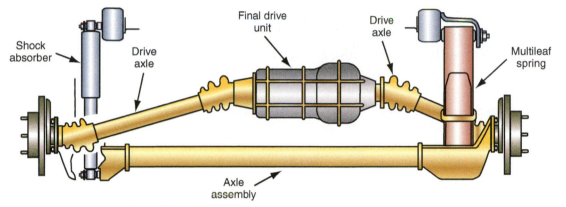

FIGURE 8-41 The front suspension on a 4WD Ford Expedition.

The rear axle used on AWD applications is mounted on a multileaf spring and does not use isolator bushings between the rear axle and the leaf springs. A track bar is not used on AWD models (Figure 8-40).

When a normally RWD vehicle is equipped with 4WD, the front suspension must be modified to support the weight of, and make room for, the drive axle. Let us do another comparison to show how different the suspensions on the same vehicle can be. A Ford Expedition is available in both 2WD and 4WD. Besides modifying the steering knuckles to accept half-shafts on the 4WD model, the major change in suspension design is the use of torsion bars on 4WD models (Figure 8-41), instead of coil springs. This change requires unique mounting and supporting hardware and, as a result, the suspensions of the two models are very different.

SUMMARY

- With 4WD, engine power can flow to all four wheels, which can greatly increase a vehicle's traction when traveling in adverse conditions and can also improve handling as side forces generated by the turning of a vehicle or by wind gusts will have less of an effect on a vehicle that has power applied to the road on four wheels.
- FWD cars are modified by adding a transfer case, a rear drive shaft, and a rear axle with a differential. Some are equipped with a center differential.

SUMMARY

- 4WD systems use a separate transfer case that allows the driver to choose or feature automatic controls that select the transfer of the engine's power to either two or four wheels.

- The front wheel hubs on some 4WD vehicles must be locked or unlocked by the driver to provide for efficient operation in 2WD or 4WD. The rear wheel hubs are always engaged.

- Integrated full-time 4WD systems use computer controls to enhance full-time operation, adjusting the torque split depending on which wheels have traction.

- On-demand 4WD systems power a second axle only after the first begins to slip.

- Many 4WD units are equipped to allow the driver to select in and out of 4WD. The systems capable of operating in both 2WD and 4WD are called part-time 4WD systems. Full-time 4WD systems cannot be selected out of 4WD. Some units feature the option of selecting part-time or full-time capabilities.

- During turns, the front wheels travel a greater distance than the rear wheels. This is because the front wheels move through a wider arc than the rear wheels.

- Part-time 4WD systems are designed to be used only when driving off the road or on slippery surfaces. When 4WD is engaged, a part-time system locks the front and rear axles together.

- A transfer case is equipped with a gear or chain drive to transmit power to one or both of the drive axles. Some transfer cases are equipped with two speeds: a high range for normal driving and a low range for especially difficult terrain.

- Two-speed transfer cases are controlled by a shift lever that typically has four positions: 2WD-High, which engages only the rear axle and is used for all dry-road driving; 4WD-High, which engages both axles and is used at any speed on slippery surfaces; Neutral, which disengages both axles; and 4WD-Low, which engages both axles and lowers the ratios of the entire driveline.

- The purpose of a transfer case is to transfer torque from the output of the transmission to the vehicle's front and rear axles.

- Some transfer cases use a spur or helical gear set to provide for the speed ranges and for engagement and disengagement of 4WD. The front axle is engaged by shifting a sliding or clutching gear into engagement with the driven gear on the drive shaft for the front wheels inside the transfer case. The sliding gears and clutches are driven through splines on the shafts.

- Some "shift-on-the-fly" systems use a magnetic clutch in the transfer case to bring the front drive shaft, differential, and drive axles to the same speed as the transmission. When the speeds are synchronized, an electric motor in the transfer case completes the shift.

- Drive chains are often used to link the input and output shafts in a transfer case and serve as a link that does not influence gear ratios.

- A simple planetary gear set is made up of three gears. In the center of the gear set is the sun gear around which all other gears in the set revolve. Meshing with the sun gear are three or four planetary pinion gears whose pinions are held together by the planetary pinion carrier, or planetary carrier. The carrier holds the gears in place while allowing them to rotate around their shafts. On the outside of the planetary pinions is the ring gear that has teeth around its inside circumference that mesh with the teeth of the planetary pinions. Each planetary pinion gear is mounted to the planetary carrier by a pin or shaft.

- Low- and high-range speeds are available from the planetary gear set. Speed reduction provides the low range and direct drive provides the high range. Speed reduction is provided when the ring gear is held stationary. When the transfer case's shift lever is moved to low range, the planetary gear assembly slides forward on its shaft and engages with a locking plate that is bolted to the case and engages with the teeth of the ring gear to hold it stationary.

- A locking hub is a type of clutch that disengages the outer ends of the axle shafts from the wheel hub.
- Automatic hubs lock when the driver shifts into 4WD and normally, automatically unlock when the vehicle is driven in reverse for a few feet. Automatic hubs are engaged by the rotational force of the axle shafts whenever the transfer case is in 4WD.
- 4WD vehicles equipped with an axle disconnect do not need locking hubs. Axle disconnects can be vacuum, cable, or motor activated. A sliding collar connects two sections of one front axle.
- Locking hubs are not needed with full-time 4WD. The wheels and hubs are always engaged with the axle shafts. The interaxle differential or transfer case prevents damage to and undue wear of the parts of the powertrain.
- Driveline windup occurs when the drive axles are rotating at different speeds on dry pavements or when the vehicle is turning a corner.
- The suspension components necessary for 4WD vehicles are basically the same as for FWD and RWD vehicles.
- The solid front 4WD axle is basically a rear axle turned so that the pinion gear faces the center of the vehicle.
- An independent front suspension (IFS) axle has the differential housing mounted solidly to the vehicle's frame. Short drive shafts or half shafts connect the differential to the wheels. These half shafts are suspended by springs and are able to respond independently to road surfaces.
- The wheel bearings in front and rear axle assemblies of 4WD vehicles are typically located so that they do not support the weight of the vehicle and are held in the hub assembly in the same way as other full-floating axle bearings.

REVIEW QUESTIONS

Short Answer Essays

1. What are the advantages of 4WD?

2. What are the main differences between 4WD and AWD?

3. What is the primary purpose of a transfer case?

4. Describe the operation of front locking wheel hubs.

5. What are the differences between an integrated 4WD system and an on-demand 4WD system?

6. What are the primary differences between a full-time and part-time 4WD system?

7. Name the three main driveline components that are added to a 2WD vehicle to make it a 4WD vehicle.

8. What is the primary purpose of an interaxle differential?

9. Why are chain drives used in many transfer cases?

10. Briefly explain the operation of a simple planetary gear set.

Fill-in-the-Blanks

1. A typical car has _____ wheels, _____ of which can deliver power; therefore, they could be designated as 4×2 vehicles. A 4WD vehicle is designated as a 4×4 because it has _____ wheels and power can be applied to all _____ wheels.

2. A simple planetary gear set is made up of three gears. In the center of the gear set is the _____ gear. Meshing with this gear are three or four _____ gears. These are held together by the _____. This holds the gears in place while allowing them to rotate around their shafts. On the outside of the planetary gear set is the _____ gear. This gear has teeth around its inside circumference.

3. In a typical planetary gear set, speed reduction is provided when the _____ gear is held stationary.

4. _____ occurs when the drive axles are rotating at different speeds on dry pavements or when the vehicle is turning a corner.

5. To provide for _____ _____ suspension, some vehicles have one half shaft and one solid axle for the front-drive axle.

6. Common ways of connecting and disconnecting the front axles on a 4WD vehicle include having locking hubs, _____ motors, _____ motors, and mechanical _____.

7. Two-speed transfer cases typically have four positions: _____, which engages only the rear axle and is used for all dry-road driving; _____, which engages both axles and is used at any speed on slippery surfaces; _____, which disengages both axles; and _____, which engages both axles and lowers the ratios of the entire driveline, and should only be used at low speeds and on very demanding terrain.

8. Some vehicles use a planetary gear set to couple the front and rear axles. This gear set functions as a _____.

9. _____ _____ are not used on full-time 4WD systems because the axles and hubs are always driven by the transfer case.

10. An IFS 4WD axle has the _____ housing mounted solidly to the vehicle's frame and uses short _____ _____ to connect the differential to the wheels.

ASE-Style Review Questions

1. When discussing the purpose of a transfer case,
   Technician A says that it transfers power to an additional drive axle.
   Technician B says that most have two speeds, which affect the overall gear ratio of the vehicle.
   Who is correct?
   A. A only
   B. B only
   C. Both A and B
   D. Neither A nor B

2. When discussing 4WD systems,
   Technician A says that 4WD vehicles with a front-axle disconnect use automatic locking hubs.
   Technician B says that some transfer cases use an electric switch to engage both drive axles.
   Who is correct?
   A. A only
   B. B only
   C. Both A and B
   D. Neither A nor B

3. When discussing the gear sets of a transfer case,
   Technician A says that most use a planetary gear set to provide for the different gear range selections.
   Technician B says that a chain and sprocket assembly is used to provide 2WD operation.
   Who is correct?
   A. A only
   B. B only
   C. Both A and B
   D. Neither A nor B

4. When discussing the various gear positions of a transfer case,
   Technician A says that when the transfer case is in low, the overall gear ratio is numerically increased.
   Technician B says that when the transfer case is in the high position, the vehicle operates in an overdrive mode due to the decrease in torque multiplication.
   Who is correct?
   A. A only
   B. B only
   C. Both A and B
   D. Neither A nor B

5. When discussing the different range positions of a transfer case,
   Technician A says that 4WD High engages both axles and is used at any speed on slippery surfaces.
   Technician B says that 4WD Low engages both axles and lowers the ratios of the entire driveline.
   Who is correct?
   A. A only
   B. B only
   C. Both A and B
   D. Neither A nor B

6. When discussing wheel speed differences during cornering,
   Technician A says that when a vehicle is turning a corner, the rear wheels travel a greater distance than the front wheels.
   Technician B says that the front wheels move through a wider arc than the rear wheels when the vehicle is making a turn.
   Who is correct?
   A. A only
   B. B only
   C. Both A and B
   D. Neither A nor B

7. When discussing differences between 4WD and AWD systems,
   Technician A says that part-time 4WD systems are designed to be used only when driving off the road or on slippery surfaces.
   Technician B says that AWD systems are intended for off-road use only.
   Who is correct?
   A. A only
   B. B only
   C. Both A and B
   D. Neither A nor B

8. What results from having different axle ratios on the front and rear of a 4WD vehicle when it is operating in 4WD?

   A. Poor handling on dry surfaces
   B. Driveline windup
   C. Mechanical damage to the driveline
   D. All of the above

9. While discussing AWD systems,

   Technician A says that some systems are electronically controlled and use an electromagnetic clutch to engage and disengage the drive axles.

   Technician B says that some cars do not use a transfer case, but are equipped with a third differential unit.

   Who is correct?

   A. A only   C. Both A and B
   B. B only   D. Neither A nor B

10. When discussing front-axle disconnect systems,

    Technician A says that an indicator switch at a front-axle disconnect is used to inform the driver that 4WD is engaged.

    Technician B says that some front-axle disconnects are operated by a magnetic clutch.

    Who is correct?

    A. A only   C. Both A and B
    B. B only   D. Neither A nor B

ADVANCED FOUR-WHEEL-DRIVE SYSTEMS

UPON COMPLETION AND REVIEW OF THIS CHAPTER, YOU SHOULD BE ABLE TO:

- Use the correct terminology when discussing 4WD systems.
- Describe the different designs of 4WD systems and their applications.
- Compare and contrast the components of part- and full-time 4WD systems.

- Discuss the purpose of an interaxle differential and the design variations used by the industry.
- Discuss the purpose, operation, and application of a viscous coupling in 4WD systems.
- Explain the operation of some common AWD systems.

A BIT OF HISTORY

In 1912, the United States Army tested the viability of trucks by hosting a 1,509-mile race that took the four competing trucks off- and on-road to complete the test. The winner was a 4WD truck.

Shop Manual
Chapter 9, page 414

INTRODUCTION

4WD and AWD systems have consistently been modified and enhanced to provide improved safety, reliability, handling, and fuel economy. Most of the changes allow the driver to have more control over the system or to allow the system to have more control of the vehicle. Most systems are integrated into the vehicle's electronic network circuit. This means the action of the 4WD or AWD system can respond to the vehicle's operating conditions very quickly (Figure 9-1). Although many of the advances made with these systems are accomplished through electronics, new differential and coupling designs also have an important part.

FIGURE 9-1 This Audi crossover has AWD.

FOUR-WHEEL-DRIVE SYSTEMS

Many late-model vehicles have automatic four-wheel-drive systems. These systems have the ability to automatically switch from 2WD to 4WD and back, without driver involvement. Often these systems are referred to as full time, but this is misleading because the vehicle does not always operate in 4WD. Normally an electromagnetic clutch in the transfer case controls the torque split between the front and rear axles. The clutch is controlled by a computer or control module in the vehicle. The driver can also select a desired operational mode.

During turns, the front wheels travel a greater distance than the rear wheels. This is because the front wheels move through a wider arc than the rear wheels (Figure 9-2). With full-time 4WD, an open-center differential allows the front wheels to travel farther or rotate faster than the rear wheels without slipping.

4WD Operational Modes

The names of a 4WD's operational modes will vary by manufacturer; it would be difficult to list them all. The following explanation uses the most common names for the most commonly available operational modes. This example is based on a vehicle with a RWD bias, that is, it was designed to use the rear wheels for primary motivation and the front wheels are driven as commanded by the system or driver.

The 4WD-Auto mode provides 4WD with full power delivered to the rear axle and to the front axle as required for increased traction. This is appropriate for normal on-road operating conditions, such as dry road surfaces, wet pavement, snow, and gravel.

The 4WD-High mode provides 4WD with full power to both axles. It is intended for severe winter or off-road conditions, such as deep snow and ice (where no dry or wet pavement remains uncovered) and shallow sand.

The 4WD-Low mode supplies 4WD with full power to both axles and includes a lower gear ratio for low speed. It is intended only for off-road applications that require extra power including deep sand, steep grades, and pulling a boat and trailer out of the water.

The vehicle should not be operated in 4WD-High and 4WD-Low on dry or merely wet pavement. Doing so will produce excessive noise, increase tire wear, and may damage driveline components. These modes are intended for use only on consistently slippery or loose surfaces.

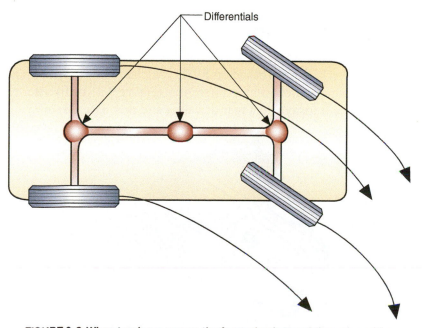

FIGURE 9-2 When turning a corner, the front wheels travel through a wider arc than the rear wheels.

When a viscous coupling directly connects two shafts, it is called a viscous transmission. Although most full-time systems cannot be selected out of 4WD, some, such as Jeep's full-time system, called Selec-Trac, can be disengaged for better fuel economy and less drivetrain wear in dry conditions.

Certain procedures must be followed when selecting the operational modes. This is important, regardless of how the system is advertised. Always follow the instructions given for the particular make and model vehicle. Failure to do so may result in system damage.

Shifting between 4WD-Auto and 4WD-High. When the control is moved to 4WD-High, the indicator light will illuminate in the instrument cluster. When the control is moved to 4WD-Auto, the indicator light will turn off. Either shift can be done at a stop or while driving at any speed.

Shifting from 4WD-Auto or 4WD-High to 4WD-Low. When shifting into Low, most vehicles must first be brought to a complete stop. Then depress the brake pedal. Place the automatic transmission gearshift into neutral, or if the vehicle has a manual transmission, depress the clutch pedal. Move the 4WD control to the 4WD-Low position. If the vehicle is equipped with an electronic shift 4WD system, and the control is moved to 4WD-Low while the vehicle is moving, the system will not engage and no damage will occur to the 4WD system. The 4WD-Low indicator lamp will come on momentarily when the ignition is turned on and when 4WD-Low is selected. It will flash if the system requires service.

Shifting from 4WD-Low to 4WD-Auto or 4WD-High. When shifting out of Low, the vehicle must be at a complete stop. Depress the brake pedal. Place the automatic transmission gearshift into neutral, or if the vehicle has a manual transmission, depress the clutch pedal. Move the 4WD control to the 4WD-Auto or 4WD-High position. The 4WD-High indicator lamp will come on momentarily when the ignition is turned on and when 4WD-High is selected. It will flash if the system requires service.

> **AUTHOR'S NOTE:** It seems that every time there is a new model car or truck that offers 4WD, there is a new 4WD system. Each of these systems has some feature that it did not have before or one the competition does not have. It is hard to predict what will happen next. I highly recommend that you look at the owner's manual and any labeling that may be around the 4WD controls before doing anything with the vehicle (including driving). If you do not know how it is supposed to work, how can you tell if it is working the way it should?

ALL-WHEEL-DRIVE SYSTEMS

Shop Manual
Chapter 9, page 419

AWD systems are basically the same as 4WD systems, except there is no low-gear option (Figure 9-3). Many AWD vehicles are based on FWD vehicles. They may not have a separate transfer case; rather, a viscous clutch, center differential (Figure 9-4), or transfer clutch is used to transfer power from the transaxle to a rear driveline and rear axle assembly.

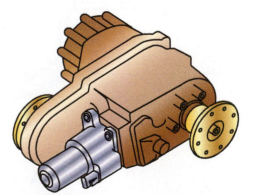

FIGURE 9-3 A transfer case for a full-time 4WD system.

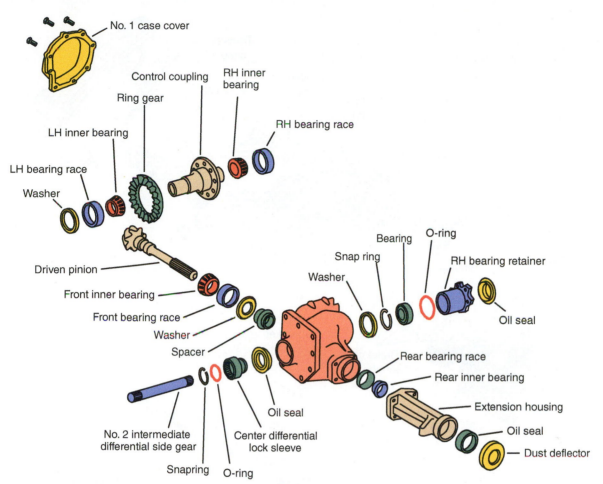

FIGURE 9-4 An exploded view of a center differential assembly for a full-time AWD vehicle.

Full-time 4WD and AWD systems require a device that allows the front and rear axles to rotate at different speeds. These units work to eliminate driveline windup. The most common setup has an interaxle differential located in or in place of the transfer case. The front and rear drivelines are connected to the interaxle differential. Just as a drive axle differential allows for different left and right drive axle shaft speeds, the interaxle differential allows for different front and rear driveline shaft speeds.

While the interaxle differential solves the problem of driveline windup during turns, it also lowers performance in poor traction conditions. This is because the interaxle differential will tend to deliver more power to the wheels with the least traction. The result is increased slippage, which is the exact opposite of what is desired.

To counteract this problem, some interaxle differentials are designed much like a limited-slip differential. They use a multiple-disc clutch pack to maintain a predetermined amount of torque transfer before the differential action begins to take effect. Other systems such as the one shown in Figure 9-5 use a cone-braking system rather than a clutch pack, but the end result is the same. Power is supplied to both axles regardless of the traction encountered.

Many systems give the driver the option of locking the interaxle differential in certain operating modes. This eliminates differential action altogether. However, this differential should only be locked while driving in slippery conditions and can only be activated at low speeds.

Full-Time AWD

These systems are similar to full-time 4WD and power the four wheels at all times. They are not designed for off-road driving; rather, they enhance a vehicle's stability and performance during normal conditions. The engine's torque is divided according to operating conditions.

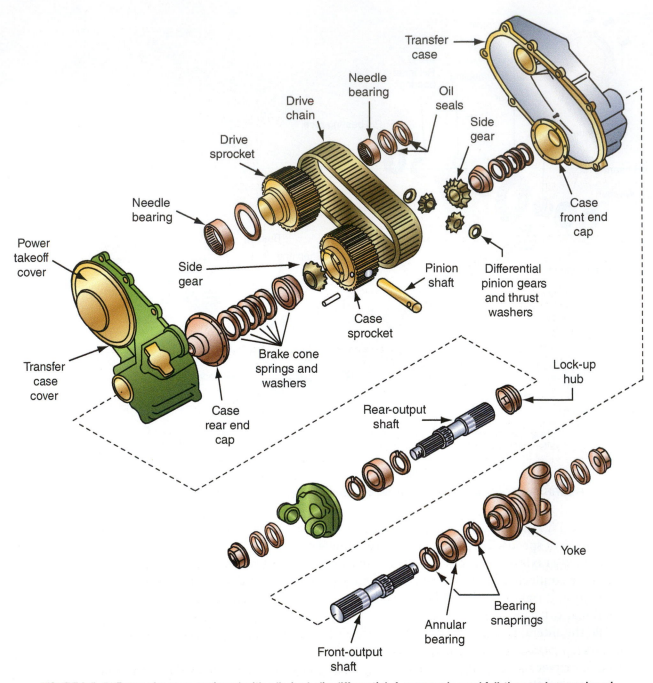

Transfer
case

Needle
bearing

Oil
seals

Drive
chain

Side
gear

Drive
sprocket

Needle
bearing

Case
front end
cap

Power
takeoff
cover

Side
gear

Pinion
shaft

Differential
pinion gears
and thrust
washers

Transfer
case
cover

Case
sprocket

Brake cone
springs and
washers

Lock-up
hub

Rear-output
shaft

Case
rear end
cap

Yoke

Front-output
shaft

Annular
bearing

Bearing
snaprings

FIGURE 9-5 4WD transfer case equipped with a limited-slip differential. A commonly used full-time system equipped with a limited-slip differential in the transfer case was called the Quadra-Trac. It was built by the Warner Gear Division of the Borg-Warner Corporation.

When the vehicle is moving straight on a level road, each wheel receives the same amount of torque. During a turn, the front wheels receive less torque; this prevents wheel slip.

Some AWD vehicles have an electromagnetic clutch in the transfer case or center differential. During normal driving, the control module keeps the clutch working at a minimum level. This allows for a slight difference between speeds of the front and rear drive shafts, enabling the vehicle to negotiate turns. When the system detects wheel slippage, the control module fully activates the clutch to send more torque to the front or rear wheels (Figure 9-6). Wheel slip is monitored by inputs from wheel speed sensors that are available on the CAN bus.

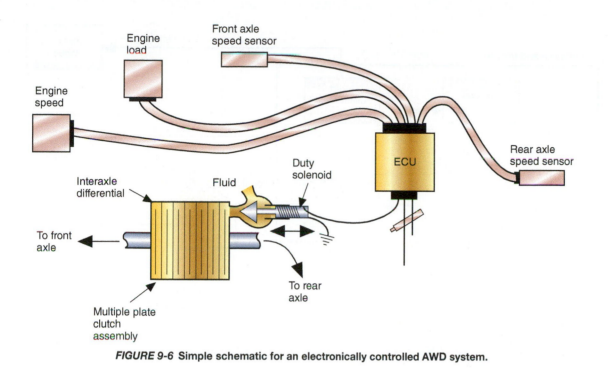

FIGURE 9-6 Simple schematic for an electronically controlled AWD system.

Some systems allow the driver to select an operating mode that locks the front and rear drive shafts. This causes the drive shafts to rotate at the same speed and torque. This mode is intended for very slippery conditions and should not be used on dry payment.

Automatic AWD

These systems operate in 2WD most of the time. Power to four wheels only occurs when the conditions dictate the need. These systems were designed to enhance vehicle stability and safety and are not intended for off-road usage.

During normal driving, one axle receives all of the output from the transmission. When that axle experiences some slippage, the AWD control unit allows up to 50 percent of the torque to move to the other axle. This power split can be accomplished hydraulically, mechanically, or electrically, depending on the system. As soon as the axle that was slipping is no longer slipping, all torque is sent to that axle. Under normal conditions one axle gets 100 percent of the torque—meaning you are driving in 2WD.

Many AWD systems are fitted with a viscous coupling in addition to or in place of a center differential (Figure 9-7). The viscous coupling allows limited slip between the front and rear drive wheels. This provides power to all four wheels but prevents driveline windup.

On-demand 4WD systems are electronically controlled. To split the power between the axles, a multiple-disc clutch is used. This clutch serves as an interaxle differential and permits a speed difference between the front and rear drive axles. Sensors monitor front- and rear-axle speeds, engine speed, and load on the engine and driveline. An electronic control unit (often called the transmission control unit [TCU]) receives information from the sensors and controls a solenoid that operates on a duty cycle to control the fluid flow that engages the transfer clutch (Figure 9-8). The duty solenoid pulses, cycling on and off very rapidly, which develops a controlled slip condition. As a result, the transfer clutch operates like an interaxle differential and allows for a power split from 95 percent FWD and 5 percent RWD to 50 percent FWD and 50 percent RWD. This power split takes place so rapidly that the driver is unaware of the traction problem.

Shop Manual
Chapter 9, page 431

Shop Manual
Chapter 9, page 430

Duty cycle is also called jitter cycle.

The center or interaxle differential is incorporated into the transfer case and splits power, in some proportion, to the front and rear drive axles.

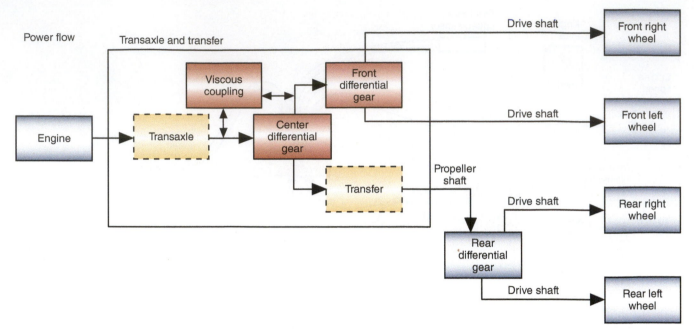

FIGURE 9-7 A chart showing the power-flow through a viscous clutch-type center differential.

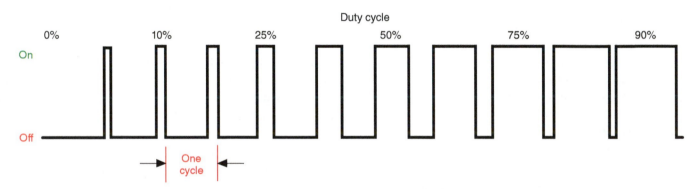

FIGURE 9-8 Duty cycle is the percentage of on-time per cycle. Duty cycle can be changed; however, total cycle time remains constant.

VISCOUS COUPLINGS

Viscous couplings are often referred to as viscous clutches because they engage and disengage power-flow.

The viscous clutch is used in the driveline of vehicles to drive the axle with low tractive effort, taking the place of the interaxle differential. In existence for several years, the viscous clutch is installed to improve the mobility factor under difficult driving conditions. It is similar in action to the viscous clutch described for the cooling system fan. The viscous clutch in AWD is a self-contained unit. When it malfunctions, it is simply replaced as an assembly. The viscous clutch assembly is very compact, permitting installation within a front transaxle housing, center interaxle assembly, or rear axle assembly. Viscous clutches operate automatically without external controls while constantly transmitting power to the axle assembly as soon as it becomes necessary to improve driving wheel traction. This action is also known as biasing driving torque to the axle with tractive effort. The viscous clutch assembly is designed similarly to a multiple-disc clutch with alternating driven and driving plates.

Shop Manual
Chapter 9, page 432

AUTHOR'S NOTE: Keep in mind that an open differential sends power to the axle or wheel that has the *least* traction. Viscous coupling differentials send power to the axle or wheel that has the *most* traction.

The handling characteristics of 4WD vehicles are excellent in all driving conditions except when turning on a dry pavement. When turning a corner, the front wheels cannot rotate at the same speed as the rear wheels. This causes one set of wheels to scuff along the pavement. To overcome this tendency, many full-time 4WD and AWD vehicles use a viscous coupling between the front and rear axles.

A viscous coupling (Figure 9-9) is basically a drum with some thick fluid in it that houses several multiple closely fitted, thin steel discs. One set of plates is connected to the front wheels and the other to the rear. The advantage of the viscous coupling is that it splits the engine torque according to the needs of each axle.

Viscous couplings also may be used in the center and the front or rear axle differentials. They provide a holding force between two shafts as they rotate under force. Like a limited-slip differential, a viscous coupling transfers torque to the opposite drive wheel when the other wheel has less traction. A viscous clutch often takes the place of the interaxle differential. Viscous clutches operate automatically as soon as it becomes necessary to improve wheel traction.

> When a viscous coupling directly connects two shafts, it is called a viscous transmission.

High-performance full-time AWD vehicles use a viscous coupling in the center and rear differentials to improve cornering and maneuvering at higher speeds. The center differential sends power to all wheels and allows front/rear tire speed to differ on corners. If excessive wheel spin occurs on one axle, the viscous coupling locks the differential and torque is transferred to wheels that have traction (Figure 9-10). The combination of a center differential viscous coupling with open front and rear differentials improves the vehicle's distribution of braking forces and is compatible with antilock braking systems.

Examples of AWD vehicles that used a viscous coupling, rather than a center differential, are several Volkswagens. These vehicles had automatic AWD systems with a viscous coupling and a freewheel mechanism to disconnect the driven axle during braking. The freewheel mechanism is inside the rear differential. It allows the rear wheels to rotate faster than the front wheels without locking the viscous coupling and preventing ABS from applying the brakes to each wheel independently.

During normal conditions, the front axle receives 95 percent of the torque. The remaining 5 percent is sent to the rear axle. This is done to reduce the activation time of the viscous coupling. Viscous couplings require some time for the fluid to heat. When the discs inside the coupling are not turning, the fluid stays cold. The slight torque transfer to the rear axle allows the discs to rotate and the fluid becomes warm and ready to engage the clutch when needed. Once the front axle begins to slip, up to 50 percent of the torque is automatically transferred to the rear axle.

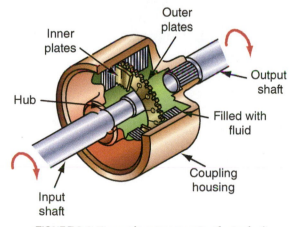

FIGURE 9-9 The main components of a typical viscous coupling.

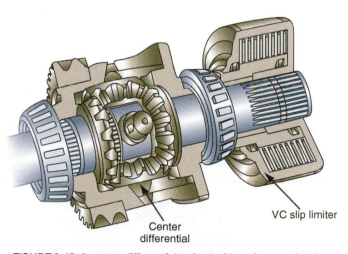

FIGURE 9-10 A center differential paired with a viscous clutch.

Operation of a Viscous Coupling

A viscous coupling consists of an input shaft connected to a set of splined drive plates, an additional set of driven plates splined internally to the drum, and an output shaft splined to the drum. The drive plates and driven plates are alternately stacked and evenly spaced (Figure 9-11). The drum contains an exact amount of thick silicone oil. **Silicone oil** is used because it has high shearing resistance and volumetric change when its temperature increases. There is very little silicone fluid in a viscous coupling case. About one-quarter of the case contains fluid at rest. The fluid expands rapidly to fill the case, and most couplings are designed with air bleeds to permit this to happen. There are various designs of couplings, all of which have different viscosity characteristics.

Based on practical experience, vehicles operating with this clutch transmit power automatically, smoothly, and with the added benefit of the fluid being capable of dampening driveline shocks. When a difference in speed of 8 percent exists between the input shaft driven by the driving axle with tractive effort and the other axle, the clutch plates begin shearing (cutting) the special silicone fluid. The shearing action causes heat to build within the housing very rapidly, which results in the silicone fluid stiffening. The stiffening action causes a locking action between the clutch plates to take place within approximately one-tenth of a second. The locking action results from the stiff silicone fluid becoming very hard for the plates to shear. The stiff silicone fluid transfers power-flow from the driving to the driven plates. The driving shaft is then connected to the driven shaft through the clutch plates and stiff silicone fluid.

The viscous clutch has a self-regulating control. When the clutch assembly locks up, there is very little, if any, relative movement between the clutch plates. Because of this, silicone fluid temperature drops, which reduces pressure within the clutch housing. But as speed fluctuates between the driving and driven members, heat increases, causing the silicone fluid to stiffen. Speed differences between the driving and driven members regulate the amount of slip in a viscous clutch driveline. When used in an automatic AWD system, the viscous clutch takes the place of the interaxle differential, biasing driving torque to the normally undriven axle during difficult driving conditions.

A viscous coupling is a speed-sensitive limited slip. When it rotates very slowly, very little torque is transferred to the fixed side. As the speed difference increases, so does the torque

Shearing means to cut through.

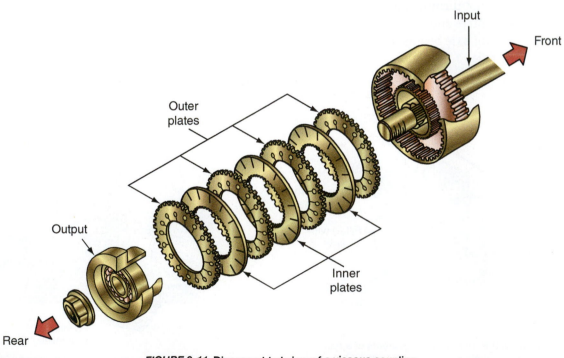

FIGURE 9-11 Disassembled view of a viscous coupling.

transferred. A viscous clutch is most often used as, or part of, the center differential in an AWD system. With a center viscous clutch, nearly all of the power is normally transferred to one axle until more transfer is needed. Normally 90 percent of the available torque is sent to one axle and the remaining 10 percent to the other. If the system is based on a FWD vehicle and the front wheels begin to spin, the viscous clutch will transfer torque to the rear axle. If the viscous clutch is connected to a center differential, power is normally equally split between the front and rear axles.

Haldex Clutch

Some AWD vehicles are equipped with a Haldex clutch, which serves as a center differential. This clutch unit distributes the drive force variably between two axles. In a typical application, the Haldex unit mounts in front of the rear differential and receives torque from the front axle (Figure 9-12).

The Haldex unit has three main parts: the hydraulic pump driven by the slip between the axles or wheels, a wet multidisc clutch, and an electronically controlled valve (Figure 9-13). The unit is much like a hydraulic pump in which the housing and a piston are connected to one shaft and a piston actuator connected to the other.

When a front wheel slips, the input shaft to the Haldex unit spins faster than its output shaft. This causes the pump to immediately generate oil flow. The oil flow and pressure engages the multidisc clutch to send power to the rear wheels. This happens extremely quickly because an electric pump and accumulator keep the circuit primed.

The oil from the pump flows to the clutch's piston to compress the clutch pack. The oil returns to the reservoir through a controllable valve, which adjusts the oil pressure and the force on the clutch pack. An electronic control module controls the valve and also determines when to decouple the axles to prevent the rear brakes from braking the front axle.

In high slip conditions, a high pressure is delivered to the clutch pack; in tight curves or at high speeds, a much lower pressure is provided. When there is no difference in speed between the front and rear axles, the pump does not supply pressure to the clutch pack.

Volkswagen 4MOTION Haldex System. The Volkswagen 4MOTION AWD system uses a Haldex coupling as a center differential. The coupling is mounted in front of the rear axle differential and is part of the rear differential case. The input shaft of the Haldex center differential

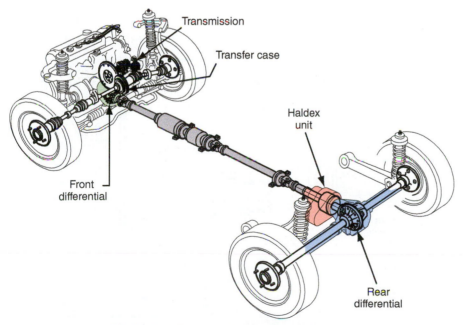

FIGURE 9-12 The typical location of a Haldex unit.

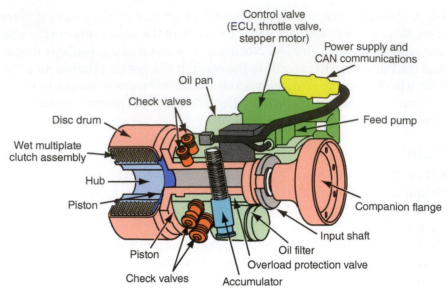

FIGURE 9-13 A Haldex clutch assembly.

is driven by a drive shaft from the transaxle. In a Haldex differential, the input shaft is totally separated from the output shaft, which is connected to the rear final drive gears. Therefore, power is only sent to the rear wheels when the Haldex clutch is engaged.

The Haldex unit is operated by an engine control unit (ECU) that receives inputs from a variety of sensors. This means the system can respond to other driving conditions and not just wheel slip. When there is no slip, understeer, or oversteer, the vehicle operates as a 2WD vehicle. It only distributes power to the rear axle when it is needed.

The pump of a Haldex differential is an axial piston pump. The system also has an accumulator that stores pressurized oil. Between the pump and the accumulator is a filter that prevents particles from reaching the control valve. This control valve is controlled by the system's control module and directs pressurized oil to the clutch's piston to compress the multiple-disc pack (Figure 9-14). The valve controls the pressure to the piston, which allows the unit to transfer from zero to 100 percent of the torque, near instantly.

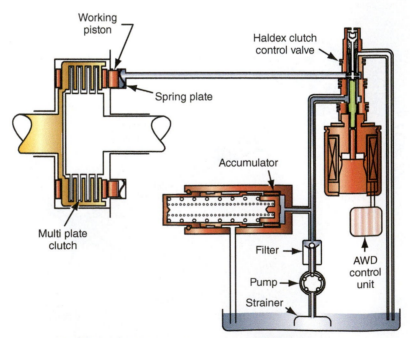

FIGURE 9-14 The hydraulic circuit for a VW 4Motion system.

The control module totally controls the action of the Haldex unit. It receives inputs from a variety of sensors through the CAN bus. All of these inputs are used to determine the exact current operating conditions. The primary inputs include engine torque, engine speed, and wheel speeds. The system also has a limp mode. This mode can be immediately selected by the control unit if an important input signal is lost. This mode does not totally disable the AWD system; rather, it attempts to provide the best operation for the conditions. The only time the system totally shuts down is when the control unit believes a fault can cause an unsafe condition.

FIFTH-GENERATION HALDEX UNITS

The Haldex differential has gone through many changes over the years. All designs are multiple-clutch based, and the third-, fourth-, and fifth-generation designs are all proactive, meaning that the controller can pretension the clutch and activate all-wheel drive before any wheel slip happens. Haldex Generation V units, now made by Borg Warner, use a new electro-hydraulic actuator and an integrated ECU (Figure 9-15). The accumulator, solenoid valve, and filter used in earlier systems are eliminated. The system is used by Volkswagen and Audi beginning in 2012, and the Volvo V70 AWD began using the system in 2013. Basic operating principles are similar to earlier Haldex units, but response time is reduced, and full locking torque is available at any time and speed. Based on the inputs, all recent Haldex generations can control wheel slip, improve overall vehicle handling, counteract oversteer and understeer, and improve stability.

Saab XWD. Saab's active AWD system is called a cross-wheel drive (XWD) system. It relies on computer-controlled center and rear differentials to control the stability of the vehicle. It is a fully automatic, on-demand system that can send up to 100 percent of the engine's torque to the front or rear wheels. It is called a cross system because it also can apply power to the opposite corner of the car from the wheel that has lost traction. It also can vary the amount of torque applied to each of the rear wheels. Adjustments to power distribution can be made in 80 milliseconds.

A three-piece drive shaft connects the power takeoff at the transaxle to the rear drive module (RDM). The RDM is comprised of a torque transfer device (TTD) and an electronically controlled limited-slip differential (eLSD). Both of these units are wet, multiple-plate Haldex clutches. The TTD regulates the amount of torque delivered to the rear drive axle. The eLSD controls the amount of torque applied to each of the rear wheels. The system's operation is controlled by a separate ECU that receives inputs from and shared information with the engine, transmission, and antilock brake system (ABS)/electronic stability program

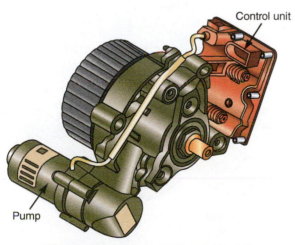

FIGURE 9-15 A fifth-generation Haldex unit.

ECUs. Based on the inputs, the system can control wheel slip, improve overall vehicle handling, counteract oversteer and understeer, and improve stability.

Torsen Units

Many manufacturers use a Torsen (a contraction of TORque SENsing) unit in the center differential, the most well known being Audi and its Quattro vehicles. Other manufacturers include Chevrolet, Lexus, Range Rover, Toyota, and Volkswagen. Audi also uses an additional Torsen unit in the rear axle of a few models. In 2012 the Ford SVT Raptor started using a Torsen front differential, and the Boss 302 a rear Torsen differential.

A Torsen is a geared, torque-sensitive mechanical limited-slip differential that uses worm gears to "sense" torque on one shaft (Figure 9-16). When a Torsen differential is used, the slower-rotating axle always receives more torque than the faster-rotating axle. When the speeds are equal, torque is normally split to provide a 50/50 distribution. When there is a variance in speed, a maximum of 75–80 percent of the engine's torque can be transferred to the front or rear axles, or right or left wheels.

A Torsen unit responds automatically to the slightest speed changes and immediately sends more torque to the shaft that is spinning slower. To do this, it relies on gear friction, not the electronic inputs necessary for the operation of electronically operated differentials. Friction in the differential will oppose motion, and that tends to slow the faster side and increase the speed on the slower side.

The most recent Torsen Type 3 differential uses a compact planetary helical gear setup capable of torque spits other than 50/50 (Figure 9-17). The compact size of this unit means it can be designed to fit most transmissions, transaxles, and differential housings with few modifications. The Torsen T-3 is used as a center differential by some Audi and Toyota SUV models.

Quaife Differentials

Quaife Automatic Torque Biasing (ATB) Differential are also used as the center differential in AWD vehicles. The Quaife differential is a gear-operated torque biasing differential with no clutch packs or preloading to transfer torque from one axle to the other, and torque transfer occurs automatically when one wheel loses traction. When quickly accelerating an AWD vehicle, the initial weight transfer to the rear causes the front wheels to spin. With a viscous clutch at the center differential, the fluid needs to heat up before it transfers more torque. With a Quaife unit, the transfer is almost immediate. Quaife differentials are available on the Dodge

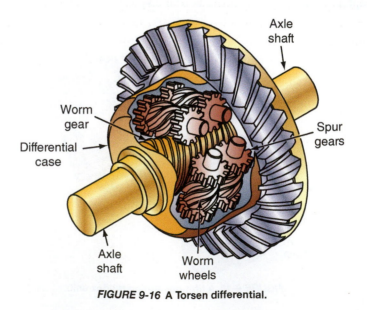

FIGURE 9-16 A Torsen differential.

FIGURE 9-17 A Torsen Type 3 planetary differential.

Viper as an option, and are original equipment on the 300 HP Ford Focus RS. Aftermarket applications include a wide number of FWD, 4WD, and RWD applications.

NEWER 4WD DESIGN VARIATIONS

There are many variations of advanced 4WD and AWD systems found on today's vehicles. We now take a look at some of them.

> **AUTHOR'S NOTE:** The electronically controlled limited-slip differential (eLSD) has become commonplace in the 4WD and AWD marketplace. eLSDs can be found in front and rear differentials and in the center differential. A wide variety of inputs are used by each system to immediately adjust clutch clamping force between front and rear or left and right. For example, the 2014 Corvette eLSD adjusts itself 83 times per second! While most people associate 4WD/AWD vehicles as being good in the traction department, modern eLSD systems contribute to driving performance and safety in many ways besides reducing wheel slip. eLSDs are integrated seamlessly into traction control and vehicle stability systems. Many performance-oriented vehicles allow various levels of driver control over these systems. When it comes to cornering, braking, and overall vehicle performance, eLSDs are among the most important components of the vehicle.

Acura and Honda's VTM-4

Late-model Honda and Acura SUVs may be equipped with a Variable Torque Management-4 Wheel Drive (VTM-4) system. This system operates the vehicle in front-wheel drive most of the time to ensure maximum fuel economy. However, the system's computer is also programmed to know when the vehicle could benefit from rear-wheel power by reading signals from the engine control system. Basically, the level of torque delivery, front to rear, is determined by the rate of acceleration and wheel slip.

Unlike most electronically controlled systems, this system does not wait until there is wheel slip to send torque to the rear wheels. It anticipates the need for all-wheel drive and engages the rear wheels whenever the vehicle is accelerating. During acceleration, the computer activates a pair of clutch packs to connect the ends of the right- and left-rear axle shafts.

The VTM-4 does not use a rear differential; rather, there is a simple ring and pinion setup with the clutches providing differential action. Each drive clutch includes an electromagnetic coil, a ball-cam device, and a set of 19 wet clutch plates, similar to those used in an automatic transmission. Nine of the clutch plates are connected to the axle shafts and 10 are connected to the ring gear. Each drive axle clutch plate is positioned between two ring gear clutch plates. When the control unit determines that torque should be transferred to the rear wheels, current is sent to the two electromagnetic coils. The magnetic field around the coils pulls on steel plates located next to each coil. Friction between those steel plates and the adjoining cam plates cause the cam plates to begin turning. As they do, the balls in the cam plates roll up curved ramps. This action creates an axial thrust against clutch-engagement plates. This thrust, in turn, compresses the clutch plate assemblies.

Depending on how much current goes to the coils, the system can vary the amount of torque distributed to each rear wheel from zero to a maximum of 55 percent. Torque sent to the rear wheels is infinitely variable and is directly proportional to the current sent to the coils by the computer.

Like other systems, the VTM-4 also kicks in when it senses wheel slip, and drivers can lock the system at the maximum rear torque setting by depressing a button on the dash. The lock mode is intended to be used in extremely slippery or stuck conditions. The lock mode only works in first, second, and reverse gears. The coils stay fully energized up to 6 mph and then gradually cut back until, at 18 mph, the system automatically shuts off.

Super Handling All-Wheel Drive System. Late-model Hondas and Acuras have a modified version of the VTM-4. This system is called the Super Handling All-Wheel Drive (SH-AWD) system. These vehicles have an electronically controlled rear drive axle. The drivetrain in these systems does include a center or rear differential. Rather, a unit is installed in front of the rear axle. This unit contains the final drive gears, planetary gears, and electromagnetic clutches (Figure 9-18). The system controls wheel spin and can transfer torque to the rear wheels during acceleration or to reduce torque steer. This system has a driver-controlled switch that engages both clutch packs to lock the system if the vehicle gets stuck.

The planetary gear set controls the torque split between the front and rear final drive units. The two electromagnetic multidisc clutch packs (one at each rear axle shaft) are controlled by an ECU that monitors the powertrain and the ABS. The action of the clutch packs is

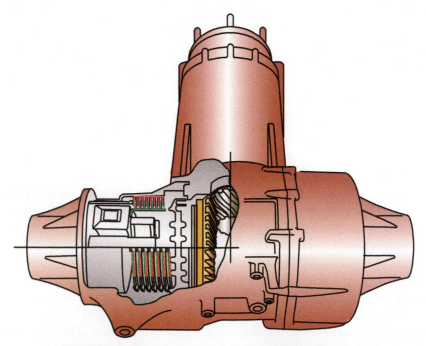

FIGURE 9-18 The rear axle unit for a Honda SH-AWD system.

controlled by varying the current sent to them. Computer-controlled electromagnetic clutches control torque distribution to each wheel of the rear drive axle. The computer relies on inputs from many sensors, including those that indicate yaw rates, steering input, brake inputs, engine speed, throttle position, wheel speeds, and front and rear g-forces. Based on these, the clutches can send power to either or both rear wheels in infinite proportions.

Audi Quattro

Audi's Quattro permanent all-wheel drive system transfers from the front to the rear and side to side as needed. On cars with longitudinally mounted engines, a Torsen differential distributes power and compensates for differences in wheel rotation when cornering. A Torsen differential relies on helical gears to send torque to the different driving wheels; if the car has a transverse engine, a Haldex clutch is used.

The center differential compensates for the speed differences between the front and rear axles and distributes engine power between the front and rear wheels. The system automatically regulates the distribution of power within milliseconds. This action is based on engine speed and torque, wheel spreads, and longitudinal and lateral acceleration rates.

The locking functions of the center differential and the electronic differential lock make sure the car can still move with only one wheel able to transfer engine power to the road.

If one of the wheels on an axle loses grip and starts spinning, power is transferred to the other wheel by the axle's differential. The electronic differential lock transfers the excess power of the spinning wheel to the other wheels with better traction.

If both wheels on an axle start spinning, the locking of the center differential sends most of the torque to the other axle.

BMW xDrive

BMW has an intelligent, permanent AWD system called the xDrive. The xDrive unit is mounted in the transfer case, and does not use a center differential. Ford also uses a similar system. These systems rely on electronics to vary the distribution of engine power between the front and rear axles and to each wheel on the rear axle. The drive system is tied to the stability control system, the active steering system, and the ABS. It helps steer the vehicle by sending torque to either of the rear wheels, which not only prevents wheel slip but also eliminates oversteer and understeer. The system is designed to anticipate slippage and handling concerns by monitoring many inputs on the CAN bus. The dynamic stability control system also brakes wheels individually to help with cornering and to regain traction.

Normally the system provides a 40/60 split of power between the front and rear drive axles but is capable of sending 100 percent of the engine's power to either the front or rear axles. At high speeds or while parking, all power goes to the rear wheels. The system is capable of responding to needs within 100 milliseconds, which is much quicker than conventional AWD systems; this is due to the advanced electronics involved.

In addition to the electronics, the system has a rear differential that contains two planetary gear sets and two electrically activated clutch packs that are capable of multiplying torque to each rear wheel (Figure 9-19). In addition, there is a power divider that splits the power from the front and rear axles; this unit also uses an electrically controlled clutch pack. The system works to counteract any oversteer or understeer during turns by moving power to the front or rear. When the possibility of understeer is detected, the system reduces power to the front wheels. When oversteering is detected, more power is sent to the front axle.

Chrysler AWD

DaimlerChrysler offers an AWD option on their Town & Country and Caravan minivans and other vehicles. The vans are FWD platforms with a rear driveline added. The rear driveline assembly consists of the rear carrier, a torque tube, an overrunning clutch assembly, a vacuum-operated dog clutch, and a viscous coupling. During normal operating conditions, the van

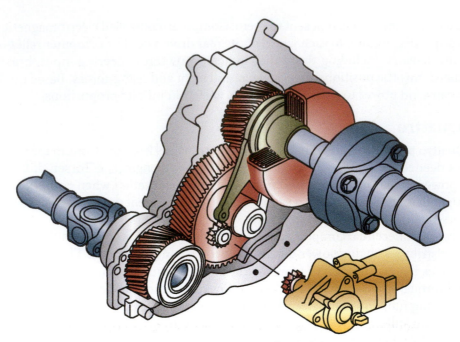

FIGURE 9-19 The transfer case for a BMW xDrive system.

operates as a FWD vehicle with 90 percent of the torque allocated to the front drive axles. As wheel slippage occurs, the viscous clutch will transmit more torque to the rear wheels. The amount of torque sent to the rear axle is determined solely by the amount of front wheel slippage. RWD-based systems work the same way.

The rear carrier houses an open differential, and half shafts connect the differential to the rear wheels. A torque tube is used to connect the transaxle to the rear carrier. Inside the torque tube is the torque shaft that does the actual transmitting of power to the rear. The torque shaft rotates on bearings located in the torque tube. The vacuum reservoir and solenoid assembly are attached to the top of the torque tube.

At the front of the torque tube is the viscous coupling, which controls the amount of torque transmitted to the rear wheels. Like most viscous couplings, this unit increases the amount of torque to the rear wheels in response to increases of front-wheel slip.

Ford's Control Trac II

This on-demand AWD system was released with Ford's small SUV, the Escape, and does not offer fully locked 4WD; rather, it sends torque to the rear wheels only when there is slippage of the front wheels. The system does not use an interaxle differential and both the front and rear differentials are open.

The system has two primary modes of operation: 4×4-Auto and 4×4-On. In the 4×4-Auto mode, all of the engine's torque is sent to the front wheels until some slip is detected. When this happens, a rotary blade coupling (Figure 9-20), which is similar to a viscous coupling, generates enough pressure to activate a multiple-disc clutch that sends up to 100 percent of the torque to the rear wheels. The rotary blade coupling relies on a three-bladed fan enclosed in a chamber filled with silicone fluid. When the front wheels slip, the fan spins through the fluid, causing the fluid's temperature to rise. This rise in temperature causes the fluid to expand and, in turn, causes the pressure of the fluid to increase. The pressure of the fluid is used to activate the clutch assembly. This is different than a viscous coupling. With a viscous coupling, the fluid transmits the torque. In Ford's setup, the fluid activates the clutch and the clutch transmits the torque. The amount of torque transmitted to the rear is proportional to the speed of the fan and is infinitely variable.

When the driver selects the 4×4-On mode, an electromagnetic clutch is activated. As this clutch is energized, it locks a ball ramp to the input shaft and exerts pressure on the

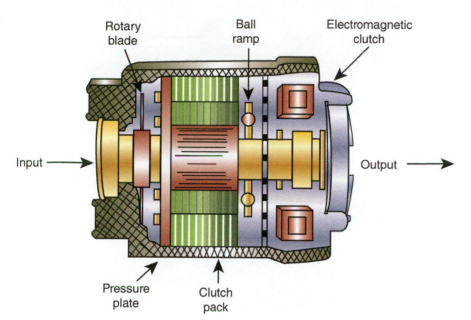

FIGURE 9-20 A rotary blade coupling as used in the **Control Trac II** system.

clutch pack. Now when the front wheels slip, the electromagnetic clutch and multiple-clutch assembly spin via the ball ramp with the input shaft and torque is transmitted to the rear wheels. This is the only condition or operational mode in which the front and rear axles are somewhat locked together.

Chrysler ITM with Front Axle Disconnect

The Chrysler Interactive Torque Management (ITM) system combines an active electronically controlled transfer case with a front-axle disconnect unit on its Chrysler 300 model AWD cars. When all-wheel drive is not needed, the system automatically disconnects the front axle to improve fuel economy. Up to 38 percent of power is sent to the front when needed. While transfer case operation is similar to many other models, the front-axle disconnect unit is unique. The unit uses a centrifugal electro-hydraulic actuator assembly to make the car essentially in RWD most of the time (Figure 9-21).

GM Versatrak

Similar to the Ford system, GM's Versatrak uses a gerotor pump in the rear-axle coupling to build oil pressure against a piston and the clutch pack to transfer torque from one side of the rear axle to the other (Figure 9-22). When the front wheels have good traction, the drive shaft

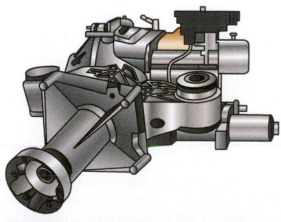

FIGURE 9-21 A Chrysler ITM front-axle disconnect unit.

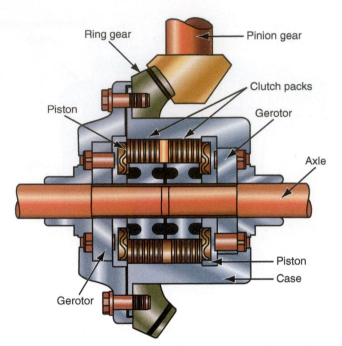

FIGURE 9-22 GM's Versatrak limited-slip rear differential.

to the rear axle and the rear ring gear spin freely. The clutches in the coupling are disengaged and no torque is transferred to the rear wheels. When the front wheels lose traction, the speed of the drive shaft and ring gear increases, and they rotate at a greater speed than the rear drive axles. This causes the coupling's oil pump to turn faster and produce a pressure to engage the clutch packs. As the clutch packs engage, they become locked to the differential housing and transfer torque to the rear drive axles.

Hyundai On-Demand

When Hyundai introduced its first SUV, the vehicle had an option for on-demand AWD. This SUV is a FWD vehicle by design. To provide AWD, the vehicles were equipped with a compact unit that contained the front differential and a power takeoff for the rear axle. This unit also contained a viscous coupling and two planetary gear sets that shared a common stationary ring gear. The sun gears of the planetary sets were connected to the front wheel axle shafts. Engine torque spun one planet carrier whereas the other drove the rear power takeoff and the viscous coupling. Slippage at either or both front wheels accelerated the power takeoff relative to the rear wheels, causing the viscous coupling to transfer torque to the rear drive axles. The planetary gears allowed a maximum of 40 percent of the engine's torque to be transferred to the rear axles.

Mitsubishi Super All Wheel Control

Mitsubishi's Super All Wheel Control (S-AWC) is a full-time 4WD system used in the Lancer Evolution. The system is integrated with the vehicle's active center differential (ACD), active yaw control (AYC), active stability control (ASC), and sports ABS components. This provides for regulation of torque and braking force at each wheel.

The ACD is an electronically controlled hydraulic multiplate clutch that limits the action of the center differential gears. The ACD regulates the torque split between the front and rear drive axles. The AYC acts like a limited-slip differential by reducing rear-wheel slip to improve traction. It controls rear-wheel torque to limit the yaw of the vehicle to improve the vehicle's cornering performance. The ASC regulates engine power and the braking force at each wheel. This system improves vehicle stability and improves traction during acceleration. The system relies on two ECUs: one controls the ACD and AYC and the other controls the ASC and ABS. The ECUs communicate to each other via the CAN bus.

Porsche AWD

Porsche offers many models with AWD; the systems vary with the different models. The manufacturer offers AWD on their rear-engined, rear-wheel-drive vehicles (these are the 4-series) and on their front-engined SUVs. The normally RWD models are fitted with a transfer case at the rear transaxle. A drive shaft extends to a front drive axle unit (Figure 9-23). These models are performance oriented and have excellent handling.

The system used in their Cayenne SUV is tied to their Porsche Traction Management (PTM) that allocates 62 percent of the engine torque to the rear wheels and 38 percent to the front. The AWD system works along with the vehicle ABS and stability control systems. When operating conditions change, the Cayenne can instantly reallocate more engine torque to the front axle or the rear, within 100 milliseconds. The system is based on a multiple-disc clutch operated by an electric motor that is electronically controlled and located in the transfer case. The motor's operation is based on many inputs, including the vehicle's speed, lateral acceleration, engine speed and torque, steering angle, and accelerator pedal position. The system also includes driver-controlled and automatic differential locks.

Subaru's Driver Control Center Differential

Subaru may have the strongest AWD reputation in the automotive world. The Driver Control Center Differential (DCCD) system brings a whole new world of vehicle/driver interface into the automotive market. The center differential is a complex unit that incorporates a torque distribution planetary gear set that maintains a 35/65 front/rear torque split. The center unit also includes an electronically controlled LSD mechanism that automatically limits slip in the wheels.

Driving conditions are determined by individual wheel speed, throttle and brake inputs, and rotational g-forces. The WRX model has a selector switch and control dial that allow the driver to select the desired DCCD involvement (Figure 9-24).

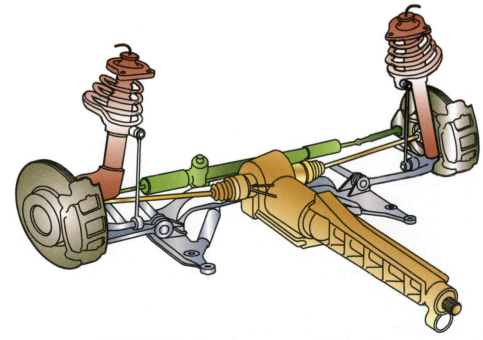

FIGURE 9-23 The front axle assembly for a AWD Porsche.

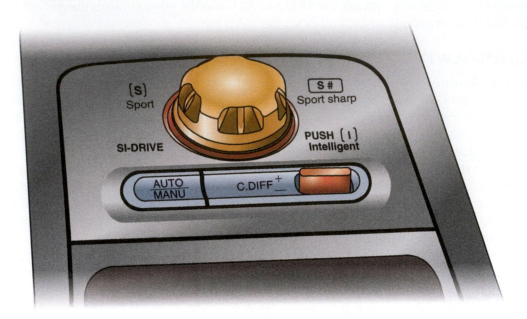

FIGURE 9-24 Subaru's DCCD system.

Subaru's Variable Torque Distribution

Most AWD Subarus send nearly all of the engine's power (90 percent) to the front wheels. This design prevents driveline windup and gives the vehicles stability on dry surfaces. When the front wheels begin to slip, more power is sent to the rear. This more-torque-to-the-front bias changed when Subaru introduced a stability control system on some models. Subaru also incorporated the AWD system into its stability control system. The result is Subaru's Variable Torque Distribution (VTD) AWD system.

The VTD system splits torque by sending 45 percent to the front axle. Traction and stability controls are part of the torque distribution system. The system uses a planetary gear-type center differential to vary the torque between the front and rear axles according to weight transfer during acceleration and deceleration as indicated by throttle position. The traction and stability controls brake the wheel or axle that is spinning to keep the vehicle heading in the correct direction.

SUMMARY

- Normally AWD systems use a center differential, viscous coupling, or transfer clutch assembly to transmit engine power to the front and rear axles.
- Integrated full-time 4WD systems use computer controls to enhance full-time operation, adjusting the torque split depending on which wheels have traction.
- Most on-demand 4WD systems power a second axle only after the first begins to slip.
- AWD vehicles are not designed for off-road operation; rather, they are designed to increase vehicle performance in poor traction situations, such as icy or snowy roads. AWD allows for maximum control by transferring a large portion of the engine's power to the axle with the most traction. Most AWD designs use a center differential to split the power between the front and rear axles.
- The interaxle differential allows for different front and rear driveline shaft speeds to prevent driveline windup but may also result in a loss of traction in very slippery conditions.

SUMMARY

- Clutches, viscous coupings, Torsen, and Quaife differentials are used in place of the differential in many AWD vehicles.
- A viscous coupling is basically a drum with some thick fluid that houses several closely fitted, thin steel discs. When used between the front and rear axle, one set of the discs is connected to the front wheels and the other to the rear.
- An open differential sends power to the axle or wheel that has the *least* traction. Viscous coupling differentials send power to the axle or wheel that has the *most* traction.
- In a typical viscous coupling, one of the two shafts that have external splines meshes with the viscous coupling housing's internal splines that also mesh with the viscous coupling plates. The other shaft rotates on seals in the housing. The plates rotate in a reservoir of silicone oil.
- The viscous coupling plates rotate at different speeds and can shear the silicone fluid with ease.

TERM TO KNOW

Silicone oil

REVIEW QUESTIONS

Short Answer Essays

1. How does a viscous clutch assembly work? Why is it used in AWD systems?

2. What are the main differences between 4WD and AWD?

3. In basic construction, what is the major difference between full-sized 4WD trucks and SUVs and smaller 4WD trucks and SUVs?

4. What types of limited-slip units are commonly used in center differentials?

5. What is a major advantage of an AWD system when compared to a conventional 4WD system?

6. What is the primary purpose of an interaxle differential?

7. Briefly explain how a Haldex clutch works.

8. What is the major difference between full-time AWD and automatic AWD?

9. Give two examples of AWD systems that rely on a Haldex clutch.

10. Briefly explain what a Torsen differential is and why it is commonly used as a center differential.

Fill-in-the-Blanks

1. Normally, AWD systems use a _____, _____ _____, or _____ assembly to transmit engine power to the front and rear axles.

2. An electronically controlled clutch-type limited-slip differential is called _____. A viscous coupling is a _____ _____ limited slip.

3. The Quaife differential is a _____ torque biasing differential.

4. The interaxle differential prevents _____ _____ but may also result is a loss of traction in very slippery conditions.

5. A _____ differential allows for different front and rear axle speeds.

6. Silicone oil is used in a viscous coupling because its _____ is not affected very much by temperature and it has high _____ resistance and _____ changes at high temperatures.

7. When the plates of a viscous coupling rotate at different speeds, the plates _____ the fluid.

8. When a viscous coupling provides holding ability for two parts of a planetary or bevel gear set, it is called a _____ _____.

9. An open differential sends power to the axle or wheel that has the _____ traction. Viscous coupling differentials send power to the axle or wheel that has the _____ traction.

10. A Torsen unit responds automatically to the slightest speed changes and immediately sends _____ torque to the shaft that is spinning faster.

ASE-Style Review Questions

1. While discussing automatic AWD systems,
   Technician A says that the systems operate in 2WD nearly all of the time.
   Technician B says that full-time AWD systems are designed for on- and off-road driving.
   Who is correct?
   A. A only
   B. B only
   C. Both A and B
   D. Neither A nor B

2. When discussing Hyundai's on-demand 4WD system,
   Technician A says that on-demand 4WD systems require the driver to engage 4WD by moving a lever.
   Technician B says that a duty cycle solenoid controls fluid flow to engage the on-demand 4WD system.
   Who is correct?
   A. A only
   B. B only
   C. Both A and B
   D. Neither A nor B

3. When discussing on-demand 4WD systems,
   Technician A says that these systems are automatically controlled.
   Technician B says that these systems drive the front and rear axles equally until some slip is detected.
   Who is correct?
   A. A only
   B. B only
   C. Both A and B
   D. Neither A nor B

4. When discussing 4WD operation,
   Technician A says that during turns the front wheels travel a greater distance than the rear wheels.
   Technician B says that when a vehicle is operating in 4WD and the front and rear axles are turning at the same speed, driveline windup can result.
   Who is correct?
   A. A only
   B. B only
   C. Both A and B
   D. Neither A nor B

5. When discussing an Acura MDX with VTM-4,
   Technician A says that the system operates the vehicle in RWD most of the time to ensure maximum fuel economy.
   Technician B says that unlike most electronically controlled systems, this system does not wait until there is wheel slip to send torque to the other axle.
   Who is correct?
   A. A only
   B. B only
   C. Both A and B
   D. Neither A nor B

6. When discussing differential operation,
   Technician A says that an open differential sends power to the axle or wheel that has the least slip.
   Technician B says that viscous coupling differentials send power to the axle or wheel that has the most traction.
   Who is correct?
   A. A only
   B. B only
   C. Both A and B
   D. Neither A nor B

7. When discussing Audi's Quattro all-wheel drive system,
   Technician A says that the system transfers power from the front to the rear and side to side as needed.
   Technician B says that Quattro-equipped cars with a transverse engine use a Haldex clutch.
   Who is correct?
   A. A only
   B. B only
   C. Both A and B
   D. Neither A nor B

8. When discussing AWD systems,
   Technician A says that they normally use a center differential, viscous coupling, or transfer clutch assembly to transmit and split engine power to the front and rear axles.
   Technician B says that nearly all AWD vehicles are designed for off-road operation and extremely poor road conditions.
   Who is correct?
   A. A only
   B. B only
   C. Both A and B
   D. Neither A nor B

9. When discussing the Subaru DCCD AWD system,
   Technician A says that all operation modes are automatic and do not provide for manual control.
   Technician B says the front/rear torque split is provided by a planetary gear set in the center differential.
   Who is correct?
   A. A only
   B. B only
   C. Both A and B
   D. Neither A nor B

10. When discussing viscous clutches,
    Technician A says that the driver activates the viscous clutch through a selector switch to improve driving wheel traction.
    Technician B says that the gears inside the clutch assembly rotate in the viscous fluid.
    Who is correct?
    A. A only
    B. B only
    C. Both A and B
    D. Neither A nor B

Chapter 10

DRIVETRAIN ELECTRICAL AND ELECTRONIC SYSTEMS

UPON COMPLETION AND REVIEW OF THIS CHAPTER, YOU SHOULD BE ABLE TO:

- Understand and explain the basic principles of electricity.
- Understand and explain the basic difference between electricity and electronics.
- Understand and define the terms *voltage*, *current*, and *resistance*.
- Name the various electrical components and their uses in electrical circuits.
- Understand and describe the purpose and operation of a clutch safety switch.

- Understand and describe the purpose and operation of a reverse lamp switch.
- Understand and describe the purpose and operation of an upshift light and a high gear switch.
- Understand and describe the location and operation of ABS speed sensor circuits.
- Understand and describe the purpose and operation of a shift blocking circuit.

INTRODUCTION

Although a manual transmission is not typically electrically operated or controlled, a few accessories of the car are controlled by the transmission. The driveline is also fitted with sensors that give vital information to computers that control other systems of the automobile. There are a few transmissions that have their shifting controlled or limited by electronics. To understand the operation of the accessories, sensors, and controls, you must have a good understanding of electricity and electronics.

There is often confusion concerning the terms *electrical* and *electronic*. In this book, electrical and electrical systems will refer to wiring and electrical parts. Electronics will mean computers and other black box–type items used to control engine and vehicle systems.

A basic understanding of electrical principles is important for proper diagnosis of any system that is monitored, controlled, or operated by electricity. Although the subject is normally covered in a separate course, a quick overview of electricity and its principles is presented here.

BASIC ELECTRICITY

All things are made up of atoms. An atom is the smallest particle of something. Atoms are very small and cannot be seen with the naked eye. This may be the reason many technicians struggle to understand electricity. The basics of electricity focus on atoms. Understanding the structure of the atom is the first step to understanding how electricity works. The following principles describe atoms, which are the building blocks of all materials.

- In the center of every atom is the nucleus.
- The nucleus contains positively charged particles called protons and particles called neutrons that have no charge.
- Negatively charged particles called electrons orbit around every nucleus.
- Every type of atom has a different number of protons and electrons.

In all atoms, the electrons are arranged in different orbits, called shells. Each shell contains a specific or certain number of electrons. For example, a copper atom (Figure 10-1) has 29 electrons and 29 protons. The 29 electrons are arranged in shells. The outer shell has only one electron. This outer shell needs 32 electrons to be completely full. This means that the one electron in the outer shell is loosely tied to the atom and can be easily removed.

The looseness or tightness of the electrons in orbit around the nucleus of an atom explains the behavior of electricity. Electricity is caused by the flow of electrons from one atom to another. The release of energy as one electron leaves the orbit of one atom and jumps into the orbit of another is **electricity** (Figure 10-2). The key behind creating electricity is to give a reason for the electrons to move.

There is a natural attraction of electrons to protons. Electrons have a negative charge and are attracted to something with a positive charge. When an electron leaves the orbit of an atom, the atom then has a positive charge. An electron moves from one atom to another because the atom next to it appears to be more positive than the one around which it is orbiting. An electrical power source provides for a more positive charge and in order to allow for a continuous flow of electricity, it supplies free electrons. In order to have a continuous flow of electricity, three things must be present: an excess of electrons in one place, a lack of electrons in another place, and a path between the two places.

Two power or energy sources are used in an automobile's electrical system. These are based on a chemical reaction and on magnetism. A car's battery is a source of chemical energy. A chemical reaction in the battery provides for an excess of electrons in one place and a lack of electrons in another place. Batteries have two terminals, a positive and a negative. Basically, the negative terminal is the outlet for the electrons, and the positive terminal is the inlet for the electrons to get to the protons. The chemical reaction in a battery causes a lack of electrons at the positive (+) terminal and an excess at the negative (−) terminal. This creates

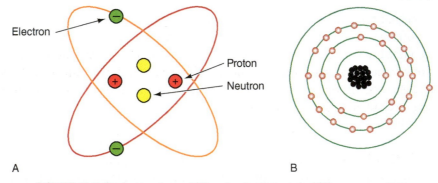

FIGURE 10-1 Basic structure of (A) a simple atom and of (B) a copper atom.

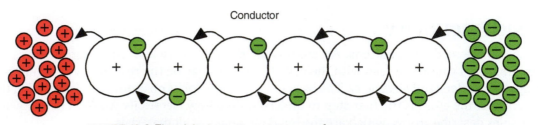

FIGURE 10-2 Electricity is the flow of electrons from one atom to another.

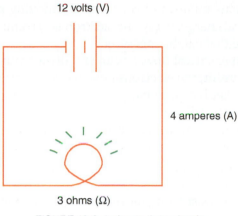

12 volts (V)

4 amperes (A)

3 ohms (Ω)

FIGURE 10-3 A simple light circuit.

an electrical imbalance, causing the electrons to flow through the path provided by a wire. A simple example of this process is shown in the battery and light arrangement in Figure 10-3.

The chemical process in the battery continues to provide electrons until the chemicals become weak. At that time, either the battery has run out of electrons or all of the protons are matched with an electron. When this happens, there is no longer a reason for the electrons to want to move to the positive side of the battery. It no longer looks more positive. Fortunately, the vehicle's charging system restores the battery's supply of electrons. This allows the chemical reaction in the battery to continue almost indefinitely. Electricity and magnetism are interrelated. One can be used to produce the other. Moving a wire (a conductor) through an already existing magnetic field can produce electricity. This process of producing electricity through magnetism is called *induction*. In a generator, a coil of wire is moved through a magnetic field. In an alternator, a magnetic field is moved through a coil of wire. In both cases, electricity is produced. The amount of electricity that is produced depends on a number of factors, including the strength of the magnetic field, the number of wires that pass through the field, and the speed at which the wire moves through the magnetic field.

Measuring Electricity

Electrical **current** is a term used to describe the movement or flow of electricity. The greater the number of electrons flowing past a given point in a given amount of time, the more current the circuit has. This current, like the flow of water or any other substance, can be measured. **Voltage** is electrical pressure (Figure 10-4). Voltage is the force developed by the attraction of the electrons to the protons. The more positive one side of the circuit is, the more voltage is present in the circuit. Voltage does not flow; rather, it is the pressure that causes current flow. When any substance flows, it meets resistance. The resistance to electrical flow can be measured.

> **Voltage** is the force that causes electrons to move.

Electrical Flow (Current)

The unit for measuring electrical current is the **ampere (amp)**, usually called an amp. The instrument used to measure electrical current flow in a circuit is called an **ammeter**.

> The **ampere** is the unit for measuring electrical current. One ampere means that 6.25 billion electrons are flowing past a given point in 1 second.

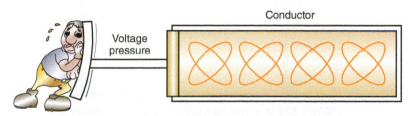

Conductor

Voltage pressure

FIGURE 10-4 Voltage is the force that causes electrons to move.

In the flow of electricity, millions of electrons are moving past any given point at the speed of light. The electrical charge of any one electron is extremely small. It takes millions of electrons to make a charge that can be measured.

There are two types of electrical flow, or current: **direct current (DC)** and **alternating current (AC)**. In direct current, the electrons flow in one direction only. The example of the battery and light shown earlier is based on direct current. In alternating current, the electrons change direction at a fixed rate. Most automobile circuits operate on DC, whereas the current in homes and buildings is AC. Generators and alternators produce AC volts, but the AC voltage is converted to DC before it is released into the rest of the electrical system.

Resistance

In every atom, the electrons resist being moved out of their shell. The amount of resistance depends on the type of atom. As explained earlier, in some atoms (such as those in copper) there is very little resistance to electron flow because the outer electron is loosely held. In other substances there is more resistance to flow because the outer electrons are tightly held.

The resistance to current flow produces heat. This heat can be measured to determine the amount of resistance. A unit of measured resistance is called an **ohm**. Resistance can be measured by an instrument called an **ohmmeter**.

Pressure

In electrical flow, some force is needed to move the electrons between atoms. This force is the pressure that exists between a positive and less positive point within an electrical circuit. This force, also called electromotive force (EMF), is measured in units called **volts**. One volt is the amount of pressure (force) required to move 1 ampere of current through a resistance of 1 ohm. Voltage is measured by an instrument called a **voltmeter**.

Circuits

When electrons are able to flow along a path (wire) between two points, an electrical circuit is formed. An electrical circuit is considered complete when there is a path that connects the positive and negative terminals of the electrical power source. Somewhere in the circuit there must be a load or resistance to control the amount of current in the circuit. Most automotive electrical circuits use the chassis as the path to the negative side of the battery (Figure 10-5). Electrical components have a lead that connects them to the chassis. These are called the chassis ground connections. In a complete circuit, the flow of electricity can be controlled and applied to do useful work, such as light a headlamp or turn over a starter motor. Components that use electrical power put a load on the circuit and consume electrical energy.

Although electrons in a circuit actually flow from negative to positive (the electron theory of current flow), automotive electrical diagrams continue to assume that current flow is from positive to negative (the conventional theory of current flow).

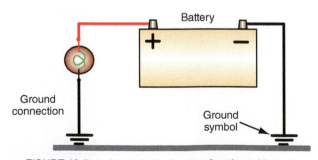

FIGURE 10-5 A simple light circuit using the vehicle as the negative conductor for the circuit.

The amount of current that flows in a circuit is determined by the resistance in that circuit. As resistance goes up, the current goes down. The energy used by a load is measured in volts. Amperage stays constant in a circuit but the voltage is dropped as it powers a load. Measuring voltage drop determines the amount of energy consumed by the load.

Ohm's Law

To understand the relationship between current, voltage, and resistance in a circuit, you need to know **Ohm's law**, the basic law of electricity. This law states that it takes 1 volt of electrical pressure to push 1 ampere of electrical current through 1 ohm of resistance. As such, the law provides a mathematical formula for determining the amount of current, voltage, or resistance in a circuit when two of these are known. The basic formula is: Voltage = Current × Resistance.

Although the basic premise of this formula is calculating unknown values in an electrical circuit (Figure 10-6), it also helps to define the behaviors of electrical circuits. A knowledge of these behaviors is important to an automotive technician.

If voltage does not change, but there is a change in the resistance of the circuit, the current will change. If resistance increases, current decreases. If resistance decreases, current will increase. If voltage changes, so must the current or resistance. If the resistance stays the same and current decreases, so will voltage. Likewise, if current increases, so will the voltage.

Ohm's law, the basic law of electricity states that it takes 1 volt of electrical pressure to push 1 ampere of electrical current through 1 ohm of resistance. The basic formula is: Voltage = Current × Resistance.

AUTHOR'S NOTE: The electrical property you will most often measure is voltage. Understanding Ohm's law and how conditions will affect voltage is your key to diagnosis. Whenever you get a voltage measurement that is not what you expected or what the specification calls for, think of resistance! For example, if you measure the voltage at both ends of a wire, you would expect the readings to be the same. However, if the readings aren't alike, you know there is some resistance between those two points. The higher the difference, the higher the resistance!

Electrical Circuits

A complete electrical circuit exists when electrons flow along a path between two points. In a complete circuit, resistance must be low enough to allow the available voltage to push electrons between the two points. Most automotive circuits contain five basic parts.

1. Power sources, such as the battery or alternator, that provide the energy needed to create electron flow.
2. **Conductors**, such as copper wires, that provide a path for current flow.

Voltage (E) = Current (I) times Resistance (R), therefore

$$E = I \times R.$$

Current (I) = Voltage (E) divided by Resistance (R), therefore

$$I = E/R.$$

Resistance (R) = Voltage (E) divided by Current (I), therefore

$$R = E/I.$$

FIGURE 10-6 Ohm's law.

FIGURE 10-7 A typical fuse and relay panel for a 4WD vehicle.

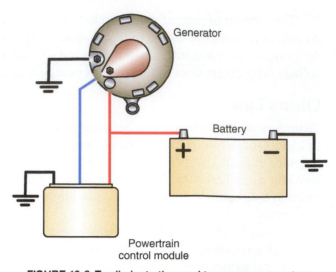

FIGURE 10-8 To eliminate the need to run separate return wires, many accessories and automotive components are grounded to the vehicle's chassis.

Shop Manual
Chapter 10, page 450

3. Loads, which are devices that use electricity to perform work, such as light bulbs, electric motors, or resistors.
4. **Controllers**, such as switches or relays, that direct the flow of electrons.
5. **Protection devices**, such as fuses, circuit breakers, and fusible links.

A complete circuit must have a complete path from the power source to the load and back to the source. With the many circuits on an automobile, this would require hundreds of wires connected to both sides of the battery. To avoid this, automobiles are equipped with power distribution centers or fuse blocks that distribute battery voltage to various circuits. The positive side of the battery is connected to the fuse block, and power is distributed from there (Figure 10-7).

As a common return circuit, auto manufacturers use a wiring style that involves using the vehicle's metal frame (Figure 10-8) as part of the return circuit. The load is often grounded directly to the metal frame, which then acts as the return wire in the circuit. Current passes from the battery, through the load, and into the frame. The frame is connected to the negative terminal of the battery through the battery's ground wire. This completes the circuit.

An electrical component, such as an AC generator, is often mounted directly to the engine block, transmission case, or frame. This direct mounting effectively grounds the component without the use of a separate ground wire. In other cases, however, a separate ground wire must be run from the component to the frame or another metal part to ensure a sound return path. The increased use of plastics and other nonmetallic materials in body panels and engine parts has made electrical grounding more difficult. To ensure good grounding back to the battery, some manufacturers now use a network of common grounding terminals and wires.

Circuit Components

Automotive electrical circuits contain a number of different types of electrical devices. The more common components are outlined in the following sections.

Resistors are used to limit current flow (and thereby voltage) in circuits where full current flow and voltage are not needed. Resistors are devices specially constructed to introduce a measured amount of electrical resistance into a circuit (Figure 10-9). In addition, some other components use resistance to produce heat and even light. An electric window defroster is a specialized type of resistor that produces heat. Electric lights are resistors that get so hot they produce light.

Resistors in common use in automotive circuits are of three types: fixed value, stepped or tapped, and variable.

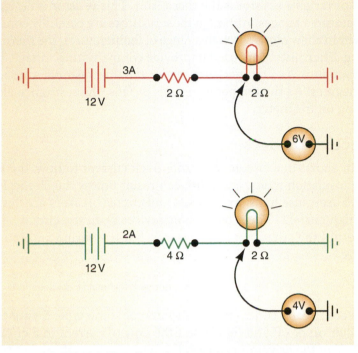

FIGURE 10-9 The effect of resistors in electrical circuits.

Fixed value resistors are designed to have only one rating, which should not change. These resistors are used to control voltage, such as in an automotive ignition system.

Tapped or stepped resistors are designed to have two or more fixed values, available by connecting wires to the several taps of the resistor. Heater motor resistor packs, which provide for different fan speeds, are an example of this type of resistor.

Variable resistors are designed to have a range of resistances available through two or more taps and a control. Two examples of this type of resistor are rheostats and potentiometers. **Rheostats** have two connections (Figure 10-10), one to the fixed end of a resistor and one to a sliding contact with the resistor. Turning the control moves the sliding contact away from or toward the fixed end tap, increasing or decreasing the resistance. **Potentiometers** have three connections (Figure 10-11), one at each end of the resistance and one connected to a sliding contact with the resistor; turning the control moves the sliding contact away from one end of the resistance, but toward the other end.

Rheostats are variable resistors with two connections that are used to change current flow through a circuit.

Potentiometers are variable resistors with three connections that are typically used to change voltage.

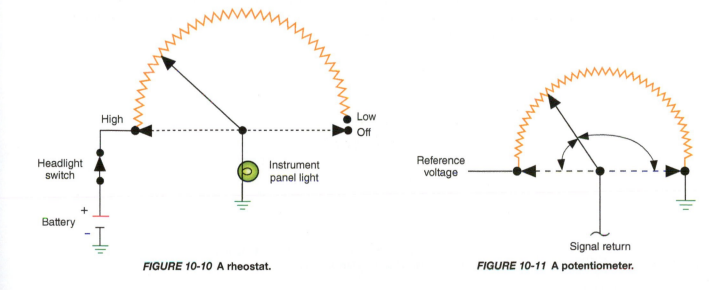

FIGURE 10-10 A rheostat.

FIGURE 10-11 A potentiometer.

Another type of variable resistor is the thermistor. This resistor is designed to change in values as its temperature changes. Although most resistors are carefully constructed to maintain their rating within a few ohms through a range of temperatures, the thermistor is designed to change its rating. Thermistors are used to provide compensating voltage in components or to determine temperature. As a temperature sender, the thermistor is connected to a voltmeter calibrated in degrees. As the temperature rises or falls, the resistance also changes. This changes the reading on the meter.

Circuit Protective Devices

Shop Manual
Chapter 10, page 476

When overloads or shorts in a circuit cause too much current to flow, the wiring in the circuit heats up, the insulation melts, and a fire can result unless the circuit has some kind of protective device. Fuses, fuse links, maxi-fuses, and circuit breakers are designed to provide protection from high current. These protection devices open the circuit when high current is present. As a result, the circuit no longer works, but the wiring and the components are saved from damage.

Switches

Shop Manual
Chapter 10, page 478

Electrical circuits are usually controlled by a switch of some type. Switches do two things. They turn the circuit on or off and they direct the flow of current in a circuit. Switches can be under the control of the driver or can be self-operating through a condition of the circuit, the vehicle, or the environment.

Contacts in a switch can be of several types, each named for the job they do or the sequence in which they work. A hinged-pawl switch is the simplest type of switch. It either makes or breaks the current in a single conductor or circuit. It is a single-pole, single-throw (SPST) switch. The throw refers to the number of output circuits, and the pole refers to the number of input circuits made by the switch.

Another type of SPST switch is the momentary contact switch. The spring-loaded contact on this switch keeps it from making the circuit except when pressure is being applied to the button. A horn switch is of this type. Because the spring holds the contacts open, the switch has a further designation: normally open. In the case in which the contacts are held closed except when the button is pressed, the switch is designated normally closed.

Single-pole, double-throw switches have one wire in and two wires out. This type of switch allows the driver to select between two circuits, such as high-beam or low-beam headlights.

Automatic transmission–equipped vehicles have a neutral safety switch that only allows the starter to work when park or neutral is selected.

Switches can be designed with a great number of poles and throws. The transmission neutral start switch may have two poles and six throws and is referred to as a multiple-pole, multiple-throw (MPMT) switch. It contains two movable wipers that move in unison across two sets of terminals. The dotted line shows that the wipers are mechanically linked, or ganged. The switch closes a circuit to the starter in either P (park) or N (neutral) and to the back-up lights in R (reverse) (Figure 10-12).

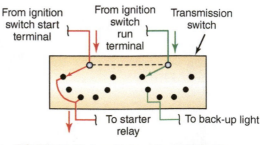

From ignition switch start terminal
From ignition switch run terminal
Transmission switch
To starter relay
To back-up light

FIGURE 10-12 A multiple-pole, multiple-throw neutral start safety switch.

Most switches are combinations of hinged-pawl and push–pull switches, with different numbers of poles and throws. Some special switches are required, however, to satisfy the circuits of modern automobiles. A mercury switch is sometimes used to detect motion in a component, such as the one used in the engine compartment to turn on the compartment light.

A temperature-sensitive switch usually contains a bimetallic element heated either electrically or by some component in which the switch is used as a sensor. When engine coolant is below or at normal operating temperature, the engine coolant temperature sensor is in its normally open condition. If the coolant exceeds the temperature limit, the bimetallic element bends the two contacts together, and the switch to the indicator or the instrument panel is closed. Other applications for heat-sensitive switches are time-delay switches and flashers.

Relays

A **relay** (Figure 10-13) is an electric switch that allows a small amount of current to control a much larger one. It consists of a control circuit. When the control circuit switch is open, no current flows to the coil, so the windings are deenergized. When the switch is closed, the coil is energized, turning the soft iron core into an electromagnet and drawing the armature down. This closes the power circuit contacts, connecting power to the load circuit. When the control switch is opened, the current stops flowing in the coil, the electromagnet disappears, and the armature is released, which breaks the power circuit contacts.

Solenoids

Solenoids (Figure 10-14) are also electromagnets with movable cores used to translate electrical current flow into mechanical movement. The movement of the core causes something else, such as a lever, to move. They also can close electrical contacts, acting as a relay at the same time.

Shop Manual
Chapter 10, page 482

Electromagnetism Basics

Electricity and magnetism are related. One can be used to create the other. Current flowing through a wire creates a magnetic field around the wire. Moving a wire through a magnetic field creates current flow in the wire.

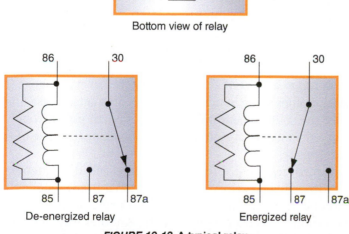

FIGURE 10-13 A typical relay.

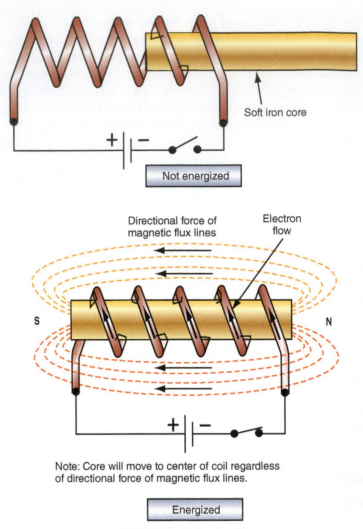

FIGURE 10-14 A solenoid.

Many automotive components, such as alternators, ignition coils, starter solenoids, and magnetic pulse generators operate using principles of electromagnetism.

Although almost everyone has seen magnets at work, a simple review of magnetic principles is in order to ensure a clear understanding of electromagnetism.

A substance is said to be a magnet if it has the property of magnetism—the ability to attract such substances as iron, steel, nickel, or cobalt. These are called magnetic materials.

A magnet has two points of maximum attraction, one at each end of the magnet. These points are called poles, with one being designated the north pole and the other the south pole. When two magnets are brought together, opposite poles attract, whereas similar poles repel.

A magnetic field, called a **field of flux**, exists around every magnet (Figure 10-15). The field consists of invisible lines along which a magnetic force acts. These lines emerge from the north pole and enter the south pole, returning to the north pole through the magnet itself. All lines of force leave the magnet at right angles to the magnet. None of the lines cross each other. All lines are complete.

Magnets can occur naturally in the form of a mineral called magnetite. Artificial magnets can also be made by inserting a bar of magnetic material inside a coil of insulated wire and passing a heavy direct current through the coil. This principle is very important in understanding certain automotive electrical components. Another way of creating a magnet is by stroking the magnetic material with a bar magnet. Both methods force the randomly arranged molecules of the magnetic material to align themselves along north and south poles.

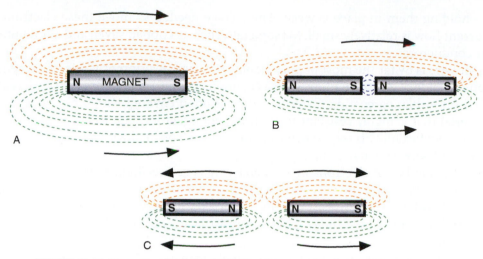

FIGURE 10-15 Magnetic principles: (A) field of flux around a magnet, (B) unlike poles attract each other, and (C) like poles repel.

Artificial magnets can be either temporary or permanent. Temporary magnets are usually made of soft iron. They are easy to magnetize but quickly lose their magnetism when the magnetizing force is removed. Permanent magnets are difficult to magnetize, but once magnetized they retain this property for very long periods.

Induced Voltage

Now that we have explained how current can be used to generate a magnetic field, it is time to examine the opposite effect of how magnetic fields can produce electricity. Consider a straight piece of conducting wire with the terminals of a voltmeter attached to both ends. If the wire is moved across a magnetic field, the voltmeter registers a small voltage reading (Figure 10-16). A voltage has been induced in the wire.

It is important to remember that the conducting wire must cut across the flux lines to induce a voltage. Moving the wire parallel to the lines of flux does not induce voltage. The wire need not be the moving component in this setup. Holding the conducting wire still and moving the magnetic field at right angles to it also induces voltage in the wire.

The wire or conductor becomes a source of electricity and has a polarity or distinct positive and negative end. However, this polarity can be switched depending on the relative direction of movement between the wire and magnetic field. This is why an AC generator produces alternating current.

Conductors and Insulators

Controlling and routing the flow of electricity requires the use of materials known as conductors and insulators. Conductors are materials with a low resistance to the flow of current. If the number of electrons in the outer shell or ring of an atom is less than four,

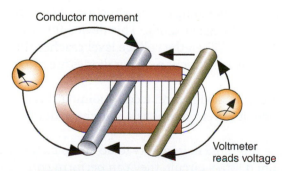

FIGURE 10-16 Moving a conductor through a magnetic field induces a voltage.

the force holding them in place is weak. The voltage needed to move these electrons and create current flow is relatively small. Most metals, such as copper, silver, and aluminum, are excellent conductors.

Copper wire is by far the most popular conductor used in automotive electrical systems. Wire wound inside of electrical units, such as ignition coils and generators, usually has a very thin baked-on insulating coating. External wiring is often covered with a plastic-type insulating material that is highly resistant to environmental factors such as heat, vibration, and moisture. Where flexibility is required, the copper wire will be made of a large number of very small strands of wire woven together.

When the number of electrons in the outer ring is greater than four, the force holding them in orbit is very strong and very high voltages are needed to move them. These materials are known as **insulators**. They resist the flow of current. Thermal plastics are the most common electrical insulators used today. They can resist heat, moisture, and corrosion without breaking down.

When there are four electrons in the outer ring, the material is a semiconductor. A semiconductor is neither a conductor nor an insulator.

BASICS OF ELECTRONICS

Computerized engine controls and other features of today's cars would not be possible if it were not for electronics. For purposes of clarity, let us define electronics as the technology of controlling electricity. Electronics has become a special technology beyond electricity. Transistors, diodes, semiconductors, integrated circuits, and solid-state devices are all considered to be part of electronics rather than just electrical devices. But keep in mind that all the basic laws of electricity apply to electronic controls.

Semiconductors

A **semiconductor** is a material or device that can function as either a conductor or an insulator, depending on how its structure is arranged. Semiconductor materials have less resistance than an insulator, but more resistance than a conductor. Some common semiconductor materials include silicon (Si) and germanium (Ge).

In semiconductor applications, materials have a crystal structure. This means that their atoms do not lose and gain electrons as the atoms in conductors do. Instead, the atoms in these semiconductor materials share outer electrons with each other. In this type of atomic structure, the electrons are tightly held and the element is stable.

Because the electrons are not free, crystals cannot conduct current. These materials are called electrically inert materials. A small amount of trace element must be added for these to function as semiconductors. The addition of these traces, called impurities, allows the material to function as a semiconductor. The type of impurity added determines what type of semiconductor will be produced.

The **diode** is the simplest semiconductor device. A diode allows current to flow in one direction, but not in the opposite direction. Therefore, it can function as a switch, acting as either a conductor or insulator, depending on the direction of current flow.

A variation of the diode is the zener diode. This device functions like a standard diode until a certain voltage is reached. When the voltage level reaches this point, the zener diode allows current to flow in the reverse direction. Zener diodes are often used in electronic voltage regulators.

A **transistor** is an electronic device produced by joining three sections of semiconductor materials. Like the diode, it is very useful as a switching device, functioning as either a conductor or an insulator.

One transistor or diode is limited in its ability to do complex tasks. However, when many semiconductors are combined into a circuit, they can perform complex functions.

An **integrated circuit (IC)** is simply a large number of diodes, transistors, and other electronic components, such as resistors and capacitors, all mounted on a single piece of semiconductor material. This type of circuit has a tremendous size advantage. It is extremely small. Circuitry that used to take up entire rooms can now fit into a pocket. The principles of semiconductor operation remain the same in integrated circuits—only the size has changed.

The increasingly small size of integrated circuits is very important to automobiles. This means that electronics is no longer confined to simple tasks, such as rectifying alternator current. Enough transistors, diodes, and other solid-state components can be installed in a car to make logical decisions and issue commands to other areas of the engine. This is the foundation of computerized control systems.

The computer (Figure 10-17) has taken over many of the tasks in cars and trucks that were formerly performed by vacuum, electromechanical, or mechanical devices. When properly programmed, they can carry out explicit instructions with blinding speed and almost flawless consistency.

A computer is called a PCM, ECM, central processor, or microprocessor.

A typical electronic control system is made up of sensors, actuators, and related wiring that is tied to a computer.

Sensors

All sensors perform the same basic function. They detect a mechanical condition (movement or position), chemical state, or temperature condition and change it into an electrical signal that can be used by the computer to make decisions. The computer makes decisions based on information it receives from sensors. Each sensor used in a particular system has a specific job to do. Together these sensors provide enough information to help the computer form a complete picture of vehicle operation. Even though there are a variety of different sensor designs, they all fall under one of two operating categories: reference voltage sensors or voltage-generating sensors.

Shop Manual
Chapter 10, page 487

Reference voltage (Vref) sensors provide input to the computer by modifying or controlling a constant, predetermined voltage signal (Figure 10-18). This signal, which can have a reference value from 5 to 9 volts, is generated and sent out to many sensors by a reference voltage regulator located inside the processor. The term *processor* is used to describe the actual metal box that houses the computer and its related components. Because the computer knows that a certain voltage value has been sent out, it can indirectly interpret things like motion, temperature, and component position, based on what comes back. For example, consider the operation of the throttle position sensor (TP sensor). During acceleration

FIGURE 10-17 A typical automotive computer.

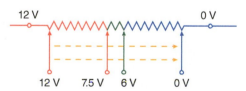

FIGURE 10-18 The voltage present across a potentiometer as the wiper moves. It alters the reference voltage in response to the movement of the wiper and sends a signal back to the voltage-sensing circuit of the computer.

(from idle to wide-open throttle), the computer monitors throttle plate movement based on the changing reference voltage signal returned by the TP sensor. (The TP sensor is a type of variable resistor known as a rotary potentiometer that changes circuit resistance based on throttle shaft rotation.) As TP sensor resistance varies, the computer is programmed to respond in a specific manner (e.g., increase fuel delivery or alter spark timing) to each corresponding voltage change.

Most sensors presently in use are variable resistors (potentiometers). They modify a voltage to or from the computer, indicating a constantly changing status that can be calculated, compensated for, and modified. That is, most sensors simply control a voltage signal from the computer. When varying internal resistance of the sensor allows more or less voltage to ground, the computer senses a voltage change on a monitored signal line. The monitored signal line may be the output signal from the computer to the sensor (one- and two-wire sensors), or the computer may use a separate return line from the sensor to monitor voltage changes (three-wire sensors).

While most sensors are variable reference voltage, there is another category of sensor—the **voltage-generating devices** (Figure 10-19). These sensors include components like the Hall-effect switch, oxygen sensor (zirconium dioxide), and knock sensor (piezoelectric), which are capable of producing their own input voltage signals. This varying voltage signal, when received by the computer, enables the computer to monitor and adjust for changes in the computerized engine control system.

In addition to variable resistors, two other commonly used reference voltage sensors are switches and thermistors. Switches provide the necessary voltage information to the computer so that vehicles can maintain the proper performance and driveability. Thermistors are special types of resistors that convert temperature into a voltage. Regardless of the type of sensors used in electronic control systems, the computer is incapable of functioning properly without input-signal voltage from sensors.

Communication Signals

Analog is a type of signal that carries a constant or changing voltage.

Most input sensors are designed to produce a voltage signal that varies within a given range (from high to low, including all points in between). A signal of this type is called an **analog** signal. Unfortunately, the computer does not understand analog signals. It can only read a **digital** binary signal, which is a signal that has only two values—on or off (Figure 10-20).

Digital is a type of signal that is caused by switching the circuit on and off.

To overcome this communication problem, all analog voltage signals are converted to a digital format by a device known as an analog-to-digital converter (A/D converter). The A/D converter is located in a section of the processor that receives the input signal. Some sensors like the Hall-effect switch produce a digital or square wave signal that can go directly to the computer as input (Figure 10-21).

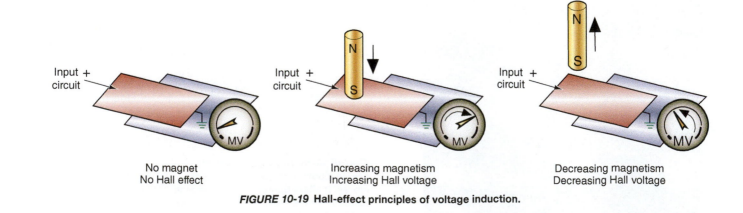

No magnet
No Hall effect

Increasing magnetism
Increasing Hall voltage

Decreasing magnetism
Decreasing Hall voltage

FIGURE 10-19 Hall-effect principles of voltage induction.

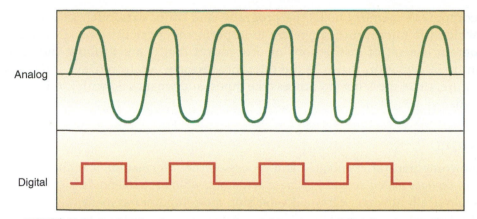

FIGURE 10-20 Analog signals are constantly variable, whereas digital signals are either on or off, or high or low.

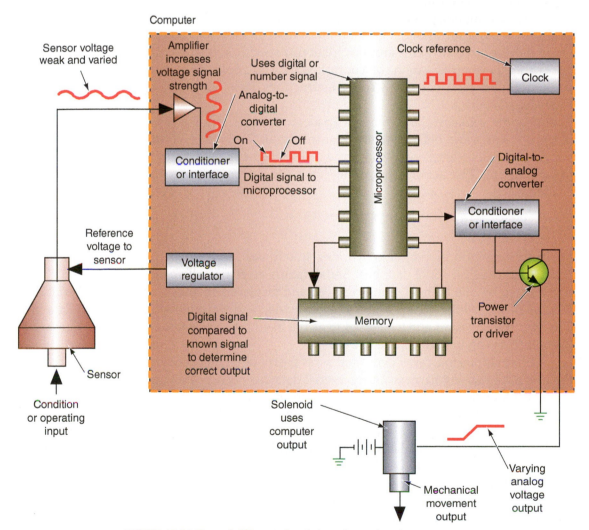

FIGURE 10-21 Flow of different signals in and out of a microprocessor.

A computer's memory holds the programs and other data, such as vehicle calibrations, which the microprocessor refers to in performing calculations. To the computer, the program is a set of instructions or procedures that it must follow. Included in the program is information that tells the microprocessor when to retrieve input (based on temperature, time, etc.), how to process the input, and what to do with it once it has been processed.

Actuators

After the computer has assimilated and processed the information, it sends output signals to control devices called actuators. These actuators, which are solenoids, switches, relays, or motors, physically act on or carry out a decision the computer has made.

Actuators are electromechanical devices that convert an electrical current into mechanical action. This mechanical action can then be used to open and close valves, control vacuum to other components, or open and close switches. When the microcomputer receives an input signal indicating a change in one or more of the operating conditions, the microcomputer determines the best strategy for handling the conditions. The microcomputer then controls a set of actuators to achieve a desired effect or strategy goal. In order for the computer to control an actuator, it must rely on a component called an **output driver**.

Output drivers are also located in the processor (along with the input conditioners, microprocessor, and memory) and operate by the digital commands issued by the microcomputer. Basically, the output driver is nothing more than an electronic on and off switch that the computer uses to control the ground circuit of a specific actuator.

For actuators that cannot be controlled by a solenoid, relay, switches, or motors, the computer must turn its digitally coded instructions back into an analog format via a digital-to-analog converter.

On some vehicles, the actuator is an electromagnetic clutch. These clutches are found in transfer cases and older clutchless transmissions. This clutch consists of a wired coil and a plate. When current flows through the coil, a magnetic field is formed that pulls the plate toward the coil. This action engages the clutch. Once current flow to the coil stops, the clutch is disengaged.

The **output driver** is basically an electronic on and off switch that the computer uses to control the ground circuit of a specific actuator.

Multiplexing

Multiplexing is an in-vehicle networking system used to transfer data between electronic modules through a serial data bus. Serial data is electronically coded information that is transmitted by one computer and received and displayed by another computer. Serial data (Figure 10-22) is information that is digitally coded and transmitted in a series of data bits. The data transmission rate is referred to as the baud rate. Baud rate refers to the number of data bits that can be transmitted in a second.

With multiplexing, fewer dedicated wires are required for each function, and this reduces the size of the wiring harness. Using a serial data bus reduces the number of wires by combining the signals on a single wire through time division multiplexing. Information is sent to individual control modules that control each function, such as antilock braking, turn signals, power windows, and the instrument panel.

Multiplexing also eliminates the need for redundant sensors because the data from one sensor is available to all electronic modules. Multiplexing also allows for greater vehicle content flexibility because functions can be added through software changes, rather than adding another module or modifying an existing one.

The common multiplex system is called the CAN or controller area network. It is used to interconnect a network of electronic control modules.

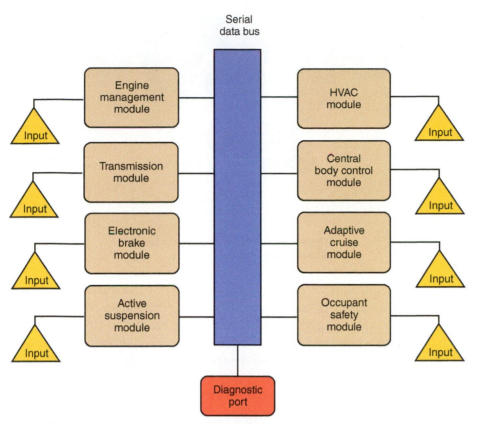

Serial
data bus

Engine
management
module

Input

Transmission
module

Input

Electronic
brake
module

Input

Active
suspension
module

Input

HVAC
module

Input

Central
body control
module

Input

Adaptive
cruise
module

Input

Occupant
safety
module

Input

Diagnostic
port

FIGURE 10-22 The serial data bus (normally CAN) carries many of the input signals from the vehicle and is used by many different control units.

CLUTCH SAFETY SWITCH

The **clutch safety switch** (Figure 10-23) is connected into the starting circuit. The switch prevents the starting of the engine unless the clutch pedal is fully depressed. The switch is normally open when the clutch pedal is released. When the clutch pedal is depressed, the switch closes and completes the circuit between the ignition switch and the starter solenoid. Sometimes a neutral switch is also used, and this switch bypasses the clutch switch when the transmission is shifted into neutral.

Shop Manual
Chapter 10, page 478

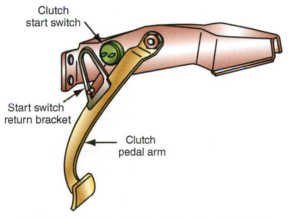

Clutch
start switch

Start switch
return bracket

Clutch
pedal arm

FIGURE 10-23 The clutch switch responds to the movement of the clutch pedal.

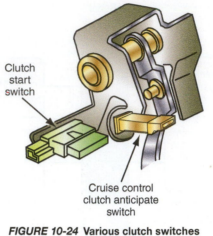

FIGURE 10-24 Various clutch switches used with cruise control systems.

Disengaging the clutch also reduces the current draw of the starter when the engine is being started in cold weather. By disengaging the clutch, the starter motor only rotates the engine. When the clutch is engaged, the starter motor must also turn the input shaft of the transmission even when the transmission is in neutral. This can provide quite a load on the starter, especially when it is cold and the transmission oil is thick.

Clutch switches are also used in cruise control circuits (Figure 10-24). When the driver depresses the clutch pedal, power for the cruise control system is shut off. This prevents high throttle operation when there is no load on the engine.

REVERSE LAMP SWITCH

Shop Manual
Chapter 10, page 480

All vehicles sold in the United States after 1971 are required to have back-up (reverse) lights. Back-up lights illuminate the area behind the vehicle and warn other drivers and pedestrians that the vehicle is moving in reverse (Figure 10-25). Typically power for the lamps is supplied through the ignition switch when it is in the on position. When the driver shifts the transmission into reverse, the contacts of the back-up light switch are closed; this completes the light circuit. Most manual transmissions are equipped with a separate switch located on the transmission (Figure 10-26) but can be mounted to the shift linkage away from the transmission. If the switch is mounted in the transmission, the shifting fork will close the switch and

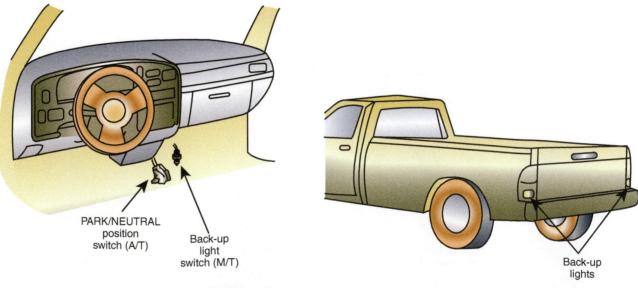

PARK/NEUTRAL position switch (A/T)

Back-up light switch (M/T)

Back-up lights

FIGURE 10-25 A typical back-up light system.

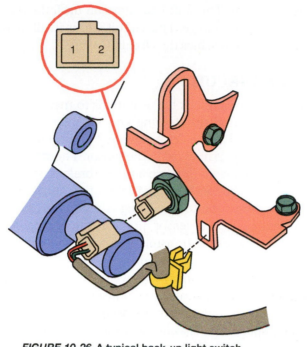

FIGURE 10-26 A typical back-up light switch mounted to a transaxle.

Some late-model Corvettes have a redundant reverse switch to ensure reliability of the keyless entry and starting system. The car must be in reverse gear to start and to lock after it is shut off.

complete the circuit whenever the transmission is shifted into reverse gear. If the switch is mounted on the linkage, the switch is closed directly by the linkage.

UPSHIFT LAMP CIRCUIT

Upshift and shift lamps inform the driver when to shift into the next gear in order to maximize fuel economy. These lights are controlled by the PCM, which activates the light according to engine speed, engine load, and vehicle speed (Figure 10-27). Basically these light circuits operate like a vacuum gauge. When engine load is low, engine vacuum is high. And when engine load is high, vacuum is low. The shift light will come on whenever there is high vacuum. The shift lamp is lit at those engine speeds and loads in which engine vacuum is high and the

Engine load is determined by the throttle position sensor and the manifold absolute pressure (MAP) sensor.

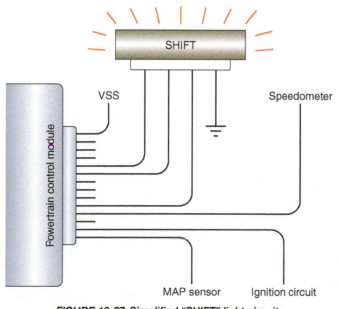

FIGURE 10-27 Simplified "SHIFT" light circuit.

transmission is in a forward gear. The shift light stays on until the transmission is shifted or the engine's operating conditions change. This circuit works in all forward gears except high gear, in which a high gear switch disables the circuit.

ABS SPEED SENSOR CIRCUITS

Wheel speed sensors are used in antilock brake systems to measure the speed of the wheels. The tip of the sensor is located on the steering knuckle near a toothed ring or rotor. The toothed ring is typically attached to the drive axle shaft and rotates at the same speed as the wheel. As the ring spins, a voltage is induced in the sensor. The strength and frequency of this voltage varies with the wheel's speed. In some antilock brake systems, the wheel speed sensor is mounted at each wheel. The toothed ring is part of the outer CV joint or axle assembly (Figure 10-28). In other systems, the toothed ring is mounted next to the differential ring gear (Figure 10-29). In these cases, one sensor monitors the speed of the entire axle assembly instead of the individual wheels. Some transmissions are fitted with a sensor (Figure 10-30), which is typically used to monitor vehicle speed.

Speed sensors are critical inputs for many four-wheel-drive (4WD), all-wheel-drive (AWD), and automatic shift manual transmissions. Electronically controlled transfer cases, differentials, and transmissions must be able to monitor vehicle speed and the speed of individual wheels. Vehicle and wheel speed data is available on the CAN and is used by many systems.

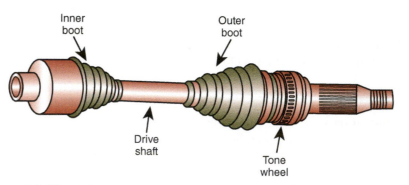

FIGURE 10-28 Location of speed sensor toothed ring (tone wheel) on a half shaft.

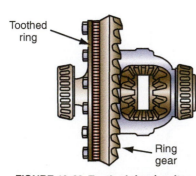

FIGURE 10-29 Toothed ring (excitor ring) mounted to a differential's ring gear.

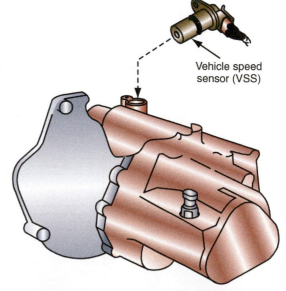

FIGURE 10-30 Typical location of a VSS mounted in a transaxle.

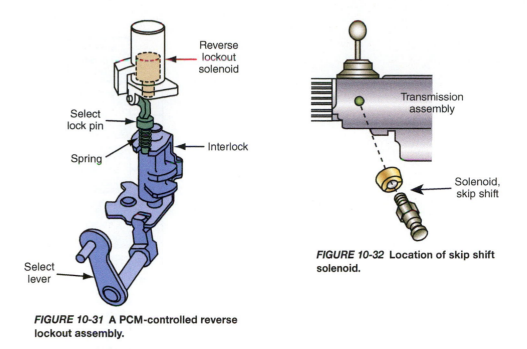

FIGURE 10-31 A PCM-controlled reverse lockout assembly.

FIGURE 10-32 Location of skip shift solenoid.

SHIFT BLOCKING

Some six-speed transmissions have a feature called shift blocking or *skip shift*. This prevents the driver from shifting into second or third gears from first gear, when the coolant temperature is above 50°C (122°F), the speed of the vehicle is between 12 and 22 mph (20 and 29 km/h), and the throttle is opened 35 percent or less. Shift blocking occurs to ensure good fuel economy and keeps the vehicle in compliance with federal fuel economy standards. These transmissions are equipped with reverse lockout (Figure 10-31), as are some others. This feature prevents the engagement of reverse whenever the vehicle is moving forward.

Shift blocking is controlled by the PCM (Figure 10-32). A solenoid is used to block off the shift pattern from first gear to second or third. The driver moves the gearshift from its up position to a lower position, as if shifting into second, and fourth gear is selected. The solenoid does not impede downshifting.

Shop Manual
Chapter 10, page 487

ELECTRICAL CLUTCHES

Some older vehicles were equipped with a clutchless manual transmission. These offered the driver control of the transmission without needing to depress the clutch during gear changes. These transmissions promised to offer the advantages of an automatic transmission with those of a manual transmission. In short, they were designed to work the clutch according to needs, instead of the driver's foot. Often the clutch is misused and performance, fuel economy, and durability suffer because a driver does not use the clutch wisely. One of the most commonly found clutchless transmissions was the Volkswagen "Automatic Stickshift." Although these systems were clever and convenient, they were dropped because they were very complicated, expensive to produce, and gear shifting felt very abrupt. With the advances made in the electronics world, these systems are once again under development.

This manual transmission has only two pedals, with no clutch pedal. To change gears, the driver merely moves the shifter. Sensors monitor the position of the shifter and throttle pedal (Figure 10-33). The system then automatically operates the electromagnetic clutch and throttle to match revs and engage the clutch again. Some transmissions with this system will automatically shift gears at the right points and others require the driver to manually select the desired gear. Also, many of these transmissions operate in a sequential mode where the driver can only upshift or downshift one gear at a time.

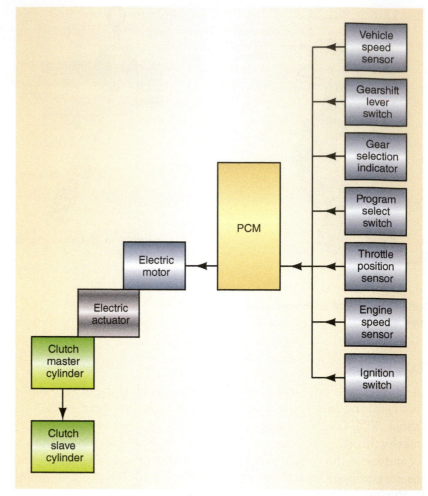

FIGURE 10-33 Simplified electric clutch circuit.

Shop Manual

Chapter 10, page 483

The clutch is released by a small electric motor that is controlled by a computer. The computer responds to sensors that send information about engine speed, transmission speed, throttle position, gearshift position, and driver intent (Figure 10-32). A gearshift lever switch responds only to fore and aft movements and is load sensitive. When the driver moves the shifter to change gears, the control module actuates the hydraulic clutch system. If the driver quickly and forcibly moves the shifter, the clutch will be disengaged and engaged quickly. If the gearshift is moved slowly, the clutch action will also be slower. A switch allowing the driver to select the harshness and quickness of the shifts allows the computer to control the clutch according to the intent of the driver, as well as to the movement of the shifter.

When the vehicle is stopped and in gear, the clutch is controlled by throttle movement. The clutch remains disengaged until the driver applies pressure on the throttle; then the clutch engages to match the movement of the throttle as well as the driver's intent.

OTHER ELECTRONIC SYSTEMS

Shop Manual

Chapter 10, page 482

As mentioned in the previous chapter, some 4WD vehicles use an electric switch to control the transfer case. Some models use an electronic control unit to control the engagement and disengagement of 4WD. These systems use a clutch pedal position switch, an electric shift motor, a shift position sensor, and a speed sensor mounted to the transfer case, or speed sensors at each wheel. A few systems use a vacuum motor in place of the electric shift motor. These vacuum units are controlled by solenoids. The selector switch tells the computer when the clutch pedal has been fully depressed to allow the transfer case to be shifted from 4WD-High to 4WD-Low. The shift position sensor informs the computer as to what gear range the transfer case is in. The speed sensor tells the control unit when the vehicle is stopped, allowing shifts from High to Low to occur.

The rear axle differential, on some model vehicles, is equipped with an electromagnetic clutch for lockup of the differential. This clutch is the same as used in 4WD systems to lock the center differential and may be controlled by a switch or by a traction control computer. When one rear wheel loses traction and speeds up, the computer energizes the clutch, which locks up the differential and provides equal power to both rear wheels.

The electromagnetic clutch consists of an actuator coil, an armature, and a stack of steel clutch plates. When the clutch is energized, a magnetic field is produced to compress the clutch; this locks the differential.

Some models are equipped with a transmission fluid–level indicator (Figure 10-34). This type of warning light is normally a simple circuit consisting of a sensor or sending unit, wires, and the lamp. The sending unit provides a path for ground when the fluid is low, thereby lighting the warning lamp. When the fluid level is high enough to keep the switch open, the lamp remains off.

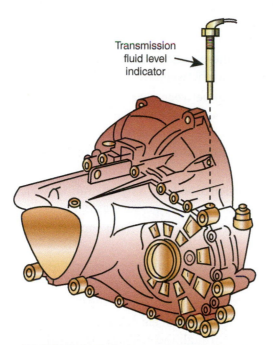

Transmission fluid level indicator

FIGURE 10-34 Typical transmission fluid–level indicator.

SUMMARY

- A basic understanding of electrical principles is important for effective electrical diagnosis and for the proper operation and interpretation of diagnostic tools.
- The release of energy as one electron leaves the orbit of one atom and jumps into the orbit of another is electricity.
- Two power or energy sources are used in an automobile's electrical system; these are based on a chemical reaction and on magnetism.
- A car's battery is a source of chemical energy. A chemical reaction in the battery provides for an excess of electrons in one place and a lack of electrons in another place.
- The flow of electricity is called current and is measured in amperes. There are two types of electrical flow: direct current (DC) and alternating current (AC).

SUMMARY

- Resistance to current flow produces heat. The amount of resistance is measured in ohms.
- In a complete electrical circuit, the flow of electricity is controlled and applied to perform tasks, such as lighting headlights and turning over the starter motor. Circuit testers are used to identify shorted and open circuits.
- Voltage is electrical pressure and is measured in volts.
- For electrical flow to occur, there must be an excess of electrons in one place, a lack of electrons in another, and a path between the two places.
- Ohm's law states that it takes 1 volt of electrical pressure to push 1 ampere of electrical current through 1 ohm of resistance.
- If voltage does not change, but there is a change in the resistance of the circuit, the current will change. If resistance increases, current decreases. If resistance decreases, current increases. If voltage changes, so must the current or resistance. If the resistance stays the same and the current decreases, so will the voltage. Likewise, if the current increases, so will the voltage.
- Voltage is electrical pressure and is measured in volts.
- For electrical flow to occur, there must be an excess of electrons in one place, a lack of electrons in another, and a path between the two places.
- The mathematical relationship between current, resistance, and voltage is expressed in Ohm's law, E = IR, where voltage is measured in volts, current in amperes, and resistance in ohms.
- Resistors in common use in automotive circuits are of three types: fixed value, stepped or tapped, and variable.
- Fixed-value resistors are designed to have only one rating, which should not change. These resistors are used to control voltage, such as in an automotive ignition system.
- Rheostats have two connections, one to the fixed end of a resistor and one to a sliding contact with the resistor.
- Potentiometers have three connections, one at each end of the resistance and one connected to a sliding contact with the resistor. Turning the control moves the sliding contact away from one end of the resistance, but toward the other end.
- Electrical schematics are diagrams with electrical symbols that show the parts and how electrical current flows through the vehicle's electrical circuits. They are used in troubleshooting.
- The strength of an electromagnet depends on the number of current-carrying conductors and what is in the core of the coil. Inducing a voltage requires a magnetic field producing lines of force, conductors that can be moved, and movement between the conductors and the magnetic field so that the lines of force are cut.
- Fuses, fuse links, maxi-fuses, and circuit breakers protect circuits against overloads. Switches control on and off and direct current flow in a circuit. A relay is an electric switch. A solenoid is an electromagnet that translates current flow into mechanical movement. Resistors limit current flow.
- A semiconductor is a material or device that can function as either a conductor or an insulator, depending on how its structure is arranged.
- An integrated circuit is simply a large number of diodes, transistors, and other electronic components, such as resistors and capacitors, all mounted on a single piece of semiconductor material.
- The diode allows current to flow in one direction but not in the opposite direction.

TERMS TO KNOW

Actuator

Alternating current (AC)

Ammeter

Ampere (amp)

Analog

Clutch safety switch

268

- Transistors are used as switching devices.
- Computers are electronic decision-making centers. Input devices called sensors feed information to the computer. The computer processes this information and sends signals to controlling devices. A typical electronic control system is made up of sensors, actuators, a microcomputer, and related wiring.
- Most input sensors are variable resistance or reference types, switches, and thermistors.
- All sensors detect a mechanical condition, chemical state, or temperature condition and change it into an electrical signal that can be used by the computer to make decisions.
- All sensors are either reference voltage sensors or voltage-generating sensors.
- Most input sensors are designed to produce a voltage signal that varies within a given range called an analog signal.
- A computer does not understand analog signals. It can only read a digital binary signal, which is a signal that has only two values—on or off.
- After the computer has assimilated and processed the information, it sends output signals to control devices called actuators, which are solenoids, switches, relays, or motors that physically act on or carry out a decision the computer has made.
- The clutch start switch prevents starting of the engine unless the clutch pedal is fully depressed.
- Most manual transmissions are equipped with a clutch safety switch located on the transmission, but they can be mounted to the shift linkage away from the transmission.
- Upshift and shift lamps inform the driver when to shift into the next gear in order to maximize fuel economy. These lights are controlled by the PCM, which activates the light according to engine speed, engine load, and vehicle speed.
- Wheel speed sensors are used in antilock brake systems to measure the speed of the wheels. The tip of the sensor is located on the steering knuckle near a toothed ring or rotor. The toothed ring is typically attached to the drive axle shaft and rotates at the same speed as the wheel.
- Some transmissions are fitted with a sensor, which is typically used to monitor vehicle speed.
- Shift blocking prevents the driver from shifting into second or third gears from first gear, when the coolant temperature is above 50°C (122°F), the speed of the vehicle is between 12 and 22 mph (20 and 29 km/h), and the throttle is opened 35 percent or less.
- Shift blocking is controlled by the PCM. A solenoid is used to block off the shift pattern from first gear to second or third.
- Some transmissions are equipped with a transmission fluid–level indicator.

REVIEW QUESTIONS

Short Answer Essays

1. Name the two energy sources used in automobile electrical systems.

2. For electrical flow to occur, what must be present?

3. Define *electricity*.

4. What is the difference between voltage and current?

5. What must be present in a circuit in order to cause electricity to flow?

6. State Ohm's law.

7. Describe the differences between a rheostat and a potentiometer.

8. What is the difference between a fixed resistor and a variable resistor?

9. What types of sensors are typically used in an automotive computer system?

10. What is the purpose of shift blocking?

Fill-in-the-Blanks

1. The two power or energy sources used in an automobile's electrical system are based on a _____ _____ and on _____.

2. Current is measured in _____; electrical voltage is measured in _____; and electrical resistance is measured in _____.

3. Back-up lights are normally controlled by a switch located in the _____ or controlled by the movement of the _____.

4. _____, _____, _____ _____, and _____ _____ are used to protect circuits against current overloads.

5. _____ control on and off and direct current flow in a circuit.

6. The center of an atom is known as the _____.

7. Resistance is measured in _____.

8. A _____ is a material or device that can function as either a conductor or an insulator, depending on how its structure is arranged.

9. A computerized circuit depends on two types of signals: _____ and _____.

10. Wheel speed sensors are used in antilock brake systems to measure the speed of the wheels. The toothed ring of the sensor is typically attached to the _____ _____, _____, or is pressed onto the _____ of the differential.

ASE-Style Review Questions

1. When discussing sources of electrical energy in an automobile,
   Technician A says that magnetism is a source of electrical energy in an automobile.
   Technician B says that chemical reaction is a source of electrical energy in an automobile.
   Who is correct?
   A. A only
   B. B only
   C. Both A and B
   D. Neither A nor B

2. When discussing the behavior of electricity,
   Technician A says that if voltage does not change, but there is a change in the resistance of the circuit, the current will change.
   Technician B says that if resistance increases, current decreases.
   Who is correct?
   A. A only
   B. B only
   C. Both A and B
   D. Neither A nor B

3. When discussing rheostats and potentiometers,
   Technician A says that rheostats have three connections, one at each end of the resistance and one connected to a sliding contact with the resistor.
   Technician B says that potentiometers have two connections, one at the power end of the resistance and one connected to a sliding contact with the resistor.
   Who is correct?
   A. A only
   B. B only
   C. Both A and B
   D. Neither A nor B

4. When discussing electrical resistance,
   Technician A says that electrical resistance is the pressure that causes current to flow in a circuit.
   Technician B says that if there is zero resistance in a circuit, a maximum amount of current will flow in the circuit.
   Who is correct?
   A. A only
   B. B only
   C. Both A and B
   D. Neither A nor B

5. When discussing electrical devices,
   Technician A says that a diode allows current to flow in one direction but not in the opposite direction.
   Technician B says that transistors are used as switching devices.
   Who is correct?
   A. A only
   B. B only
   C. Both A and B
   D. Neither A nor B

6. When discussing the flow of electricity,
   Technician A says that the flow of electricity is called current.
   Technician B says that the flow of electricity is measured in amperes.
   Who is correct?
   A. A only
   B. B only
   C. Both A and B
   D. Neither A nor B

7. When discussing Ohm's Law,
   Technician A says that if the resistance in a circuit changes, so must the voltage or current.
   Technician B says that if voltage changes, so must the current or resistance.
   Who is correct?
   A. A only
   B. B only
   C. Both A and B
   D. Neither A nor B

8. When discussing control of upshift and shift lamps,

Technician A says that upshift and shift lamps are controlled by vehicle speed.

Technician B says that upshift and shift lamps are controlled by throttle position.

Who is correct?

A. A only
B. B only
C. Both A and B
D. Neither A nor B

9. When discussing batteries,

Technician A says that a chemical reaction in the battery provides for an excess of electrons in one part of the battery.

Technician B says that a chemical reaction in the battery provides for a lack of electrons in one part of the battery.

Who is correct?

A. A only
B. B only
C. Both A and B
D. Neither A nor B

10. When discussing electrical flow,

Technician A says that there are two types of electrical flow: direct current (DC) and alternating current (AC).

Technician B says that DC is used to operate most automotive electrical circuits.

Who is correct?

A. A only
B. B only
C. Both A and B
D. Neither A nor B

Chapter 11

ELECTRONICALLY CONTROLLED AND AUTOMATED TRANSMISSIONS

UPON COMPLETION AND REVIEW OF THIS CHAPTER, YOU SHOULD BE ABLE TO:

- Understand and explain how a typical CVT is controlled and how it provides various gear ratios.

- Describe how the various transmissions used in hybrid vehicles allow for hybrid activity.

- Understand and explain how a typical sequential manual transmission works.

- List some examples of the use of sequential transmissions.

- Understand and explain the basic operation of a dual-clutch transmission.

- List some examples of a dual-clutch transmission.

INTRODUCTION

Through the years, transmissions have been the focus of much research and development. This is true not only for automatic transmissions but also for manual transmissions. One variant of transmission design is a continuously variable transmission (CVT), which does not directly fall into the category of an automatic or a manual transmission. These transmission designs, as well as the transmissions used in most hybrid vehicles, are discussed in this chapter. Also included are sequential and dual shaft (or clutch) transmissions.

CONTINUOUSLY VARIABLE TRANSMISSIONS

A CVT may be regarded as an automatic transmission in that the driver does nothing to change forward gear ratios. However, the design is also rather more mechanical than most automatic transmissions; therefore, it is sort of a manual transmission. A CVT can automatically select any desired drive ratio within its operating range. It automatically and continuously selects the best overall ratio for the operating conditions. During the drive ratio changes, there is no perceptible shift unless a manufacturer engineers individual shift feel into the CVT operating program. The controls of this type of transmission attempt to keep the engine operating at its most efficient speed. This decreases fuel consumption and exhaust emissions. When driving a vehicle with a typical CVT, shifts are unnoticeable and the transmissions are referred to as "stepless" or "nonstepped."

During maximum acceleration, the drive ratio is adjusted to maintain peak engine horsepower. At a constant vehicle speed, the drive ratio is set to obtain maximum fuel mileage while maintaining good driveability.

In a typical CVT, one pulley is the driven member and the other is the drive member (Figure 11-1). Each pulley has a moveable face and a fixed face. When the moveable face

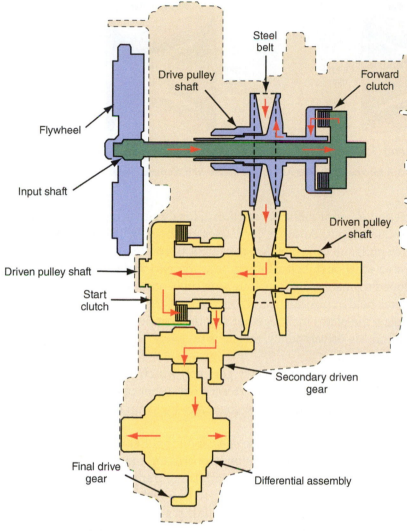

FIGURE 11-1 Components of a CVT transmission.

Labels on Figure 11-1:
Steel belt · Drive pulley shaft · Forward clutch · Flywheel · Input shaft · Driven pulley shaft · Driven pulley shaft · Start clutch · Secondary driven gear · Final drive gear · Differential assembly

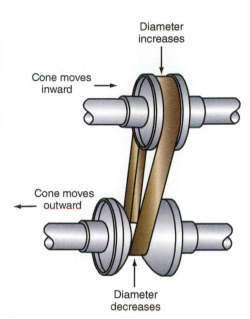

FIGURE 11-2 The action of the pulleys in a CVT.

Labels on Figure 11-2:
Diameter increases · Cone moves inward · Cone moves outward · Diameter decreases

moves, the effective diameter of the pulley changes (Figure 11-2). The change in effective diameter changes the effective pulley (gear) ratio. A steel push belt transfers power between the driven and drive pulleys.

Many late-model CVTs are equipped with a feature that stimulates the activity of a manual shifting automatic transmission. Transmissions with this feature are typically called stepped CVTs. They have five or six predetermined areas that the pulleys stop in when the driver selects manual control. While in manual, these steps provide the feel and shift effect of distinct gear ratios.

To change the direction of the vehicle, two wet clutches are used, one for forward and one for reverse. A planetary gear set is used in conjunction with the clutches to change direction. These clutches are applied when the shifter is placed into drive or reverse.

CVT Designs

Several designs have been used over the years. The torodial design uses a moveable roller between two curved metal plates instead of a chain and pulley setup. One plate is the input and the other plate is the output. By changing the angle and position where the roller touches the plates, the drive ratio is varied (Figure 11-3). Actual transmission of power is through a film of special oil that separates the roller and plates. This film is less than $\frac{1}{1,000}$ mm thick. The toroidal CVT has found many applications in the outdoor power equipment market, but

A BIT OF HISTORY

The basic design of CVTs is based on Leonardo da Vinci's concept from the fifteenth century. Hubertus Van Doone took the concept and added pulleys and belts to successfully create a working CVT.

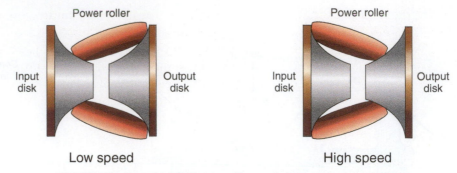

Low speed   High speed

FIGURE 11-3 A toroidal CVT uses rollers and plates with a special fluid.

only Nissan has used this design in automobile production. This design is typically used with engines that have a high torque output. Another CVT uses a rubber belt that connects the two pulleys. The drive pulley pulls the driven pulley, causing it to rotate.

Nissan's latest CVT uses a steel push belt running between two variable-width pulleys. This technology was developed to enable CVTs to mate with high-output engines. The belt is made of a series of small plates held in position by a cable. When torque is applied to the belt as it rotates off the drive pulley, the plates lock together and the belt becomes a solid link. As the belt begins to rotate around the driven pulley, torque is no longer applied to that section of the belt and the belt becomes flexible.

This CVT has one clutch and a simple planetary gear set, which are used to allow for neutral and reverse. Nissan has added a torque converter for smoother operation at very low speeds. Once the vehicle begins to move, the torque converter locks up to allow the CVT's belt and pulleys to provide all of the drive ratios.

To help in the activity of the belt, these CVTs use special oil that helps lock the steel belt to the pulleys and lubricates and cools the transmission. Controlling the temperature of the fluid is important to this CVT; therefore, it is equipped with three transmission coolers.

Audi's stepless Multitronic CVT (Figure 11-4) is based on two variable pulleys and a steel belt. However, the chain-like belt is not a conventional multiplate chain with a free pivot pin design. Instead, this CVT uses adjacent rows of plates linked together with cradle-type pressure pieces (oval-shaped pins). The cradle-type pressure pieces are jammed between the

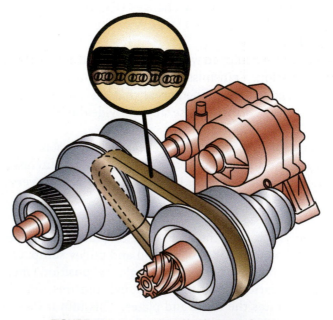

FIGURE 11-4 An Audi Multitronic transmission.

tapered sides of the pulleys as they are pressed together. Torque is transmitted only by the frictional force between the ends of the cradle pieces and the contact faces of the pulleys.

Some CVTs are based on planetary gear sets and not pulleys. These are discussed later, as they are mostly used in hybrid vehicles. However, some hybrids, such as those made by Honda, use a conventional CVT with the addition of an electric motor sandwiched between the engine and transmission.

CVT CONTROLS

The size of the pulleys is electronically controlled to select the best overall drive ratio based on throttle position, vehicle speed, and engine speed. Different speed ratios are available any time the vehicle is moving. Since the size of the drive and driven pulleys can vary greatly, vehicle loads and speeds can be responded to without changing the engine's speed.

The electronic control system for a typical CVT consists of a TCM, various sensors, linear solenoids, and an inhibitor solenoid. Input from the various sensors determines which linear solenoid the TCM will activate (Figure 11-5). Activating the shift control solenoid changes the shift control valve pressure, causing the shift valve to move. This changes the pressures applied to the driven and drive pulleys, which changes the effective pulley ratio.

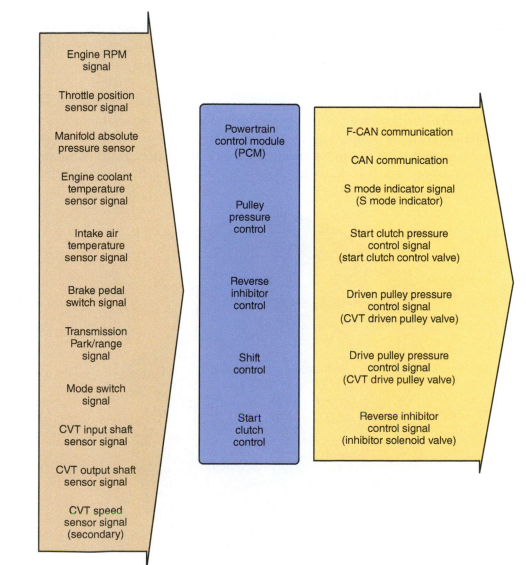

FIGURE 11-5 The electronic control system for a Honda CVT.

The start clutch allows for smooth starting. Since this transaxle does not have a torque converter, the start clutch is designed to slip just enough to get the car moving without stalling or straining the engine. The slippage is controlled by the hydraulic pressure applied to the start clutch. To compensate for engine loads, the TCM monitors the engine's vacuum and compares it to the measured vacuum of the engine while the transaxle was in park or neutral.

Activating the start clutch control solenoid moves the start clutch valve. This valve allows or disallows pressure to the start clutch assembly. When pressure is applied to the clutch, power is transmitted from the pulleys to the final drive gear set.

If the car is continuously driven at full-throttle acceleration, the TCM causes an increase in pulley ratio. This reduces engine speed and maintains normal engine temperature while not adversely affecting acceleration. After the car has been driven at a lower speed or not accelerated for a while, the TCM lowers the pulley ratio. When the gear selector is placed into reverse, the TCM sends a signal to the PCM. The PCM then turns off the car's air conditioning and causes a slight increase in engine speed.

HYBRID TRANSMISSIONS

Most hybrids are equipped with automatic shifting transmissions. This section is included to show how most hybrid transmissions are unique and are not like typical automatic or manual transmissions.

Perhaps the most complex transmissions are those used in most hybrid vehicles. The transmissions are fitted with electric motors that not only help propel the vehicle but also provide a constantly variable drive ratio. These CVTs do not rely on belts and pulleys; rather, it is the electric motors that change the drive gear ratios. It is important to note that some hybrid vehicles rely on conventional CVTs and conventional manual or automatic transmissions. This section covers the common nontraditional hybrid transmissions.

Honda Hybrid Models

A modified version of the five-speed automatic transaxle or CVT is used in Honda hybrid vehicles. The transaxles are more compact so they can fit behind the Integrated Motor Assist (IMA) electric motor mounted at the rear of the engine (Figure 11-6) and occupy the same amount of space as the transaxle in a nonhybrid vehicle. The five-speed unit is fitted with an electric oil

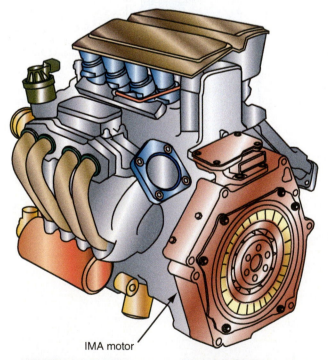

IMA motor

FIGURE 11-6 The IMA motor in Honda hybrids is mounted between the engine and the transaxle.

pump and different gear ratios to provide for better acceleration, fuel economy, and regenerative braking. These transaxles operate in the same way as other Honda units.

Toyota and Lexus Hybrids

The power-split device used in Toyota and Lexus hybrids operates as a continuously variable transaxle, although it does not use the belts and pulleys. The variability of this transaxle depends on the action of a motor/generator, referred to as MG1, and the torque supplied by another motor/generator, referred to as MG2, and/or the engine. The transaxle assembly contains:

- Differential assembly
- Reduction unit
- Motor generator 1 (MG1)
- Motor generator 2 (MG2)
- Transaxle damper
- Planetary gear set

A conventional differential unit is used to allow for good handling when the car is making a turn. The reduction unit increases the final drive ratio so that ample torque is available to the drive wheels. MG1, which generates energy and serves as the engine's starter, is connected to the planetary gear set. So is MG2, which is also connected to the differential unit by a drive chain. This transaxle does not have a torque converter or clutch. Rather, a damper is used to cushion engine vibration and the power surges that result from the sudden engagement of power to the transaxle.

The engine, MG1, and MG2 are mechanically connected at the planetary gear set. The gear set can transfer power between the engine, MG1, MG2, and drive wheels in nearly any combination of these. The unit splits power from the engine to different paths: to drive MG1, drive the car's wheels, or both. MG2 can drive the wheels or be driven by them.

Figure 11-7 summarizes the results of various gear combinations in a simple planetary gear set.

In the **power-split device**, the sun gear is attached to MG1. The ring gear is connected to MG2 and the final drive unit in the transaxle. The planetary carrier is connected to the engine's output shaft (Figure 11-8). The key to understanding how this system splits power is to realize that when there are two sources of input power, they rotate in the same direction but not at the same speed. Therefore, one can assist the rotation of the other, slow down the rotation of the other, or work together. Also, keep in mind the rotational speed of MG2 largely depends on the power generated by MG1. Therefore, MG1 basically controls the CVT function of the transaxle.

> Using a motor as a generator or vice versa is accomplished by changing the electrical connections to the rotor and stator.

Sun Gear	Carrier	Ring Gear	Speed	Torque	Direction
1. Input	Output	Held	Maximum reduction	Increase	Same as input
2. Held	Output	Input	Minimum reduction	Increase	Same as input
3. Output	Input	Held	Maximum increase	Reduction	Same as input
4. Held	Input	Output	Minimum increase	Reduction	Same as input
5. Input	Held	Output	Reduction	Increase	Reverse of input
6. Output	Held	Input	Increase	Reduction	Reverse of input
7. When any two members are held together, speed and direction are the same as input. Direct 1:1 drive occurs					
8. When no member is held or locked together, output cannot occur. The result is a neutral condition.					

FIGURE 11-7 The basic laws of simple planetary gear operation.

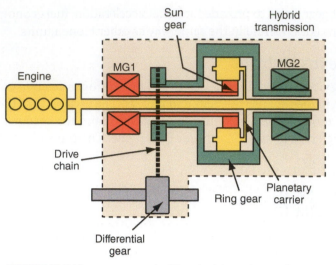

FIGURE 11-8 The arrangement of the electric motors and planetary gears in Toyota's power-split transaxle.

Here is a summary of the action of the planetary gear set during different operating conditions.

- To start the engine, MG1 is energized and the sun gear becomes the drive member of the gear set (Figure 11-9). Current is sent to MG2 to lock or hold the ring gear. The carrier is driven by the sun gear and walks around the inside of the ring gear to crank the engine at a speed higher than that of the sun gear.
- After the engine is started, MG1 becomes a generator. The ring gear remains locked by MG2 and the carrier now drives the sun gear, which spins MG1.
- When the car is driven solely by MG2 (Figure 11-10), the carrier is held because the engine is not running. MG2 rotates the ring gear and drives the sun gear in the opposite direction. This causes MG1 to spin in the opposite direction without generating electricity.
- If more torque is needed while operating on MG2 only, MG1 is activated to start the engine. There are now two inputs to the gear set, the ring gear (MG2) and the sun gear (MG1). The carrier is driven by the sun gear and walks around the inside of the rotating ring gear. This cranks the engine at a faster speed than when the ring gear is held.

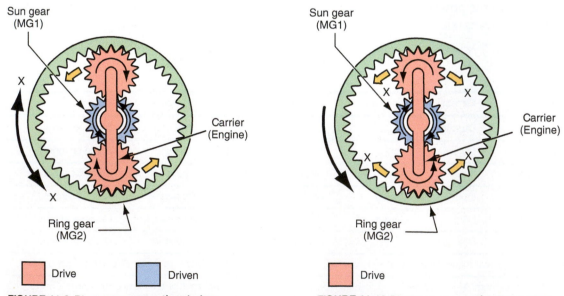

FIGURE 11-9 Planetary gear action during engine startup.

FIGURE 11-10 Planetary gear action when MG2 is propelling the vehicle.

- After the engine is started, MG2 continues to rotate the ring gear and the engine rotates the carrier to drive the sun gear and MG1, which is now a generator.
- When the car is operating under light acceleration and the engine is running, some engine power is used to drive the sun gear and MG1 and the rest is sent to the ring gear to move the car (Figure 11-11). The energy produced by MG1 is fed to MG2. MG2 is also rotating the ring gear, and the power of the engine and MG2 combine to move the vehicle.
- This condition continues until the load on the engine or the condition of the battery changes. When the load decreases, such as during low-speed cruising, the HV ECU increases the generation ability of MG1, which now supplies more energy to MG2. The increased power at the ring gear allows the engine to do less work while driving the car's wheels and to do more work driving the sun gear and MG1.
- During full-throttle acceleration, battery power is sent to MG2, in addition to the power generated by MG1. This additional electrical energy allows MG2 to produce more torque. This torque is added to the high output of the engine at the carrier.
- When the car is decelerating and the transmission is in DRIVE, the engine is shut off, which effectively holds the carrier (Figure 11-12). MG2 is now driven by the wheels and acts as a generator to charge the battery pack. The sun gear rotates in the opposite direction and MG1 does not generate electricity. If the car is decelerating from a high speed, the engine is kept running to prevent damage to the gear set. The engine, however, merely keeps the carrier rotating within the ring gear.
- When the car is decelerating and the transmission is moved into the B range, MG2 acts as a generator to charge the battery pack and to supply energy to MG1. MG1 rotates the engine, which is not running at this time, to offer some engine braking.
- During normal deceleration with the brake pedal depressed, the engine is off and the skid-control ECU calculates the required amount of regenerative brake force and sends a signal to the HV ECU. The HV ECU, in turn, controls the generative action of MG2 to provide a load on the ring gear. This load helps to slow down and stop the car. The hydraulic brake system does the rest of the braking.
- When reverse gear is selected, only MG2 powers the car. MG2 and the ring gear rotate in the reverse direction. Since the engine is not running, the carrier is effectively being held. The sun gear is rotating in its normal rotational direction but slowly and MG1 is not acting as a generator. Therefore, the only load on MG2 is the drive wheels.

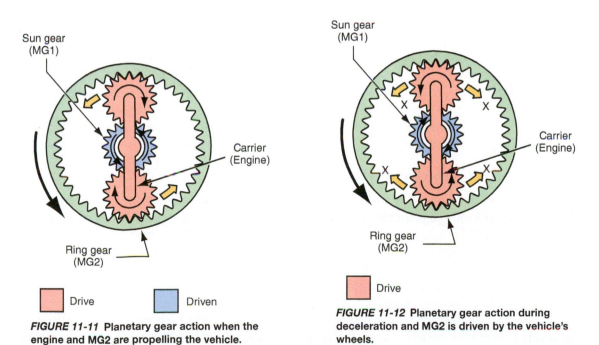

FIGURE 11-11 Planetary gear action when the engine and MG2 are propelling the vehicle.

FIGURE 11-12 Planetary gear action during deceleration and MG2 is driven by the vehicle's wheels.

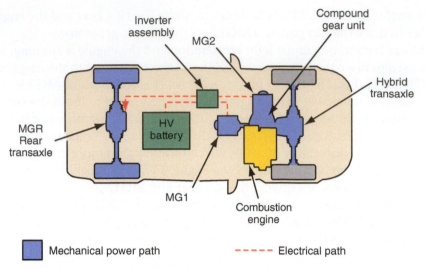

FIGURE 11-13 The layout of components for a Toyota hybrid 4WD vehicle.

It is important to remember that at any time the car is powered only by MG2, the engine may be started to correct an unsatisfactory condition, such as low-battery SOC, high-battery temperature, and heavy electrical loads.

4WD Hybrids. On Toyota and Lexus hybrid SUVs, the front transaxle assembly has been modified to include a speed reduction unit. This unit is a planetary gear set coupled to the power-split planetary gear set. This compound gear set has a common or shared ring gear that drives the vehicle's wheels. The sun gear of the power-split unit is driven by MG1 and the carrier is driven by the engine. In the reduction gear set, the carrier is held and the sun gear is driven by MG2. Because the sun gear is driving a larger gear, the ring gear, its output speed is reduced and its torque output is increased proportionally. High torque is available because MG2 can rotate at very high speeds. The rotational speed of MG1 essentially controls the overall gear ratio of the transaxle. The torque of the engine and MG2 flows to a common ring gear and to the final drive gear and differential unit.

The transaxle has three distinct shafts: a main shaft that turns with MG1, MG2, and the compound gear unit; a shaft for the counter-driven gear and final gear; and a third shaft for the differential. Since a clutch or torque converter is not used, a coil spring damper is used to absorb torque shocks from the engine and the initiation of MG2 to the driveline.

At the rear axle, an additional motor/generator (MGR) is placed in its own transaxle assembly to rotate the rear drive wheels. Unlike conventional four-wheel-drive (4WD) vehicles, there is no physical connection between the front and rear axles (Figure 11-13). The aluminum housing of the rear transaxle contains the MGR, a counter-drive gear, counter-driven gear, and a differential. The unit has three shafts: MGR and the counter-drive gear are located on the main shaft (MGR drives the counter drive gear), the counter-driven gear and the differential drive pinion gear are located on the second shaft, and the third shaft holds the differential.

Ford Motor Company Hybrids

Ford's hybrids are equipped with an electronically controlled continuously variable transmission. Based on a simple planetary gear set, like in a Toyota, the overall gear ratios are determined by the MGR. Ford's transaxle is different in construction from that found in a Toyota. In a Ford transaxle, the traction motor is not directly connected to the ring gear of the gear set. Rather, it is connected to a transfer gear assembly (Figure 11-14). The transfer gear assembly has three gears, one connected to the ring gear of the planetary set, a counter gear, and the drive gear of the traction motor.

The effective gear ratios are determined by the speed of the members in the planetary gear set. This means the speed of the MGR, engine, and traction motor determines the torque

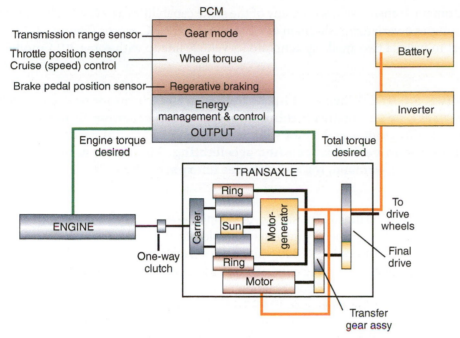

FIGURE 11-14 Layout of the Ford Escape hybrid powertrain.

that moves to the final drive unit in the transaxle. The operation of these three is controlled by the TCM. Based on information from a variety of inputs, the TCM calculates the amount of torque required for the current operating conditions. A MGR control unit then sends commands to the inverter. The inverter, in turn, sends phased AC to the stator of the motors. The timing of the phased AC is critical to the operation of the motors, as is the amount of voltage applied to each stator winding.

Angle sensors (resolvers) at the motors' stator track the position of the rotor within the stator. The signals from the resolvers are also used for the calculation of rotor speed. These calculations are shared with other control modules through CAN communications. The TCM also monitors the temperature of the inverter and transaxle fluid.

4WD Hybrids. Unlike Toyotas with 4WD, the Escape and Mariner do not have a separate motor to drive the rear wheels. Rather, these wheels are driven in a conventional way with a transfer case, rear drive shaft, and a rear axle assembly. This 4WD system is fully automatic and has a computer-controlled clutch that engages the rear axle when traction and power at the rear is needed. The system relies on inputs from sensors located at each wheel and the accelerator pedal. By monitoring these inputs, the control unit can predict and react to wheel slippage. It can also make adjustments to torque distribution when the vehicle is making a tight turn; this eliminates any driveline shutter that can occur when a 4WD vehicle is making a turn.

The acronym "CAN" stands for **controller area network**. CAN is a network protocol used to interconnect a network of electronic control modules.

Two-Mode GM, DCX, and BMW Hybrids

A two-mode hybrid vehicle can rely on electric motor power, engine power, or both. At low speed and light load, the vehicle can move using only the electric motor(s), only the internal combustion engine (ICE), or a combination of the two. In the second mode at higher speeds and loads, the ICE is always running, and electric motor assistance is brought in as controlled by the system. GM, BMW, and DaimlerChrysler have developed a two-mode full hybrid system that can be used with gasoline or diesel engines. The system can be used with gasoline or diesel engines, and with a variety of different transmissions. The **two-mode hybrid system** relies on advanced hybrid, transmission, and electronic technologies to improve fuel economy and overall vehicle performance. It is claimed that the fuel consumption of a full-size car, truck, or SUV is decreased up to 25 percent when it is equipped with this hybrid system. This transmission is an automatic transmission with many of the same capabilities as a CVT. The current Nissan unit combines a variator/push-chain CVT with a 20 HP AC motor (Figure 11-15). Honda is the latest to bring a two-mode system into its vehicle lineup with some 2014 models.

> **AUTHOR'S NOTE:** When we think of adding an electric motor or two to a vehicle's powertrain, we automatically think of increased fuel economy. But drivers of Mercedes, BMW, Dodge, and other vehicles with electric motor assist find a surprising increase in performance when accelerating. The motor assist adds instant horsepower on demand, for a sometimes unexpected benefit!

The GM/BMW/DaimlerChrysler co-developed system fits into a standard transmission housing and has three planetary gear sets coupled to two electronically controlled electric motors, which are powered by a 300-volt battery pack (Figure 11-16). Both motors can be used to help accelerate the vehicle, or can be switched to generator mode to help recharge the battery (Figure 11-17). This combination results in four forward speeds plus continuously variable gear ratios at low speeds and MGRs for hybrid operation.

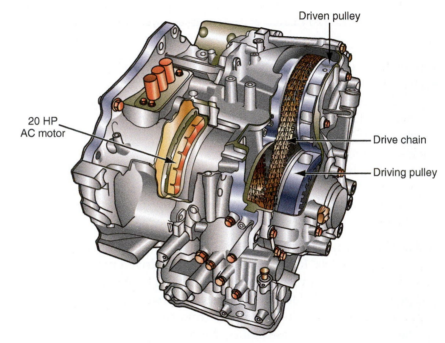

FIGURE 11-15 The Nissan hybrid transaxle uses a chain CVT and a 20 HP AC motor.

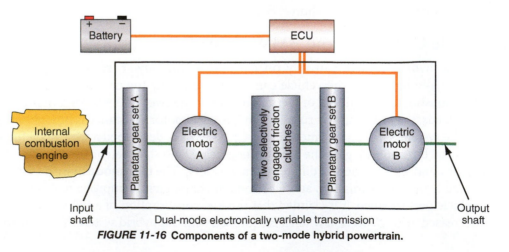

FIGURE 11-16 Components of a two-mode hybrid powertrain.

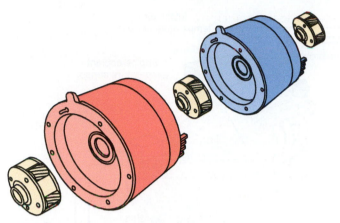

FIGURE 11-17 The electric motors and planetary gear sets in a GM, DaimlerChrysler, and BMW two-mode hybrid transmission.

SEQUENTIAL TRANSMISSIONS

Sequential manual transmissions (SMTs) are currently available on some passenger cars and are used in Formula One and other race cars. These transmissions work like typical manual transmissions except electronic or hydraulic actuators shift the gears and work the clutch (Figure 11-18). They can be set in automatic mode where the software and hardware does the shifting, or in a manual mode that allows the driver to shift sequentially, as there is no way to skip gears. The driver shifts the gears using buttons or paddles on the steering wheel and/or a shifter in the console.

Nearly all SMTs are manual transmissions with a computer-controlled actuator connected to the shift forks and a clutch actuator (Figure 11-19). When the automatic mode is selected, the computer shifts the transmission automatically at the correct time, in the correct sequence, and activates the clutch to allow for precise shifting. One type of sequential shifting uses a planetary gear set like an automatic transmission.

The driver can control gear changes by using the shifting mechanism. There is no gearshift linkage or cable; instead, a sensor at the shifter sends a signal to the computer. The computer, in turn, commands the actuators to engage or disengage the clutch and the gears with very fast response times. Engine torque is controlled during the shift by directly controlling the throttle or ignition/fuel injection system to provide smooth shifts.

Most sequential transmissions feature speed-matching downshifts. The system orders slight little blips of the throttle to decrease the wear on the transmission's synchronizers, as well as the entire drivetrain.

Operation

In many cars equipped with a SMT, the shift lever is pulled backward to upshift or pushed forward to downshift. In Formula One cars, there are two paddles on the sides of the steering wheel, instead of a shift lever. The left paddle upshifts, while the right paddle downshifts.

Like a traditional manual transmission, a set of gear selector forks move collars to engage gears. However, the driver cannot select a particular gear unless it is the next one in the normal sequence. Cars with SMTs do not have a clutch pedal; the clutch is automatically engaged and disengaged.

Typically, the movement of the shifter moves a ratcheting drum one complete increment. Each increment is assigned a particular speed gear. The drum has grooves cut into it that either move the shift control rods or move the gear selector forks directly (Figure 11-20). Therefore, the rotation of the drum causes a change in gears. The drum can be rotated by the shifter, or solenoids, pneumatics, or hydraulics that are activated electronically.

Ferrari Transmissions. The most famous application of a sequential transmission in road cars is their use in some Ferraris since the mid-1990s. These systems are based on a traditional

It is important to realize that SMTs are not automatic transmissions with manual controls! An example of this type of transmission is the Porsche Tiptronic transmission. This may duplicate the shift-lever motion of a sequential transmission, but it is an automatic transmission and has the torque converter and usually does not shift as quickly.

A BIT OF HISTORY

In 1989, Ferrari used an automated manual transmission in their Formula One race car. Then in 1998, they introduced a paddle-shift automated manual transmission. Since then, all F1 cars are equipped with a design of a sequential manual transmission.

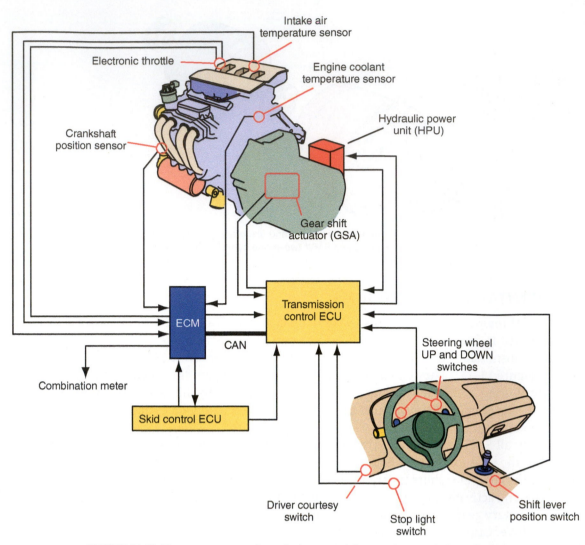

FIGURE 11-18 The components of a typical sequential manual transmission system.

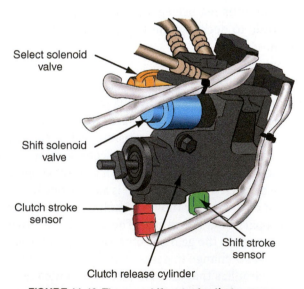

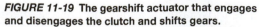

FIGURE 11-19 The gearshift actuator that engages
and disengages the clutch and shifts gears.

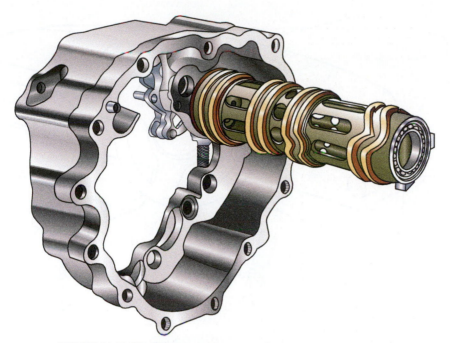

FIGURE 11-20 **The SMT ratcheting drum that moves the shift forks.**

six-speed manual transmission with the addition of an electronic clutch and a high-pressure hydraulic shift actuator.

The transmission has three different operating modes. In the fully automatic mode, the computer makes gear changes automatically according to engine speed, load, and throttle position. In the sport mode, gear selection is done by the driver via paddles mounted on the steering column. There is one paddle for upshifts and another for downshifts. During hard acceleration, upshifts will be made at over 8,000 rpm and shifting will be completed in less than 0.15 seconds. The other operating mode is a medium semi-automatic mode. While in this mode, the transmission will change gears only at 6,000 rpm. This provides slower acceleration but the gear changes are smoother.

BMW M-Sequential (SMG/SMG II). BMW's **sequential M gearbox (SMG)** is six-speed transmission that uses separate electronically controlled actuators on each of the three selector rails. The SMG does not have a clutch pedal but it does have a clutch. The transmission has both automatic and manual shift modes. Within the modes are different programs, with six settings to control the upshift/downshift speed for manual operation, and five settings for automatic mode. While operating in the S6 mode (the mode offering the best performance), gearshifts take place in 80 milliseconds.

Gear change is accomplished by paddles on the steering column or the gearshift lever. However, the actual shifting is completed by computers, solenoids, hydraulics, and linkages. Shift timing and quality is based on inputs from a variety of sensors (Figure 11-21), driver-selected programs, and the programmed logic in the control module. Basically, the control module interrupts the engine's power for just milliseconds and causes gear changes while it opens and closes the clutch.

An additional feature is the *climbing assistant.* This permits pulling away on forward slopes without rolling back. It can be used both in manual and automatic modes and for forward and reverse travel. All the driver needs to do is to press the brake pedal with the car stationary and to pull the rocker switch on the steering wheel for a short period. When the brake is released again, the car is ready to drive away within 2 seconds without first rolling away in an uncontrolled manner.

Another feature, the *acceleration assistant,* offers maximum acceleration. To activate this program, the driver selects the S6 driving program, pushes the selector lever forward, and keeps it in that position while the car is stationary. The driver now fully depresses the

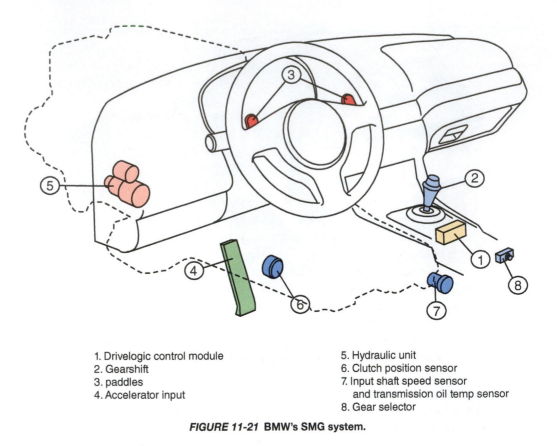

1. Drivelogic control module
2. Gearshift
3. paddles
4. Accelerator input
5. Hydraulic unit
6. Clutch position sensor
7. Input shaft speed sensor
   and transmission oil temp sensor
8. Gear selector

FIGURE 11-21 BMW's SMG system.

accelerator pedal and the optimum engine speed for maximum acceleration is automatically set. When the driver releases the selector lever, the car accelerates as quickly as possible with a minimum amount of wheel slip. This feature is now found on a variety of automated manual transmissions, and is typically called *launch control*.

DUAL-CLUTCH TRANSMISSION

Dual- or twin-clutch manual transmissions are the latest designs in transmissions. The dual clutch is essentially a fully automated, manual transmission with a computer-controlled clutch. This transmission can change gears faster than any other geared transmission. **Dual-clutch transmissions (DCTs)** were first put into production by Borg Warner. This transmission was called a "DualTronic" unit. Volkswagen licensed the technology and uses Borg Warner clutches and control modules for their dual-clutch transmissions. Dual-clutch transmissions initially appeared in the Audi TT 3.2 with the name **direct-shift gearbox (DSG)**. Since then, it has been available in several other Audi and Volkswagen models. In addition to Volkswagen and Audi, DCTs are now offered by BMW, Bugatti, Ford, Mitsubishi, Nissan, Porsche, and other manufacturers.

AUTHOR'S NOTE: The DCT transmission has been a staple of the high-performance market for years, but an examination of today's automotive offerings shows that the popularity of this concept has extended to vehicles at every level. From Volvos to the Honda Jazz to the Ford Fiesta, DCTs are everywhere due to their efficiency, relatively light weight, and reasonable cost. This new popularity has led to the use of the term "automatic" by some manufacturers when they refer to a DCT. They are free to use whatever description they feel necessary, but the technician must remember that inside the box are two manual transmissions in a single unit.

Due to the design of a dual-clutch transmission, gearshifts are made without an interruption of power flow. This results in improved performance and increased fuel efficiency. It is claimed that dual-clutch transmissions can provide a 4–12 percent improvement in fuel mileage. They also deliver more power and control than a traditional automatic transmission.

Although the unit relies on two separate clutches, there is no clutch pedal. The clutches are engaged and disengaged by electro-hydraulic actuators. The actual shifting of gears is completed by computer-controlled solenoids and hydraulics. Dual-clutch transmissions can be shifted either manually or automatically. In the automatic mode, the computer selects the proper gear for the conditions. To operate the unit manually, the driver changes gears with buttons, paddles, or a shifter. Most dual-clutch transmissions have a traditional P-R-N-D-S shift pattern. The transmission can shift automatically in either Normal (D) or Sport (S) modes. While in Normal mode, the DCT shifts into the higher gears early in order to minimize engine noise and maximize fuel economy. When Sport mode is selected, the transmission holds the lower gears longer for improved performance. The driver can also elect to operate the transmission manually by either sliding the shift lever to the side or pulling one of the paddles on the steering wheel.

This transmission is actually two separate transmissions, each with its own clutch and built as a single unit. One transmission contains the even-numbered gears, while the other contains the odd-numbered gears. In addition, the outer clutch drives the odd-numbered gears, while the inner clutch drives the even-numbered gears. For example, in a six-speed transmission, one clutch and transmission assembly is for first, third, and fifth gears; the other clutch and transmission is for second, fourth, and sixth gears. To make this work, a two-part input shaft is used. Again each part of the shaft works with the even or odd gears. The outer input shaft is hollow and has the inner shaft running through it (Figure 11-22).

Instantaneous shifting is accomplished by complex electronic controls that change the power-flow from one clutch and transmission assembly to the other, when a change in gears is ordered (Figure 11-23). The electronic controls also have a predictive feature that preselects the next gear in the sequence. To understand this concept, consider the power-flow through a DCT. When the vehicle is at a standstill and the driver begins to accelerate, transmission number 1 is in first gear while transmission 2 is in second. Clutch number 1 engages and the

The use of a DCT in the Bugatti Veyron EB 16.4 demonstrates that this transmission design can handle large amounts of engine power. The Veyron's W16 engine is rated at over 1,000 peak horsepower.

DCTs are currently available with five, six, or seven forward gears. All DCTs have two transmission assemblies each designated for odd or even-numbered gears. Reverse gear is typically tied to the odd shaft.

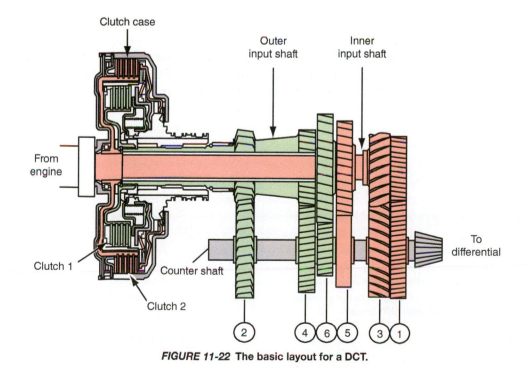

FIGURE 11-22 The basic layout for a DCT.

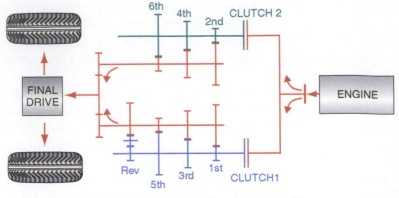

FIGURE 11-23 Power-flow through a DCT.

vehicle moves in first gear. When it is time to upshift, clutch 2 engages while clutch 1 disengages. Since second gear was already engaged, the vehicle is now operating in second gear. As soon as clutch 1 is disengaged, transmission 1 immediately shifts into third gear. This sequence continues when it is time to upshift again, clutch 1 engages and the vehicle is operating in third gear. At the same time, clutch 2 disengages and transmission 2 is shifted into fourth gear. In all cases, the preselected gear will not affect the performance of the operating gear because no input torque is reaching it.

As a result of these actions, torque is smoothly passed through without interruption. The preselection of gears allows for very quick gear changes. Most systems require only 8 milliseconds to complete an upshift. That is 10 times quicker than BMW's SMG II, which is the quickest sequential automated manual transmission currently available (Figure 11-24).

A DCT is also capable of skipping gears and will perform double-clutch downshifts when needed. Downshifting, however, is not as quick as upshifting. It takes about 600 milliseconds to complete a downshift. This is due to the time required for the throttle blip, which matches transmission speed to engine speed. Downshifting will also take longer if the driver skips some gears. This is especially true if the vehicle is operating in sixth gear and the driver wants to downshift into second. Second and sixth gears are on the same shaft and operate with same clutch. To provide this change, the transmission must first downshift into fifth gear because it is on the other shaft. Then the transmission can engage second gear and the associated clutch. Typically this sequence takes 900 milliseconds; this is not a long time but certainly longer than upshifting time.

Dual-clutch transmissions use either wet or dry clutch assemblies. These clutches are very similar; both have a compact multiplate design (Figure 11-25). Wet clutches, however, are immersed in oil and are similar to the clutches used in conventional automatic transmissions.

TRANSMISSION/VEHICLE	MINIMUM UPSHIFT TIME
DSG (VW, Audi, & Porsche)	8 ms
BMW SMG II (M3 E46)	80 ms
Ferrari F1 (Maserati 4200GT)	80 ms
Ferrari F1 (360 F1)	150 ms
Ferrari F1 (Enzo)	150 ms
Bugatti Veyron	200 ms
BMW SMG (M3 E36)	220 ms
Aston Martin Vanquish	250 ms

FIGURE 11-24 A list of some popular automated manual transmissions and their shift times.

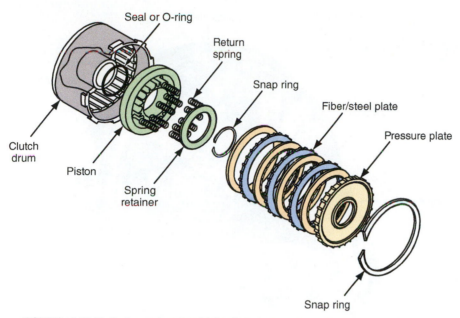

FIGURE 11-25 Both dry and wet multiple-disc clutches have the same construction.

A wet clutch comprises an inner drum connected to the gear set on the transmission's main shaft that will receive the engine's torque and an outer drum connected to the engine either by chain or gears. Both drums rotate in oil, and although they are on the same shaft, they rotate independent of each other. The two drums are linked together by clutch plates. The friction discs have internal teeth that mesh with splines in the clutch drum. Steel or fiber plates are inserted between pairs of the friction discs and are splined to the inside of the outer drum. The sets of plates are forced together by hydraulic pressure and spring tension. When the clutch is engaged, hydraulic pressure inside the clutch piston applies pressure onto the springs and the stacked clutch plates and friction discs are squeezed together against a fixed pressure plate. This action creates a linkage between the drums and they now rotate as a single unit. To disengage the clutch, the fluid pressure inside the clutch piston is reduced. This allows the springs to relax, and the force holding the clutch against the pressure plate is reduced. This action allows the drums to rotate independently, again.

The fluid in a wet clutch also serves to reduce friction and limit the production of heat, thereby improving overall performance while increasing the life of the clutch assembly. Wet clutches are capable of handling large amounts of torque.

Dry clutches are used in DCTs with engines of lower torque outputs. They are less expensive to produce and can deliver better fuel economy. The latter results from their high clamping pressures that minimize the power lost as torque flows from the engine to the transmission. The disadvantage of using a dry clutch is they will wear quickly, especially when faced with high torque demand.

Some manufacturers now use a single dry disc for each of the two input shafts (Figure 11-26). These designs are less expensive than multiple-disc setups, and are released and engaged by electromechanical actuators mounted in the bell housing.

Currently only a few vehicles offer a dual-clutch transmission. However, it is projected that the popularity of these transmissions will grow rapidly. Figure 11-22 shows how the market share for each type of transmission will change in 2014.

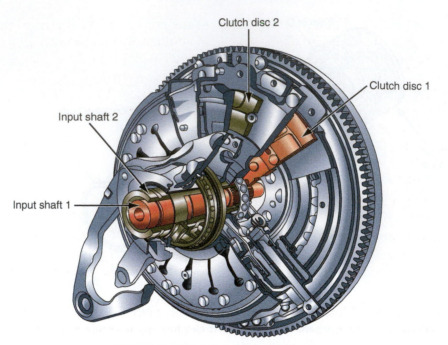

Clutch disc 2

Clutch disc 1

Input shaft 2

Input shaft 1

FIGURE 11-26 A dual dry plate clutch setup.

Volkswagen/Audi Direct Shift Gearbox

Currently, Volkswagen has more models available with a DCT than any other manufacturer. Their direct-shift gearbox (DSG) is a constant-mesh, six-speed manual transaxle with wet clutches. The DSG is also found in many Audis. They also have a dry clutch, seven-speed unit that is used only in small, low-powered vehicles.

The DSG is a dual-clutch transmission with two countershafts. All the odd-numbered gears are on one countershaft and the even-numbered gears are on the other. The inner input shaft meshes with the gears on one countershaft, and the outer input shaft meshes with the gears on the other countershaft. The DSG's electronic control unit (ECU) is mounted to the transaxle. The ECU has adaptive learning, which allows it to learn how the vehicle is typically driven and will attempt to match the shift points to the driver. The ECU controls a hydro-mechanical unit. Together these two units are called a "mechatronic" unit. The mechatronic control module is the most complex component of a DSG transaxle.

The transmission can be operated totally by the ECU or the driver can manually shift gears with a shift lever or steering wheel-mounted paddles. The ECU controls the clutches, input and output shafts, transaxle cooling, gear selection and timing, and hydraulic pressures. In addition to the various DSG-specific sensors, the ECU relies on bus data to determine the current operating conditions.

Operation in D Mode. During the D mode, all shifting is automatically completed and controlled by the ECU. Shifts times are set to obtain the best fuel economy for the conditions. This typically means shifts will occur at low speeds.

When the vehicle is in neutral and not moving, both clutches are fully disengaged and there is no input into the transmission. When the driver selects D, first gear is selected on the first shaft, and its clutch prepares to engage. At the same time, second gear is also selected, but its clutch pack remains fully disengaged. When the brake pedal is released, the computer engages the clutch for first gear and that input shaft sends torque to first gear.

As the vehicle accelerates, the computer determines when the transmission should shift into second gear. When the ideal conditions are met during intial acceleration, the second clutch is engaged while the first clutch is disengaged. Torque then flows from the engine to

The DSG is also called an S-Tronic transmission when used in late-model Audis.

the other input shaft and powers second gear. At the same time, first gear is immediately deselected, and third gear is preselected. When conditions are met to order an upshift, the first clutch engages and the second clutch disengages. The transmission is now operating in third gear. This sequence continues until the transaxle is in sixth gear.

During normal driving, the DSG will shift sequentially (1 to 2 to 3 to 4 to 5 to 6) and will reverse that sequence during downshifts. When the ECU determines the vehicle is slowing down or needs more power, it selects the appropriate gear and completes a downshift. When downshifting will occur and what gear the transmission will drop to, while in D, is determined by the ECU. However, if the vehicle is cruising at a constant speed and the driver presses the throttle to floor, the kickdown mode is activated. During this mode, the DSG can skip gears, such as dropping from sixth gear directly to third gear. The instrument panel displays the current selected gear range and the current operating gear of the transmission.

Operation in S Mode. When the driver selects the S mode, the transaxle remains to work automatically. However, shift times are altered so that all upshifts and downshifts occur at higher engine speeds. This allows for more spirited driving but has a negative effect on fuel economy.

Operation in Manual Mode. The driver can assume control over shifting by moving the shifter to the right after D has been selected. All gear changes are done sequentially; therefore, movement begins in first gear. To upshift, the lever is pushed forward, and to downshift the lever is pulled rearward. While in the manual mode, all shifts are controlled by the driver. There are exceptions to this. If the engine is close to reaching its peak speed or redline, the transmission will automatically upshift to protect the engine. Likewise, if engine speed drops to low, the transmission will automatically downshift. The driver can return to the automatic mode while driving by moving the shift lever to the left.

Some models have steering wheel-mounted paddle shifters. These shifters work in the same way as the console shifter but allow the driver to keep both hands on the steering wheel while shifting gears.

Mitsubishi's Sport Shift Transmission

Mitsubishi's twin-clutch sport shift transmission (SST) has six speeds, steering wheel shifter paddles, and a console shifter. The unit works like other DCTs. It offers three separate automatic shifting modes: normal, sport, or super sport. In the automatic mode, the system behaves much like a good automatic. The sport and super-sport modes hold each gear longer to allow for higher engine speeds and also change gears faster. When super sport is selected, the vehicle's launch control can be used, as long as stability control is disabled, the brake is depressed, and the engine is running at 5,200 rpm. The operating modes are selected by a small rocker switch located near the gear selector.

When the SST is operating in the manual mode, the transmission is under the total control of the driver. Unlike other units, it will not upshift or downshift unless the driver tells it to. The only time the transmission has a mind of its own is when the vehicle is brought to a complete stop. Then the transmission automatically drops to first gear.

Porsche Dual Klutch

Porsche was one of the early developers of a dual-clutch transmission. Recently, they introduced the Porsche Dual Klutch (PDK) transmission, which has seven speeds.

Like other DCTs, the drive unit is made up of two transmissions, two clutches, and a hydraulic control system (Figure 11-27). Two wet clutches are hydraulically controlled and use the oil for cooling and lubrication.

This transmission, which is actually a transaxle located at the rear wheels, can handle up to 333 ft.-lbs or 450 N·m of torque. This means it was designed for high-output engines. The PDK is capable of shifting 60 percent quicker than a typical automatic transmission. The result

The official name of the transmission is *Porsche-Doppelkupplung* (or Porsche double-clutch in English).

Audi and Porsche have been working on a dual-clutch transmission for many years, but until recently they were not deemed suitable for production. However, they did build race cars around the concept; in fact, in 1985, Audi made history when a Sport Quattro S1 rally car equipped with DCT won the Pikes Peak hill climb. In 1986, a Porsche 962 won the Monza 1,000-kilometer World Sports Prototype Championship race. This was the first win for a car equipped with the paddle-shifted PDK transmission. When Porsche dropped out of Group C racing in 1987, further development of the PDK was put on hold, until recently. The first DCT available in the United States was the 2004 Audi TT 3.2.

The Sports Chrono Plus package also includes a digital and analog timer, a performance display, sport and sport-plus buttons, and memory capabilities in the vehicle's communication management system.

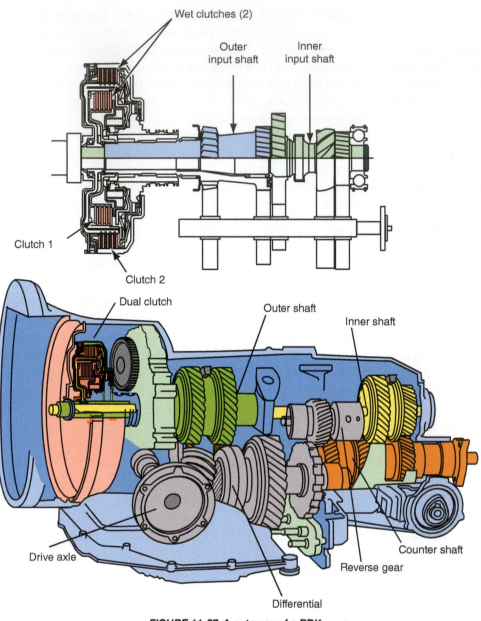

FIGURE 11-27 A cutaway of a PDK.

is drastically improved acceleration. Plus the PDK provides for a decrease in fuel consumption by as much as 15 percent. This is helped by the fact that seventh gear provides a great amount of overdrive. Top speed is achieved in sixth gear.

The transmission offers three automatic shifting modes: normal, sports, and sports plus. "Normal" controls all shifting to provide for good driveability while using the least amount of fuel. "Sports" delays the shifts and allows the engine to reach higher speeds before shifting. The "sports-plus" mode uses a motorsports-derived gearshift strategy, delaying the shifts even further and providing extremely quick gear changes. In this mode, seventh gear is not available.

The control unit relies on inputs available on the CAN bus, plus four distance sensors, two pressure sensors, and two engine-speed sensors to choose the correct gear and shift timing. Cars equipped with the optional Sports Chrono Plus package have a launch-control program that dumps the clutch automatically from 6,500 rpm, after the brake pedal is depressed, the throttle is opened completely, and the brake pedal is released. This allows the car to accelerate under full power without wheel slip or an interruption of traction. The PDK can also be operated in a full manual mode. While set in the manual mode, the transmission will not shift on its own unless engine speed exceeds a particular limit. It will also not downshift on its own unless engine speed drops too low. The system does include a kickdown switch that is activated by fully opening the throttle. This switch allows for instantaneous maximum acceleration when it is needed. An example of this is when cruising at a moderate speed in seventh gear and there is a need to quickly accelerate. When the switch is activated, the transmission will downshift into fourth, third, or second gear. When kickdown is activated, the transmission will not only downshift but it will hold the transmission in that gear until the engine redlines, then it will upshift one gear at a time as long as the throttle is fully depressed. Gear selection is done by steering wheel paddles or at the shifter.

BMW M3 Dual-Clutch

BMW's **M3 dual-clutch transmission (M-DCT)** (Figure 11-28) is fitted to a V8 that can deliver 295 ft.-lbs (400 N·m) of torque and has a red line of 8,400 rpm. The M-DCT is a seven-speed unit that is controlled by a computer system, called Drivelogic. All shifts are carried out with no interruption in power-flow. The seven speeds provide for small gear ratio changes, which provide a smooth flow of power resulting in a smooth rapid increase in vehicle speed. All gear changes are made quickly and firmly, but they are barely perceptible.

When compared with an automatic transmission connected to the same exact engine, the M-DCT provides better fuel economy and lower emissions. This is due to the seven gears and more positive connections between the engine and the speed gears.

Major Components. The transmission has wet clutches that are activated by an integrated hydraulic module. They rotate on a dual-mass flywheel that serves as the torsion damper. Lubrication for the transmission is supplied by a dry sump system. This enhances the reliability of the transmission and is maintenance free. The lubricating fluid is circulated through an oil cooler in the engine's radiator to keep it within its optimum temperature range. The oil also passes through an auxiliary oil/air cooler to prevent excessively high temperatures inside

BMW has been one of the few manufacturers that have used dual-mass flywheels consistently for many years.

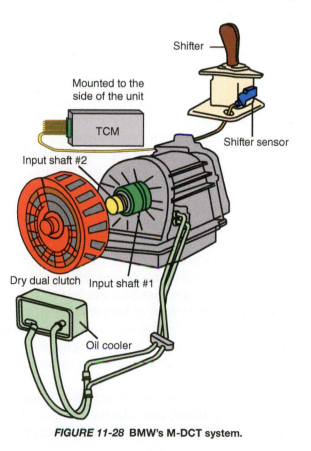

FIGURE 11-28 BMW's M-DCT system.

the transmission. Additionally, by passing the fluid through a cooler in the radiator, it can warm up quickly in cold weather. This helps reduce frictional losses while the transmission is warming up.

Operation. The transmission can be operated in an automatic or manual mode. Drivelogic provides for 11 separate operating modes: five shift programs are available in automatic and six in manual.

The automatic mode is designated as D or drive. In this mode, the transmission shifts automatically in sequence from first gear to seventh. The five available shift programs adjust the characteristics of the shifts according to the driver's desires and needs. When D1 is selected, the vehicle will initially move out in second gear and gear engagement is very soft and smooth. This range is best suited for accelerating on slippery surfaces. D2 provides the same shift quality but the vehicle begins movement in first gear. D3 through D5 provide progressively later and more positive shifts. In D5 and at full throttle, the transmission stays in one gear until the engine reaches its red line.

While operating a D range, the driver can influence shifting by slowly releasing pressure on the throttle until an upshift occurs. Moreover, the transmission has a kickdown feature that downshifts the transmission to provide maximum acceleration.

The manual or "S" (sport)-mode allows the driver to select the desired speed gear sequentially. The six manual operating modes determine the quality of the shifts. S1 provides very smooth and soft engagement, while S6 provides immediate full engagement of the gears.

To select one of the shift programs available in D and S, the shifter is placed in one of those two positions. Then the specific program is selected by depressing the Drivelogic button on the console. The driver is able to set the preferred setting into the system's memory so it does not need to be reset each time the car is driven. The selected shift program is displayed on the instrument panel along with the operating gear.

When driving in the S mode, kickdown can be activated with the shift lever or paddles. As soon as the driver fully opens the throttle and pulls the left-hand paddle or tips the gearshift lever forward, the transmission will downshift one or more gears to provide maximum acceleration.

In the S mode, a launch control feature is available. This feature allows the car to accelerate as quickly as possible with a minimum of tire slip. To initiate the launch control, the dynamic stability control (DSC) must be switched off. Then with the vehicle stopped and in S6, the shift lever is pushed and held toward the front of the vehicle. Once a starter flag symbol appears on the instrument panel, the driver fully depresses the throttle pedal to automatically reach the ideal engine speed for the launch. Slight movements of the cruise control lever can also raise or lower the engine's speed so it is exactly at the desired speed. The driver then releases pressure on the shift lever. The car will accelerate to the best of its ability all the way to its top speed, if the driver desires. When the desired speed is reached, all the driver needs to do is back off the throttle or upshift manually.

Downshifting can also be done manually or automatically. During downshifting, the control unit slowly and smoothly engages the clutches. This prevents an abrupt slowing of the rear wheels, which can cause wheel slip and tire and drivetrain problems.

The ECU determines the operation of the engine and transmission based on a variety of inputs from the CAN bus. The ECU also has adaptive learning. It not only recognizes the habits of the driver but also the current road conditions. It will adapt the shift timing and quality in response to sudden braking, hard cornering, and hard acceleration. When the vehicle seems to be out of control, the system may disengage the currently active clutch to prevent the rear of the car from swerving around due to engine drag forces suddenly acting on the wheels.

The ECU will also adjust shift timing according to the angle of the road's surface. The system monitors the road's gradient and adjusts the shift points accordingly. While moving downhill, the system operates in the lower gears to capitalize on engine braking.

FIGURE 11-29 The locations of the gearshift and shifting paddles.

While traveling uphill, the choice of gear is dictated by the load the gradient is putting on the engine.

The transmission also has a low-speed assistant feature that helps move the vehicle at slow speeds, such as during parking. This feature limits the amount of torque available to the wheels to prevent the vehicle from lurching when the throttle is opened slightly. The system also automatically engages the parking lock in the transmission when the ignition has been turned off.

Shifting. To select the D or S mode and to shift into reverse, the driver can move the shift lever or move one of the shift paddles (Figure 11-29). While in S, upshifts are selected by moving the shift lever toward the rear of the car. To downshift, the lever is pushed forward. The paddle on the right is for upshifting and the paddle on the left is for downshifting.

Shift Lights. The gear indicator always shows the current operating gear. It always displays the selected shift program. The desired engine's speed for upshifts and maximum acceleration is displayed by eight LEDs in the tachometer. These lights allow the driver to determine the ideal time to shift gears while in the S mode. Six of the LEDs are yellow and they light up progressively as engine speed increases. When the engine reaches the optimum speed for shifting, two red LEDS will light. If the engine has reached or exceeded its redline, the lights will flash.

SUMMARY

- To control the ratio of the pulleys, the electronic control system for Honda's CVT relies on a TCM, various sensors, linear solenoids, and an inhibitor solenoid.
- The power-split device used in Toyota and Lexus hybrids operates as a CVT and its operation depends on the action of MGRs and the engine.
- The controlling unit in the power-split device is a planetary gear set.
- Some 4WD hybrid vehicles have a separate electric motor at the rear wheels and there is no mechanical connection between the front and rear drive axles.
- The transaxle assembly used in Ford hybrids is similar to that used in Toyotas, but it is not entirely the same design.
- Sequential manual transmissions (SMT) work like typical manual transmissions, except electronic or hydraulic actuators shift the gears and work the clutch.
- Many SMTs have a ratcheting drum that moves with the shift linkage. The drum moves the shift fork to select the gear.
- A dual-clutch transmission (DCT) is a fully automated, manual transmission with two computer-controlled clutches.

TERMS TO KNOW

Controller area network (CAN)

Direct-shift gearbox (DSG)

Dual-clutch transmission (DCT)

M3 dual-clutch transmission (M-DCT)

Power-split device

Sequential M gearbox (SMG)

SUMMARY

TERMS TO KNOW

(continued)

Sequential manual transmission (SMT)

Two-mode hybrid system

- A DCT is actually two separate transmissions, each with its own clutch but built as a single unit. One transmission contains the even-numbered gears, while the other contains the odd-numbered gears. In addition, the outer clutch drives the odd-numbered gears, while the inner clutch drives the even-numbered gears.
- Dual-clutch transmissions use either wet or dry clutch assemblies. These clutches are very similar; both have a compact multiplate design. Wet clutches, however, are immersed in oil and are better suited for high-output engines.
- Most SMTs and DCTs can be operated manually or can be set to work automatically.

REVIEW QUESTIONS

Short Answer Essays

1. What is the basic difference between the CVT used in nonhybrid Honda vehicles and hybrid vehicles?

2. True or False: Volkswagen has DCTs as an option in many different models.

3. Describe a major difference between a sequential manual transmission and a sequential automatic transmission.

4. In a Honda variator/push-belt CVT, how do the gear ratios change?

5. Briefly describe how a double-clutch transmission works.

6. How many different modes of operation does Porsche's PDK offer? What is the difference between them?

7. What are the major parts of a Toyota hybrid transaxle?

8. Explain what happens when a Toyota hybrid is running on electric power only and there is a need for more torque.

9. Most sequential transmissions feature speed matching downshifts. How and why do they do that?

10. Describe how a gear change occurs inside a SMT.

Fill-in-the-Blanks

1. The transmission used in a GM two-mode full hybrid system is basically _____ planetary gear sets coupled to _____ electric motors, which are powered by a 300-volt battery pack.

2. When a DCT-equipped vehicle is at a standstill and the driver begins to accelerate, transmission number _____ is in first gear while transmission number _____ is in second. Clutch number 1 engages and the vehicle moves in _____ gear. When it is time to upshift, clutch number _____ engages, while clutch number _____ disengages. Since second gear was already engaged, the vehicle is now operating in second gear. As soon as clutch _____ is disengaged, transmission number _____ immediately shifts into third gear.

3. The start clutch in a CVT replaces the _____ _____. It is designed to slip just enough to move the vehicle without stalling or straining the engine. The slippage is controlled by the _____ _____ applied to the start clutch.

4. The ECU for a DSG controls the clutches, input and output shafts, transaxle cooling, gear selection and timing, and hydraulic pressures. It relies on _____ modulation valves, _____ shift valves, and numerous other valves to control the transaxle's operation, and in addition to the various DSG-specific sensors, the ECU relies on _____ _____ to determine the current operating conditions.

5. When a Toyota hybrid is operating under full-throttle acceleration, battery power is sent to _____, in addition to the power generated by _____. This additional electrical energy allows _____ to produce more torque. This torque is added to the high output of the engine at the _____ of the planetary gear set.

6. In a SMT-equipped car, there is no gearshift linkage or cable; instead, a sensor at the _____ sends a signal to the computer. The computer, in turn, commands the _____ to engage or disengage the _____ and the _____. Engine torque is controlled during the shift by controlling the _____ or ignition/fuel injection system to provide smooth shifts.

7. In DCTs, both the wet and dry clutches have a compact _____-_____ design. Wet clutches are immersed in oil and are similar to the clutches used in conventional _____ _____.

8. Since gear selection in a SMT is based on the rotation of the _____ _____, all shifts are made in _____.

9. Due to the design of a dual-clutch transmission, gearshifts are made without an interruption of power-flow. This results in improved performance and increased fuel efficiency. It is claimed that dual-clutch transmissions can provide an improvement of up to _____ to _____ percent better fuel mileage, when compared with an automatic transmission with a torque converter.

10. The fluid used in a wet clutch serves to reduce _____ and limit the production of _____. Thereby it improves overall _____ while increasing the _____ of the clutch assembly.

ASE-Style Review Questions

1. While discussing speed matching downshifts,
   Technician A says this is done to decrease the wear on the transmission's synchronizers, as well as the entire drivetrain.
   Technician B says DCT and SMT systems order slight little blips of the throttle after the downshift has occurred.
   Who is correct?
   A. A only
   B. B only
   C. Both A and B
   D. Neither A nor B

2. While discussing the PDK,
   Technician A says this transaxle is designed to handle high engine torque and horsepower.
   Technician B says the control unit relies on inputs available on the CAN bus, plus four distance sensors, two pressure sensors, and two engine-speed sensors to choose the correct gear and shift timing.
   Who is correct?
   A. A only
   B. B only
   C. Both A and B
   D. Neither A nor B

3. While discussing Toyota's hybrid system,
   Technician A says in the power-split device, the sun gear is attached to MG1. The ring gear is connected to MG2 and the final drive unit in the transaxle and the planetary carrier is connected to the engine's output shaft.
   Technician B says the rotational speed of MG1 largely depends on the power generated by MG2. Therefore, MG2 basically controls the continuously variable transmission function of the transaxle.
   Who is correct?
   A. A only
   B. B only
   C. Both A and B
   D. Neither A nor B

4. While discussing dual-clutch transmissions,
   Technician A says they are fully automated, sequential manual transmissions with computer-controlled clutches.
   Technician B says they can change gears faster than any other geared transmission.
   Who is correct?
   A. A only
   B. B only
   C. Both A and B
   D. Neither A nor B

5. While discussing SMTs,
   Technician A says all shifts are made sequentially and there is no way to skip gears.
   Technician B says the driver shifts gears using buttons or paddles on the steering wheel and/or a shifter in the console. Some units can be operated in a fully automatic mode.
   Who is correct?
   A. A only
   B. B only
   C. Both A and B
   D. Neither A nor B

6. While discussing the clutches used in a DCT,
   Technician A says wet clutches are used in most DCT applications because they are less expensive to manufacture, but they tend to wear quickly.
   Technician B says dry clutches are used with all high-performance engines, but offer less fuel savings than a wet clutch.
   Who is correct?
   A. A only
   B. B only
   C. Both A and B
   D. Neither A nor B

7. While discussing shifting a SMT,

Technician A says a conventional shifter with a conventional shift pattern is used, but there is no clutch pedal. Most systems use hydraulics that move a shift fork in response to the shifter.

Technician B says most systems have a ratcheting drum that either moves the shift control rods or the gear selector forks directly.

Who is correct?

A. A only
B. B only
C. Both A and B
D. Neither A nor B

8. While discussing the operation of DCTs,

Technician A says during upshifts the transmission engages the next gear in response to the engagement of its clutch.

Technician B says the system has two engaged gears at one time and performance is hindered until the clutch is totally disengaged, but only one is connected to the output shaft at a time.

Who is correct?

A. A only
B. B only
C. Both A and B
D. Neither A nor B

9. While discussing the Volkswagen DSGs,

Technician A says in order to keep the transmission-mounted computer within a desired temperature range, a heat exchanger is bolted to the housing.

Technician B says the ECU has adaptive learning, which allows it to learn how the vehicle is typically driven and will attempt to correct the shift points ordered by the driver.

Who is correct?

A. A only
B. B only
C. Both A and B
D. Neither A nor B

10. While discussing the sequential M gearbox (SMG) system,

Technician A says it has a feature that permits pulling away on forward slopes without rolling back.

Technician B says it has a feature that provides for maximum acceleration from a standstill.

Who is correct?

A. A only
B. B only
C. Both A and B
D. Neither A nor B

GLOSSARY
GLOSARIO

Note: **Terms are highlighted in bold**, followed by Spanish translation in color.

Acceleration An increase in velocity or speed.

Aceleración Aumento de velocidad o celeridad.

Acid A compound that has an excess of H ions and breaks into hydrogen (H^+) ions and another compound when placed in an aqueous (water) solution.

Ácido Un compuesto que tiene un exceso de iones de H y se descompone en iones de hidrógeno (H^+) y otro compuesto cuando se lo coloca en una solución acuosa (agua).

Actuator A device that carries out the instructions of a computer. Normally these devices are solenoids or motors.

Actuador Dispositivo que cumple las instrucciones de una computadora. Generalmente estos dispositivos son solenoides o motores.

Alloy A mixture of different metals such as solder, which is an alloy consisting of lead and tin.

Aleación Una mezcla de diferentes metales como por ejemplo, la soldadura blanda, que es una aleación de plomo y estaño.

All-wheel drive (AWD) Commonly refers to a full-time 4WD system.

Tracción en las cuatro ruedas Usualmente se refiere a un sistema en el cual la fuerza motriz se reparte a tiempo completo entre las cuatro ruedas.

Alternating current (AC) A type of current that changes its direction as it flows through a circuit.

Corriente alterna (AC) Tipo de corriente que cambia de dirección al fluir por un circuito.

Ammeter A meter used to measure electrical current.

Amperímetro Aparato que se usa para medir la corriente eléctrica.

Ampere (amp) The unit of measure for electrical current.

Amperio Unidad de medida para una corriente eléctrica.

Amplitude A measurement of a vibration's intensity. The height of a waveform signal strength, or the maximum measured value of a signal.

Amplitud Una medición de la intensidad de una vibración. La altura de la fuerza de una señal en forma de onda, o el valor máximo medido de una señal.

Analog A type of signal that carries a changing or constant voltage.

Análogo Un tipo de señal que lleva una voltaje que cambia o que es constante.

Annulus Another name for the ring gear of a planetary gear set.

Ánulo Otro nombre para el engranaje en forma de anillo de un conjunto de engranaje planetario.

Asbestos A material that was commonly used as a gasket material in places where temperatures are great. This material is being used less frequently today because of health hazards that are inherent to the material.

Asbesto Material que usualmente se utilizaba como material de guarnición en lugares donde las temperaturas eran muy elevadas. Hoy en día se utiliza dicho material con menos frecuencia debido a los riesgos para la salud inherentes al mismo.

Atom The smallest particle of an element in which all the chemical characteristics of the element are present.

Átomo Partícula más pequeña de un elemento en la cual están presentes todos los atributos químicos del elemento.

Axle shaft A shaft on which the road wheels are mounted.

Árbol motor Árbol sobre el cual se montan las ruedas de la carretera.

Base A solution that has an excess of OH ions, also called an alkali. Also the center layer of a bipolar transistor.

Base Una solución que tiene un exceso de iones de OH, también denominado álcali. También la capa central de un transistor bipolar.

Belleville spring A tempered spring steel cone-shaped plate used to increase the mechanical force in a pressure plate assembly.

Muelle de Belleville Lámina de muelle de acero templado en forma cónica utilizado para aumentar a la fuerza mecánica en un conjunto de placa de presión.

Bevel spur gear A gear that has teeth with a straight centerline cut on a cone.

Engranaje recto cónico Engranaje que tiene dientes con una línea central recta cortados sobre un cono.

Block synchronizers A type of transmission synchronizer assembly that uses a hub, sliding sleeve, blocking rings, insert springs, and inserts.

Sincronizadores de bloque Tipo de conjunto de sincronizador de transmisión que utiliza una maza, un manguito deslizante, aros de bloqueo, resortes insertos e insertos.

Boots Flexible protective covers fastened to an automotive component to prevent contamination and retain lubricant.

Fuelles Cobertores protectores flexibles que se sujetan a un componente automotriz para evitar la contaminación y retener el lubricante.

Bushing A cylindrical lining used as a bearing assembly and made of steel, brass, bronze, nylon, or plastic.

Buje Forro en forma cilíndrica utilizado como un conjunto de cojinete, hecho de acero, latón, bronce, nilón o plástico.

Canceling angles Opposing operating angles of two U-joints cancel the vibrations developed by the individual U-joint.

Angulos de supresión Los ángulos de funcionamiento opuestos de dos juntas universales cancelan las vibraciones producidas por la junta universal individual.

Carburizing A method used to surface-harden steel by heat or mechanical means to increase the hardness of the outer surface while leaving the core relatively soft.

Carbocementación Un método que se utiliza para endurecer la superficie del acero por medio de calor o medios mecánicos para aumentar la dureza de la superficie exterior al tiempo que el centro permanece relativamente blando.

Cardan Universal joint A nonconstant velocity U-joint consisting of two yokes with their forked ends joined by a cross. The driven yoke changes speed twice in 360 degrees of rotation.

Junta de cardán Junta universal de velocidad no constante que consiste de dos horquillas con sus extremos unidos por una cruz. El yugo accionado cambia de velocidad dos veces en una rotación de 360°.

Catalyst A substance that starts or speeds up a chemical reaction while undergoing no permanent change itself.

Catalizador Sustancia que inicia o acelera una reacción química y que, a su vez, no sufre ningún cambio permanente.

Center differential Any of several designs of differential in 4WD or AWD vehicles that allows wheel speed differences between the front and rear axles during cornering.

Diferencial central Cualquiera de los varios diseños del diferencial en vehículos 4WD o AWD que permite diferencias de velocidad entre las ruedas de los ejes delantero y trasero durante la conducción en curvas.

Centrifugal force The force acting on a rotating body that tends to move it outward and away from the center of rotation. The force increases as rotational speed increases.

Fuerza centrífuga Fuerza que acciona sobre un cuerpo giratorio y que tiende a moverlo hacia afuera y en dirección opuesta al centro de rotación. La fuerza aumenta a medida que aumenta la velocidad de rotación.

Centripetal force The force acting on a rotating body that tends to hold it inwards towards the center of the curvature of the path.

Fuerza centrípeta El fuerza que actúa sobre un cuerpo en rotación y que tiende a mantenerlo hacia adentro, hacia el centro de la curvatura de la trayectoria.

Cluster gear A common term for the counter gear assembly.

Engranajes desplazables Término común que denomina el conjunto de contraengranaje.

Clutch A device for connecting and disconnecting the engine from the transmission or for a similar purpose in other units.

Embrague Dispositivo que sirve para engranar y desengranar el motor de la transmisión o para un propósito parecido en otras unidades.

Clutch delay valve A small valve in the clutch hydraulic system designed to prevent instant engagement of the clutch to prevent damage to drivetrain components.

Válvula de demora de embrague Válvula pequeña en el sistema hidráulico de embrague diseñada para evitar la activación instantánea del embrague, a fin de evitar daños en los componentes del tren de impulsión.

Clutch (friction) disc The friction material part of the clutch assembly that fits between the flywheel and pressure plate.

Disco de embrague (fricción) La pieza material de fricción del conjunto de embrague que encaja entre el volante y la placa de presión.

Clutch gear The input shaft of a manual transmission. This shaft splines to and is driven by the clutch disc and contains the main drive gear to the counter gear.

Engranaje del embrague Árbol de transmisión de una transmisión manual. Este eje está conectado con lengüeta y es impulsado por el disco de embrague y contiene el engranaje impulsor principal de la transmisión intermedia.

Clutch housing A large aluminum or iron casting that surrounds the clutch assembly. Located between the engine and transmission, it is sometimes referred to as bell housing.

Alojamiento del embrague Pieza grande fundida en hierro o en aluminio que rodea al conjunto del embrague. Ubicado entre el motor y la transmisión, a veces se llama alojamiento de campana.

Clutch hub The splined center section of a clutch disc. Also, the component of transmission synchronizer assembly that is splined to the output shaft and synchronizer sleeve.

Maza del embrague Sección central estriada de un disco de embrague. Además, es el componente del conjunto del sincronizador de transmisión conectado con lengüeta al eje de salida y el manguito sincronizador.

Clutch pack A series of clutch discs and plates installed alternately in a housing to act as a driving or driven unit.

Paquetes del embrague Serie de discos y placas del embrague instalados de manera alternada en un alojamiento para que funcionen como una unidad de accionamiento o accionada.

Clutch release bearing A bearing assembly at the front of the transmission that is moved to release the pressure plate from the clutch disc.

Cojinete de desembrague Conjunto de cojinetes al frente de la transmisión, que se mueve para liberar la placa de presión del disco de embrague.

Clutch shaft Sometimes known as the transmission input shaft or main drive pinion. The clutch driven disc drives this shaft.

Árbol de embrague A veces llamado árbol impulsor de la transmisión o piñón principal de mando. El disco accionado del embrague acciona este árbol.

Coil spring clutch A clutch using coil springs to hold the pressure plate against the friction disc.

Embrague de muelle helicoidal Embrague que emplea muelles helicoidales para mantener la placa de presión contra el disco de fricción.

Collar shift transmission One of three types of manual transmission. The collar shift transmission uses sliding collars to connect speed gears to the output shaft. These transmissions ore not synchronized.

Transmisión tipo collar shift Uno de los tres tipos de transmisión manual. La transmisión collar shift usa collares deslizantes para conectar los engranajes de velocidad con el eje de salida. Estas transmisiones no están sincronizadas.

Combustion Rapid oxidation, with the release of energy in the form of heat and light.

Combustión Oxidación rápida, con liberación de energía en forma de calor y luz.

Concentric slave cylinder The hydraulic component that acts on the release bearing in a concentric clutch arrangement.

Cilindro secundario concéntrico Componente hidráulico que actúa en el cojinete removible en una disposición de embrague concéntrico.

Conductor A material in which electricity flows easily.

Conductor Una materia por la cual fluye fácilmente la electricidad.

Constant mesh A manual transmission design that permits the gears to be constantly enmeshed regardless of vehicle operating circumstances.

Engrane constante Diseño de transmisión manual que permite que los engranajes permanezcan siempre engranados a pesar de las condiciones de funcionamiento del vehículo.

Constant velocity joint (also called CV joint) A flexible coupling between two shafts that permits each shaft to maintain the same driving or driven speed regardless of operating angle, allowing for a smooth transfer of power. The constant velocity joint consists of an inner and outer housing with balls in between or a tripod and yoke assembly.

Cardán para velocidad constante (llamada también junta CV) Unión flexible entre dos árboles que permite que cada árbol mantenga la misma velocidad de accionamiento o accionada, a pesar del ángulo de funcionamiento, y permite que la transferencia de fuerza sea pareja. El cardán para velocidad constante consiste de un alojamiento interior y exterior entre el cual se insertan bolas, o un conjunto de trípode y yugo.

Continuously variable transmission (CVT) A transmission with no fixed forward gear ratios, typically uses pulleys and a belt rather than gears.

Transmisión variable continua (CVT) Una transmisión sin relación de engranajes delantera fija; típicamente usa poleas y una correa en lugar de de engranajes.

Contraction A reduction in mass or dimension; the opposite of expansion.

Contracción Disminución en masa o dimensión; lo opuesto de expansión.

Controller A device that switches an electrical circuit on and off or that changes the operation of the circuit.

Controlador Un dispositivo que enciende o apaga un circuito eléctrico o que cambia la operación del circuito.

Controller area network (CAN) A commonly used multiplexing protocol for serial communication. The communication wire is a twisted pair wire.

Red de área del controlador (CAN, en inglés) Un protocolo de multiplexación que se utiliza comúnmente para la comunicación serial. El cable de comunicación es un cable de par torzado.

Counter gear assembly A cluster of gears designed on one casting with short shafts supported by antifriction bearings. Closely related to the cluster assembly.

Mecanismo contador Grupo de engranajes diseñados en una sola fundición con árboles cortos apoyados por cojinetes de antifricción. Estrechamente relacionado al conjunto desplazable.

Cross groove constant velocity joint A CV joint using balls and a cage that have the grooves in the inner and outer race cut at an angle.

Junta homocinética con ranuras transversales Junta homocinética que usa bolas y una jaula con ranuras en las pistas interiores y exteriores cortadas en ángulo.

C-type retainer A steel clip shaped in the form of a "C" that fits in a machined groove in the axle and retains the axle in the housing.

Dispositivo de retención tipo C Clip de acero con forma de "C" que se ajusta en una ranura torneada del eje y retiene el eje en la carcasa.

Curb weight The weight of the vehicle when it is not loaded with passengers or cargo.

Peso en vacío El peso de un vehículo sin incluir pasajeros ni carga.

Current The flow of electricity through a circuit.

Corriente El flujo de la electricidad por un circuito.

CV joints Constant velocity joints that allow the angle of the axle shafts to change with no loss in rotational speed.

Juntas CV Juntas de velocidad constante que permiten que el ángulo de los árboles motores cambie sin que disminuya la velocidad de rotación.

Deceleration The rate of decrease in speed.

Desaceleración Disminución de la velocidad.

Density Compactness; relative mass of matter in a given volume.

Densidad Compacidad; masa relativa de materia en un volumen dado.

Detent A small depression in a shaft, rail, or rod into which a pawl or ball drops when the shaft, rail, or rod is moved. This provides a locking effect.

Retén Pequeña depresión en un árbol, una barra o una varilla sobre el/la cual cae un trinquete o una bola cuando se mueve el árbol, la barra o la varilla. Esto provee un efecto de bloqueo.

Diaphragm spring A circular disc shaped like a cone with spring tension that allows it to flex forward or backward. Often referred to as a Belleville spring.

Muelle de diafragma Disco circular en forma de cono, con tensión en el muelle que le permite flexionarse hacia adelante o hacia atrás. Conocido también como muelle de Belleville.

Differential A mechanism between drive axles that permits one wheel to run at a different speed than the other while turning.

Diferencial Mecanismo entre los ejes de mando que permite que una rueda gire a una velocidad diferente que la otra cuando el vehículo gira.

Differential case The metal unit that encases the differential side gears and pinion gears and to which the ring gear is attached.

Caja del diferencial Unidad metálica que envuelve los engranajes laterales del diferencial y los engranajes de piñón, y sobre la cual se monta la corona.

Digital A type of signal that is caused by switching the circuit on and off.

Digital Tipo de señal causada al apagar y encender el circuito.

Diode A semiconductor that allows current to flow in one direction only.

Diodo Semiconductor que permite que la corriente fluya solamente en una dirección.

Direct current (DC) A type of current that always flows in one direction, from a point of higher potential to a point of lower potential.

Corriente directa (DC) Tipo de corriente que siempre fluye en una dirección, desde el punto de energía potencial más alto hacia el punto de energía potencial más bajo.

Direct drive One turn of the input driving member compared to one complete turn of the driven member, such as when there is direct engagement between the engine and drive shaft in which the engine crankshaft and the drive shaft turn at the same rpm.

Toma directa Una vuelta de la pieza de accionamiento comparada a una vuelta completa de la pieza accionada, como por ejemplo, cuando hay un engrane directo entre el motor y el árbol de mando donde el cigüeñal del motor y el árbol de mando giran a las mismas rpm.

Direct-shift gearbox (DSG) The name given to Volkswagen's and Audi's dual clutch manual transmission.

Caja de cambio automático-secuencial (DSG, en inglés) El nombre que se da a la transmisión manual de embrague dual de Volkswagen y Audi.

Dog teeth A series of gear teeth that are part of the dog clutching action in a transmission synchronizer operation.

Diente de sierra Serie de dientes de engranaje que forman parte de la acción del embrague de garras durante una sincronización de la transmisión.

Double Cardan Universal joint A near constant velocity U-joint that consists of two Cardan Universal joints connected by a coupling yoke.

Junta doble de cardán Junta universal de velocidad casi constante compuesta de dos juntas de cardán conectadas por un yugo de acoplamiento.

Double-offset constant velocity joint Another name for the type of plunging, inner CV joint found on many GM, Ford, and Japanese FWD cars.

Junta de velocidad constante de desviación doble Otro nombre para el tipo de junta de velocidad constante interior de pistón tubular, instalada en muchos automóviles de tracción delantera japoneses, de la GM y de la Ford.

Driveline The Universal joints, drive shaft, and other parts connecting the transmission with the driving axles.

Línea de transmisión Las juntas universales, el árbol de mando y otras piezas que conectan la transmisión a los ejes motores.

Driveline windup A condition in which axles, gears, U-joints, and other components can bind or fail if the 4WD mode is used on pavement where 2WD is more suitable. This is caused by wheel speed differences between the front and rear axles.

Falla de la línea de transmisión Condición que ocurre cuando los ejes, los engranajes, las juntas universales, y otros componentes se traban o fallan si se emplea la tracción en las cuatro ruedas sobre pavimiento donde la tracción a las dos ruedas es más adecuada. Esto es causado por las diferencias de velocidad de las ruedas entre los ejes delantero y trasero.

Drive pinion flange A rim used to connect the rear of the drive shaft to the rear axle drive pinion.

Brida de piñón de mando Corona utilizada para conectar la parte trasera del árbol de mando al piñón de mando del eje trasero.

Drive shaft An assembly of one or two U-joints connected to a shaft or tube; used to transmit power from the transmission to the differential. Also called the propeller shaft.

Árbol de mando Conjunto de una o dos juntas universales conectadas a un árbol o tubo; utilizado para transmitir la fuerza motriz desde la transmisión hasta el diferencial. Llamado también árbol transmisor.

Drivetrain The components that transmit power from the engine to the drive wheels. The drivetrain consists of the clutch or torque converter, transmission, drive shaft, rear axle and differential, or transaxle and half shafts.

Tren de mando Los componentes que transmiten potencia del motor a las ruedas motrices. El tren de mando consiste en el embrague o convertidor de par de torsión, la transmisión, el arbol de mando, el eje trasero y diferencial, o el transeje y los semiejes.

Dual-clutch transmission (DCT) The name used to describe an automated manual transmission that uses two clutches and two input shafts.

Transmisión de embrague dual (DCT, en inglés) El nombre que se usa para describir una transmisión manual automatizada que utiliza dos embragues y dos ejes de entrada.

Dual-mass flywheel (DMF) A flywheel composed of two plates connected by a damper assembly. This flywheel is designed to reduce the engine vibrations that are transmitted through the transmission.

Volante de doble masa Un volante compuesto de dos placas unidas por una ensamblaje amortiguador. Este volante está diseñado para disminuir las vibraciones del motor que se transmiten a través de la transmisión.

Duty cycle The percentage of time that an actuator is energized during one complete on-off cycle during pulse-width modulation.

Ciclo de trabajo El porcentaje de tiempo que se energiza un servomotor durante un ciclo de encendido/apagado completo durante la modulación de la magnitud del impulso.

Duty solenoid A solenoid that has its duty cycle controlled by being toggled on and off by the control system.

Solenoide de trabajo Un solenoide cuyo ciclo de trabajo se controla mediante la conmutación a encendido o apagado por parte del sistema de control.

Electricity The release of energy as one electron leaves the orbit of one atom and jumps into the orbit of another.

Electricidad Liberación de energía cuando un electrón abandona la órbita de un átomo y pasa a la órbita de otro.

Electrolysis A chemical and electrical decomposition process that can damage metals such as brass, copper, and aluminum in the cooling system.

Electrólisis Un proceso de descomposición química y eléctrica que puede dañar los metales como el bronce, el cobre y el aluminio en el sistema de refrigeración.

Electrolyte A material whose atoms become ionized, or electrically charged, in solution. Automobile battery electrolyte is a mixture of sulfuric acid and water.

Electrólito Un material cuyos átomos se ionizan, o se cargan eléctricamente, cuando se coloca en una solución. El electrólito de la batería de los automóviles es una mezcla de ácido sulfúrico y agua.

Electromagnet A magnet formed by electrical flow through a conductor.

Electroimán Imán que resulta del flujo de corriente eléctrica a través de un conductor.

Electromagnetic clutch A clutch designed to connect and disconnect two adjoining components through the activation of an electromagnet.

Embrague electromagnético Embrague diseñado para conectar y desconectar dos componentes contiguos mediante la activación de un electroimán.

Element A substance with only one type of atom.

Elemento Sustancia que contiene un sólo tipo de átomo.

Ellipse A compressed form of a circle.

Elipse Forma comprimida de un círculo.

End play The amount of axial or end-to-end movement in a shaft due to clearance in the bearings.

Holgadura Amplitud de movimiento axial o movimiento de extremo a extremo en un árbol debido al espacio libre en los cojinetes.

Engine torque A turning or twisting action developed by the engine measured in foot-pounds or kilogram-meters.

Esfuerzo de rotación del motor Movimiento de giro o torción que produce el motor, medido en libras por pie o en kilogramos por metro.

Equilibrium Exists when the applied forces on an object are balanced and there is no overall resultant force.

Equilibrio Existe cuando las fuerzas aplicadas sobre un objecto son iguales y no resulta otra fuerza.

Evaporate Atoms or molecules break free from the body of the liquid to become gas particles.

Evaporar Los atomos o las moléculas se desprenden del liquido y se convierten en partículas de gas.

Expansion An increase in the size of a mass due to the movemnet of atoms and molecules as heat moves into a mass.

Expansión Aumento de tamaño de una masa a causa del movimiento de átomos y moléculas a medida que el calor se transmite al interior de la masa.

Field of flux Invisible lines around every magnet along which magnetic force acts.

Campo de flujo Líneas invisibles que rodean a los imanes. Las fuerzas magnéticas actúan a lo largo de estas líneas.

Final drive The final set of reduction gears the engine's power passes through on its way to the drive wheels.

Transmisión final Último juego de reductores por el cual pasa la fuerza del motor camino a las ruedas motrices.

Fixed-type constant velocity joint A joint that cannot telescope or plunge to compensate for suspension travel. Fixed joints are always found on the outer ends of the drive shafts of FWD cars. A fixed joint may be of either Rzeppa or tripod type.

Junta de velocidad constante de tipo fijo Junta que no es capaz de proyectarse o hundirse para compensar el movimiento de la suspensión. Siempre se encuentran juntas fijas en los extremos exteriores de los árboles de mando de vehículos de tracción delantera. Una junta fija puede ser de tipo Rzeppa o trípode.

Flange A projecting rim or collar on an object for keeping it in place.

Brida Corona proyectada o collar sobre un objeto que se usa para mantenerlo en su lugar.

Flexible clutch disc A clutch disc that contains torsional damper springs that circle the central hub.

Disco de embrague flexible Disco de embrague que contiene resortes amortiguadores de torsión que rodean la maza central.

Flexplate A lightweight flywheel used only on engines equipped with an automatic transmission. The flexplate is equipped with a starter ring gear around its outside diameter and also serves as the attachment point for the torque converter.

Placa flexible Volante liviano empleado solamente en motores equipados con transmisión automática. La placa flexible está equipada con una corona de arranque alrededor de su diámetro exterior y sirve también de punto de fijación para el convertidor del par motor.

Flywheel A heavy-metal wheel that is attached to the crankshaft and rotates with it; helps smooth out the power surges from the engine power strokes; also serves as part of the clutch and engine-cranking system.

Volante Rueda pesada de metal que se fija al cigüeñal y que gira con él; ayuda a neutralizar las sacudidas de fuerza de las carreras motrices del motor; sirve también como parte del sistema de embrague y de arranque del motor.

Force Any push or pull exerted on an object; measured in pounds and ounces, or in newtons (N) in the metric system.

Fuerza Cualquier empuje o tirón que se ejerce sobre un objeto; medido en libras y onzas, o en newtons (N) en el sistema métrico.

Four-wheel drive (4WD) On a vehicle, driving axles are found at both front and rear, so that all four wheels can be driven. 4WD is the standard abbreviation for four-wheel drive.

Tracción en las cuatro ruedas En un vehículo, los ejes motores se encuentran ubicados en las partes delantera y trasera, para que las cuatro ruedas se puedan accionar. 4WD es la abreviatura común para la tracción en las cuatro ruedas.

Frequency The number of times, or speed, at which an action occurs within a specified time interval. In electronics, frequency indicates the number of times that a signal occurs, or repeats, in cycles per second. Cycles per second are indicated by the symbol "hertz" (Hz).

Frecuencia La cantidad de veces, o la velocidad, a la cual se produce una acción en un intervalo específico. En electrónica, la frecuencia indica la cantidad de veces que se produce, o se repite, una señal en ciclos por segundo. Los ciclos por segundo se indican con el símbolo "hercio" (Hz).

Friction The resistance to motion between two bodies in contact with each other.

Fricción La resistencia al movimiento que se produce entre dos cuerpos que están en contacto el uno con el otro.

Front-wheel drive (FWD) The vehicle has all drivetrain components located at the front.

Tracción delantera Todos los componentes del tren de mando en el vehículo se encuentran en la parte delantera.

Fulcrum The support that provides a pivoting point for a lever.

Fulcro Soporte que le provee punto de apoyo a una palanca.

Fulcrum ring A circular ring over which the pressure plate diaphragm spring pivots.

Anillo de fulcro Anillo circular sobre el cual gira el muelle del diafragma de la placa de presión.

Full-floating rear axle An axle that only transmits driving force to the rear wheels. The weight of the vehicle (including payload) is supported by the axle housing.

Eje trasero enteramente flotante Eje que solamente transmite la fuerza motriz a las ruedas traseras. El puente trasero soporta el peso del vehículo (incluyendo la carga útil).

Gear A wheel with external or internal teeth that serves to transmit or change motion.

Engranaje Rueda con dientes externos o internos que sirve para transmitir o cambiar el movimiento.

Gear ratio The number of revolutions of a driving gear required to turn a driven gear through one complete revolution. For a pair of gears, the ratio is found by dividing the number of teeth on the driven gear by the number of teeth on the driving gear.

Relación de engranajes Número de revoluciones de un mecanismo de accionamiento requeridas para hacer girar un engranaje accionado una revolución completa. Para un par de engranajes se obtiene la relación al dividir el número de dientes en el engranaje accionado por el número de dientes en el mecanismo de accionamiento.

Gross weight The total weight of a vehicle plus its maximum rated payload.

Peso bruto El peso total de un vehículo más su carga útil nominal máxima.

Half shaft Either of the two drive shafts that connect the transaxle to the wheel hubs in FWD cars. Half shafts have CV joints attached to each end to allow for suspension motions and steering. The shafts may be of solid or tubular steel and may be of different lengths.

Semieje Cualquiera de los dos árboles de mando que conectan el transeje a los cubos de rueda en automóviles de tracción delantera. Los semiejes tienen juntas de velocidad constante fijadas a cada extremo para permitir el movimiento de suspensión y la dirección. Los semiejes pueden ser de acero sólido o tubular y pueden variar sus longitudes.

Hardening A process that increases the hardness of a metal, deliberately or accidentally, by hammering, rolling, carburizing, and heat treating.

Temple Un proceso que aumenta la dureza de un metal, intencional o accidentalmente, por medio del martilleo, la laminación, la carbocementación o el tratamiento térmico.

Heat A form of energy caused by the movement of atoms and molecules.

Calor Un tipo de energía que se origina con el movimiento de los átomos y las moléculas.

Heat treating Heating, followed by fast cooling, to harden metal.

Tratamiento térmico Calentamiento seguido del enfriamiento rápido, para endurecer el metal.

Helical gear Gears with the teeth cut at an angle to the axis of the gear.

Engranaje helicoidal Engranajes que tienen los dientes cortados en ángulo al pivote del engranaje.

Hertz (Hz) The unit that frequency is most often expressed and is equal to one cycle per second.

Hercio La unidad en la que se suele expresar la frecuencia y que equivale a un ciclo por segundo.

Horsepower (hp) A measure of mechanical power, or the rate at which work is done. One horsepower equals 33,000 ft.-lb. (foot-pounds) of work per minute. It is the power necessary to raise 33,000 pounds a distance of 1 foot in 1 minute.

Caballos de fuerza (hp) Medida de fuerza mecánica o velocidad a la que se realiza el trabajo. Un caballo de fuerza es equivalentea 33.000 libras-pies de trabajo por minuto. Es el esfuerzo necesario para levantar 33.000 libras una distancia de un pie por minuto.

Hotchkiss drive A type of rear suspension in which leaf springs absorb the rear axle housing torque.

Transmisión Hotchkiss Tipo de suspensión trasera en la cual unos muelles de láminas absorben el par de torsión del puente trasero.

Hub The center part of a wheel to which the wheel is attached.

Cubo Parte central de una rueda, a la cual se fija la rueda.

Hunting gear set A ring and pinion gear set in which every pinion tooth comes in contact with every ring gear tooth over the course of a number of revolutions.

Juego de engranajes con oscilación Conjunto de corona y engranaje de piñón en el que todos los dientes del piñón entran en contacto con todos los dientes de la corona durante una cierta cantidad de revoluciones.

Hypoid gear A gear that is similar in appearance to a spiral bevel gear, but the teeth are cut so that the gears match in a position where the shaft centerlines do not meet; cut in a spiral form to allow the pinion to be set below the centerline of the ring gear so that the car floor can be lower.

Engranaje hipoide Engranaje parecido a un engranaje cónico con dentado espiral, pero en el cual los dientes se cortan para que los engranajes se engranen en una posición donde las líneas centrales del árbol no se crucen; cortado en forma de espiral para permitir que el piñón se coloque debajo de la línea central de la corona para que el piso del vehículo pueda ser más bajo.

Idler gear A gear that rides between a set of gears and does not affect the ratio but does change the direction of the output gear.

Engranaje intermedio Engranaje que se coloca entre un grupo de engranajes y no afecta la relación de los engranajes, pero cambia la dirección del engranaje de salida.

Impermeable Materials that do not adsorb fluids.

Impermeable Materiales que no absorben fluídos.

Inboard constant velocity joint The inner CV joint, or the one closest to the transaxle. The inboard joint is usually a plunging-type joint that telescopes to compensate for suspension motions.

Junta de velocidad constante del interior Junta de velocidad constante interior, o la que está más cerca del transeje. La junta del interior es normalmente una junta de tipo sumergible que se extiende para compensar el movimiento de la suspensión.

Independent rear suspension (IRS) The vehicle's rear wheels move up and down independently of each other.

Suspensión trasera independiente Las ruedas traseras del vehículo suben y bajan de manera independiente la una de la otra.

Inertia The resistance to a change in motion or direction.

Inercia La resistencia a los cambios de movimiento o direción.

In-phase A condition where two events happen, one after the other. The term is used to describe a pair of Universal joints where the second joint cancels speed fluctuations introduced to the shaft by the first joint.

En fase Estado en el que ocurren dos eventos, uno después del otro. El término se utiliza para describir un par de juntas universales en el que la segunda junta elimina las fluctuaciones de velocidad introducidas al eje por la primera junta.

Input shaft The shaft carrying the driving gear by which the power is applied, as to the transmission.

Árbol impulsor Árbol que soporta el mecanismo de accionamiento a través del cual se aplica la fuerza motriz, como por ejemplo, a la transmisión.

Installation angle The measured angle of a drivetrain component expressed in the number of degrees the component is tilted from the horizontal plane.

Ángulo de instalación Ángulo medido de un componente del tren de transmisión, expresado en el número de grados de inclinación del componente con respecto al plano horizontal.

Insulator A material that does not allow electrons to flow easily through it.

Aislador Un material que no permite que los electrones lo atraviesen fácilmente.

Integral Built into, as part of the whole.

Integral Integrado, que forma parte de un todo.

Integral carrier A rear axle carrier assembly which is housed and supported by a one-piece housing that attaches to the suspension.

Portador integral Conjunto de soporte del eje sostenido y protegido por una carcasa de una sola pieza que se ajusta a la suspensión.

Integrated circuit (IC) An electrical circuit composed of many diodes and transistors. The IC is the basis for computerized systems.

Circuito integrado (IC) Un circuito eléctrico compuesto de muchos diodos y transitores. El IC es la base de los sistemas computarizados.

Interlock To prevent one action from occurring if another competing action has taken place.

Enclavamiento Para evitar que ocurra una acción si ya se ha producido otra acción que compite.

Intermediate drive shaft Located between the left and right drive shafts, it equalizes drive shaft length.

Árbol de mando intermedio Ubicado entre los árboles de mando izquierdo y derecho, ecualiza la longitud del árbol de mando.

Ion An ion is an atom or molecule that has lost or gained one or more electrons, resulting in it having a positive or negative electrical charge.

Ión Un ión es un átomo o molécula que ha obtenido o perdido uno o más electrones, lo que da como resultado que tenga una carga eléctrica positiva o negativa.

IRS A common abbreviation for independent rear suspension.

IRS Abreviatura común para la suspensión trasera independiente.

Jounce The up and down movement of the car in response to road surfaces.

Sacudida Movimiento de ascenso y descenso del vehículo causado por las superficies de la carretera.

Journal The area on a shaft that rides on the bearing.

Gorrón La sección en un árbol que se mueve sobre el cojinete.

Kinetic energy Energy in motion.

Energía cinética Energía en movimiento.

Lapping The process of fitting one surface to another by rubbing them together with an abrasive material between the two surfaces.

Lapidado Proceso de ajustar una superficie contra otra rozando la una contra la otra con un material abrasivo colocado entre las dos superficies.

Latent heat The heat released or absorbed when matter undergoes a change of phase.

Calor latente Calor liberado o absorbido cuando la materia sufre un cambio de fase.

Lever A device made up of a bar turning about a fixed pivot point, called the fulcrum, that uses a force applied at one point to move a mass on the other end of the bar.

Palanca Mecanismo formado por una barra que gira alrededor de un punto de pivot fijo, llamado el fulcro, que usa una fuerza aplicada en un punto para mover una masa ubicada en la otra punta de la barra.

Limited-slip differential (LSD) A differential designed so that when one wheel is slipping, a major portion of the drive torque is supplied to the wheel with the better traction; also called a nonslip differential.

Diferencial de deslizamiento limitado Diferencial diseñado para que cuando una rueda se desliza, la mayor parte del par de torsión de mando se transmita a la rueda que tiene mejor tracción. Llamado también diferencial antideslizante.

Live axle A shaft that transmits power from the differential to the wheels.

Eje motor Árbol que transmite la fuerza motriz del diferencial a las ruedas.

Load A term normally used to describe an electrical device that is operating in a circuit. Load also can be used to describe the relative amount of work a driveline must do.

Carga Un término que normalmente describe un dispositivo eléctrico que funciona en un circuito. La carga también puede describir la cantidad relativa de trabajo que debe efectuar una flecha motriz.

Locked differential A differential with the side and pinion gears locked together.

Diferencial trabado Diferencial en el que el engranaje lateral y el de piñón están sujetos entre sí.

Locking hubs Hub assembly at the front wheels of a 4WD vehicle that connect the axles to the hub and wheel. Locking hubs may be manual or automatic.

Mazas de enclavamiento Conjunto de masas en las ruedas delanteras de un vehículo 4WD que conecta los ejes con la maza y la rueda. Las mazas de enclavamiento pueden ser manuales o automáticas.

Lug nut The nuts that fasten the wheels to the axle hub or brake rotor. Missing lug nuts should always be replaced. Overtightening can cause warpage of the brake rotor in some cases.

Tuerca de orejetas Tuercas que sujetan las ruedas al cubo del eje o al rotor de freno. Las tuercas de orejetas faltantes deben siempre reemplazarse. En algunos casos, apretarlas demasiado puede causar la deformación del rotor de freno.

M3 dual-clutch transmission (M-DCT) The name BMW uses for its dual-clutch automated manual transmission.

Transmisión de embrague dual M3 (M-DCT, en inglés) El nombre que utiliza BMW para su transmisión manual automatizada de embrague dual.

Magneto-rheological (MR) fluid A synthetic hydrocarbon fluid containing soft particles that changes its viscosity when introduced to a magnetic field.

Fluído de imán reológica Un fluido sintético de hidrocarburo que contiene partículas blandas y que cambia de viscosidad cuando se acerca a un campo magnético.

Machinist's rule A steel ruler that is precisely graded to allow accurate measurements.

Regla técnica Regla de acero graduada con exactitud para realizar mediciones precisas.

Mass The amount of matter in an object.

Masa Cantidad de materia en un objeto.

Master cylinder The liquid-filled cylinder in the hydraulic brake system or clutch, in which hydraulic pressure is developed when the driver depresses a foot pedal.

Cilindro primario Cilindro lleno de líquido en el sistema de freno hidráulico o en el embrague, donde se produce presión hidráulica cuando el conductor oprime el pedal.

Matter Anything that occupies space.

Materia Cualquier cosa que ocupe espacio.

Mechanical advantage Based on the principles of levers, force can be increased by increasing the length of the lever or by moving the pivot or fulcrum.

Ventaja mecánica En base a los principios de las palancas, la fuerza se puede incrementar al aumentar la longitud de la palanca o al mover el pivote o el fulcro.

Meshing The mating, or engaging, of the teeth of two gears.

Engranar Emparejar, o accionar, los dientes de dos engranajes.

Mild hybrid A vehicle that uses electric motor assistance under load. The electric motor is typically used as a starter and alternator.

Semihíbrido Vehículo que utiliza la asistencia de un motor eléctrico cuando está operando bajo carga. El motor eléctrico se suele utilizar como arrancador y alternador.

Mode fork The shift fork in a transfer case that is moved to engage and disengage four-wheel drive.

Horquilla de modo Horquilla de cambio de la caja reductora que se mueve para activar o desactivar la tracción en las cuatro ruedas.

Momentum A type of mechanical energy that is the product of an object's weight times its speed.

Momento Tipo de energía que es el producto del peso de un objeto multiplicado por su velocidad.

Multiplexing A means of transmitting information between computers. It is a system in which electrical signals are transmitted by a peripheral serial bus instead of common wires, allowing several devices to share signals on a common conductor.

Multiplexación Un medio de transmitir información entre computadoras. Es un sistema en el cual las señales eléctricas se transmiten mediante un bus serial periférico en lugar de mediante cables comunes, lo que permite a varios dispositivos compartir señales en un conductor común.

Nonhunting gear set A ring and pinion gear set in which one pinion tooth contacts only some of the ring gear teeth. One ring gear revolution is required to achieve all possible contact points.

Juego de engranajes sin oscilación Juego de corona y engranaje de piñón en el que un diente de piñón entra en contacto con solo algunos dientes de la corona. Se requiere una revolución completa de la corona para alcanzar todos los puntos de contacto posibles.

Ohm A unit of measurement of electrical resistance.

Ohmio Unidad de medida de la resistencia eléctrica.

Ohm's law A statement that describes the characteristics of electricity as it flows in a circuit.

Ley de Ohm Declaración que describe las características de la electricidad al fluir dentro de un circuito.

Ohmmeter The instrument used to measure electrical resistance.

Ohmímetro Instrumento utilizado para medir la resistencia eléctrica.

On-demand 4WD Systems that power a second axle only after the first begins to slip.

Tracción a las cuatro ruedas según la demanda Sistemas que proveen fuerza motriz a un segundo eje solamente cuando el primero comienza a patinar.

Operating angle The difference between the drive shaft and transmission installation angles is the operating angle.

Ángulo de funcionamiento El ángulo de funcionamiento es la diferencia entre los ángulos de instalación del árbol de mando y de la transmisión.

Oscillation Any single swing of an object back and forth between the extremes of its travel.

Oscilación Cualquier balanceo de un objeto hacia adelante y hacia atrás entre los puntos extremos de su desplazamiento.

Outboard constant velocity joint The outer CV joint, or the one closest to the wheels. The outer joint is a fixed joint.

Junta de velocidad constante exterior Junta de velocidad constante exterior, o la que está más cerca de las ruedas. La junta exterior es una junta fija.

Out-of-round Wear of a round hole or shaft that when viewed from an end will appear egg shaped.

Deformación circunferencial Desgaste de un agujero o árbol redondo que, visto desde uno de sus extremos, parece ovalado.

Output driver The part of a circuit that controls an actuator.

Impulsor de salida La parte de un circuito que controla un servomotor.

Output shaft The shaft or gear that delivers the power from a device such as a transmission.

Árbol de rendimiento Árbol o engranje que transmite la fuerza motriz desde un mecanismo, como por ejemplo, la transmisión.

Overdrive A gear ratio whereas the output shaft of the transmission rotates faster than the input shaft. Any arrangement of gearing that produces more revolutions of the driven shaft than of the driving shaft.

Sobremultiplicación Relación de engranajes donde el árbol de rendimiento de la transmisión gira de manera más rápida que el árbol impulsor. Cualquier distribución de engranajes que produce más revoluciones del árbol accionado que del árbol de accionamiento.

Overhung-mounted pinion A differential drive pinion gear that is supported by two tapered roller bearings and does not use a third support bearing.

Piñón montado con suspensión Engranaje del piñón de mando del diferencial, sostenido por dos cilindros ahusados de cojinete y que no utiliza un tercer cojinete de soporte.

Oxidation Burning or combustion; the combining of a material with oxygen. Rusting is slow oxidation and combustion is rapid oxidation.

Oxidación Quemado o combustión; la combinación de oxígeno con otro elemento. La corrosión es una oxidación lenta, mientras que la combustión es una oxidación rápida.

Oxide A chemical compound that contains at least one oxygen atom and at least one other element.

Óxido Un compuesto químico que incluye por lo menos un átomo de oxígeno y por lo menos un elemento más.

Parallel hybrid A vehicle that uses both the electric motor and engine to power the drive wheels. The vehicle can be driven on the electric motor only under some circumstances.

Híbrido paralelo Vehículo que utiliza tanto el motor eléctrico como el motor de combustión para impulsar las ruedas. El vehículo puede utilizar el motor eléctrico solo bajo ciertas circunstancias.

Partial nonhunting gear set A ring and pinion gear set in which a pinion tooth comes in contact with only some of the ring gear teeth and more than one revolution of the ring gear is needed to achieve all possible gear contact points.

Juego parcial de engranajes sin oscilación Conjunto de corona y engranaje de piñón en el que un diente del piñón entra en contacto con solo algunos dientes de la corona y es necesaria más de una revolución de la corona para alcanzar todos los puntos de contacto del engranaje.

Part-time 4WD Systems that can be shifted in and out of 4WD.

Tracción a las cuatro ruedas a tiempo parcial Sistemas que pueden cambiarse a/o de tracción a las cuatro ruedas según sea necesario.

Permeable Materials that absorb fluids.

Permeable Materiales que absorben los fluidos.

pH scale A scale used to measure how acidic or basic a solution is.

Escala de pH Una escala que se utiliza para medir la acidez o alcalinidad de una solución.

Pin-and-rocker joint A type of drive chain construction using two convex joints that roll against one another as the chain moves.

Junta de clavija basculante Diseño de cadena de transmisión que utiliza dos juntas convexas que ruedan una contra la otra a medida que la cadena se mueve.

Pinion gear The smaller of two meshing gears.

Engranaje de piñón El más pequeño de los dos engranajes de engrane.

Planetary carrier In a planetary gear system, the carrier or bracket in a planetary system that contains the shafts on which the pinions or planet gears turn.

Portador planetario En un sistema de engranaje planetario, portador o soporte en un sistema planetario que contiene los árboles sobre los cuales giran los piñones o los engranajes planetarios.

Planetary gears The gears in a planetary gear set that connect the sun gear to the ring gear.

Engranajes planetarios Engranajes en un tren de engranaje planetario que conectan el engranaje principal a la corona.

Planetary pinions In a planetary gear system, the gears that mesh with, and revolve about, the sun gear; they also mesh with the ring gear.

Piñones planetarios En un sistema de engranaje planetario, los engranajes con que se engranan y giran entorno del engranaje principal; se engranan también con la corona.

Plasma Considered the fourth state of matter. Plasma is a partially ionized gas, in which some of the electrons are free rather than bound to an atom or molecule. The ability of the electrical charges to move somewhat independently makes the plasma electrically conductive; therefore, it responds readily to electromagnetic fields.

Plasma Se considera el cuarto estado de la materia. El plasma es un gas parcialmente ionizado, en el cual algunos de los electrones están libres en lugar de estar unidos a un átomo o una molécula. La capacidad de las cargas eléctricas de moverse con cierta independencia hace que el plasma sea eléctricamente conductivo; por lo tanto, responde fácilmente a los campos electromagnéticos.

Plunging constant velocity joint Usually the inner CV joint. The joint is designed so that it can telescope slightly to compensate for suspension motions.

Junta de velocidad constante sumergible Normalmente junta de velocidad constante interior. La junta está diseñada para que pueda extenderse ligeramente, y así compensar el movimiento de la suspensión.

Potential energy Stored energy.

Energía potencial Energía almacenada.

Potentiometer A variable resistor with three connections that is typically used to change voltage.

Potenciómetro Un resistor variable con tres conexiones, generalmente usado para cambiar el voltaje.

Power The measure of work being done.

Fuerza Medida del trabajo que se está realizando.

Power-split device The basic name for the CVT transmission used in Ford and Toyota hybrid vehicles. This device is based on planetary gears and divides the output of the engine and the electric motors to drive the wheels or the generator.

Dispositivo de división de potencia El nombre básico que se le da a la transmisión CVT que se utiliza en los vehículos híbridos de Ford y Toyota. Este dispositivo se basa en engranajes planetarios y divide la salida del motor y los motores eléctricos para impulsar las ruedas o el generador.

Pressure Force per unit area, or force divided by area. Usually measured in pounds per square inch (psi) or in kiloPascals (kPa) in the metric system.

Presión Fuerza por unidad de área, o fuerza dividida por el área. Normalmente se mide en libras por pulgada cuadrada (lpc) o en kilopascales (kPa) en el sistema métrico.

Protection device A component designed to protect an electrical circuit and electrical components. Examples include fuses, circuit breakers, and fusible links.

Dispositivo de protección Un componente diseñado para proteger un circuito eléctrico y los componente eléctricos. Son ejemplos los fusibles, los interruptores de circuito y los eslabones fusibles.

Pulley A wheel with a grooved rim in which a rope, belt, or chain runs to raise something by pulling on the other end of the rope, belt, or chain.

Polea Una rueda con un borde ranurado en el que corre una cuerda, correa, o cadena que se usa para levantar un objeto tirando del otro extremo de la cuerda, la correa, o la cadena.

Radial load A force perpendicular to the axis of rotation. Loads applied at 90 degrees to an axis of rotation.

Carga radial Fuerza perpendicular al eje de rotación. Cargas aplicadas a 90° del eje de rotación.

Ramp-type differential A design of limited-slip differential.

Diferencial tipo rampa Un diseño de diferencial con deslizamiento limitado.

Range fork The shift fork in a 4WD that is moved to engage and disengage low range.

Horquilla de rango Horquilla de cambio de 4WD que se mueve para activar o desactivar el rango bajo.

Rear-wheel drive (RWD) A term associated with a vehicle in which the engine is mounted at the front and the driving axle and driving wheels are mounted at the rear of the vehicle.

Tracción trasera Término relacionado a un vehículo donde el motor se monta en la parte delantera, y el eje motor y las ruedas motrices en la parte trasera.

Rebound The movement of the suspension system as it attempts to bring the car back to normal heights after jounce.

Rebote Movimiento que se obsrva en la suspensión cuando intenta devolver el automóvil a una altura normal después de una sacudida.

Redox The name given to a reduction–oxidation reaction. The term describes all chemical reactions that cause atoms to have their oxidation number changed.

Redox El nombre que se da a una reacción de reducción-oxidación. El término describe todas las reacciones químicas que hacen que los átomos cambien su número de oxidación.

Reduction The removal of oxygen from a molecule.

Reducción La extracción de oxígeno de una molécula.

Reference voltage (Vref) sensor A type of sensor that changes a reference voltage in response to chemical or mechanical changes.

Sensor de voltaje de referencia Un tipo de sensor que cambia un voltaje de referencia en respuesta a cambios químicos o mecánicos.

Relay An electrical device that allows a low-current circuit to control a high-current circuit.

Relé Un dispositivo eléctrico que permite que un circuito de baja corriente controle un circuito de alta corriente.

Removable carrier A type of rear axle assembly from which the axle carrier assembly can be removed for parts service and adjustment.

Alojamiento portador desmontable Tipo de puente trasero del cual se puede desmontar el conjunto del portador del eje para reparar y ajustar las piezas.

Reverse idler gear In a transmission, an additional gear that must be meshed to obtain reverse gear; a gear used only in reverse that does not transmit power when the transmission is in any other position.

Piñón de marcha atrás En una transmisión, el engranaje adicional que debe engranarse para obtener un engranaje de marcha atrás; el engranaje utilizado solamente durante la inversión de marcha, que no transmite fuerza cuando la transmisión se encuentra en cualquier otra posición.

Rheostat A variable resistor with two connections that is used to change current flow through a circuit.

Reóstato Un resistor variable de dos conexiones que sirve para cambiar el flujo de corriente a través de un circuito.

Rigid clutch disc A clutch disc that does not contain torsional damper springs.

Disco de embrague rígido Disco de embrague que no contiene resortes amortiguadores de torsión.

Ring gear A gear that surrounds or rings the sun and planet gears in a planetary system. Also the name given to the spiral bevel gear in a differential.

Corona Engranaje que rodea los engranajes planetario y principal en un sistema planetario. También es el nombre que se le da al engranaje cónico con dentado espiral en un diferencial.

Rivet A headed pin used for uniting two or more pieces by passing the shank through a hole in each piece, and securing it by forming a head on the opposite end.

Remanche Chaveta de cabeza utilizada para unir dos o más piezas insertando la espinilla en cada una de las piezas a través de un agujero. La espinilla se asegura formando una cabeza en el extremo opuesto.

Roller bearing An inner and outer race on which hardened steel rollers operate.

Cojinete de rodillos Anillo interior y exterior sobre el cual funcionan rodillos de acero templado.

Round pin A drive chain design using a single pin inserted into mating holes at each end of the link plate.

Pasador cilíndrico Una cadena de accionamiento que usa un solo pasador insertado en agujeros de acoplamiento en cada punta de la cacha de la corredera.

Rzeppa constant velocity joint The name given to the ball-type CV joint (as opposed to the tripod-type CV joint). Rzeppa joints are usually the outer joints on most FWD cars. Named after its inventor, Alfred Rzeppa, a Ford engineer.

Junta de velocidad constante Rzeppa Nombre que se le da a la junta de velocidad constante de tipo bola (en contraste con la junta de velocidad constante de tipo trípode). Las juntas Rzeppa normalmente son las juntas exteriores en la mayoría de los automóviles de tracción delantera. Nombrada por su creador, Alfred Rzeppa, ingeniero de la Ford.

Semicentrifugal pressure plate The release levers of this pressure plate are weighted to take advantage of centrifugal force to increase plate loading resulting in reduced driven disc slip.

Placa de presión semicentrífuga A las palancas de desembrague de esta placa de presión se les añade peso para aprovechar la fuerza centrífuga, y hacer que esta aumente la carga de la placa. El resultado será un deslizamiento menor del disco accionado.

Semiconductor A solid crystalline substance, such as germanium or silicon, having electrical conductivity greater than insulators but less than good conductors.

Semiconductor Sustancia cristalina sólida, como el germanio o la silicona, cuya conductividad eléctrica es mayor que los aislantes pero menor que los buenos conductores.

Semifloating rear axle An axle that supports the weight of the vehicle on the axle shaft in addition to transmitting driving forces to the rear wheels.

Eje trasero semiflotante Eje que apoya el peso del vehículo sobre el árbol motor, además de transmitir las fuerzas motrices a las ruedas traseras.

Sequential M Gearbox (SMG) BMW's name for their sequential manual transmission.

Caja de cambio de velocidades secuencial M (SMG, en inglés) El nombre que le da BMW a su transmisión manual secuencial.

Sequential manual transmission (SMT) A manual transmission that shifts automatically, or through driver controls. Gears changes are always made to the next higher or lower gear.

Transmisión manual secuencial (SMT, en inglés) Una transmisión manual que cambia automáticamente o mediante controles del conductor. Los cambios de velocidad siempre se realizan a la velocidad inmediatamente superior o inferior.

Series hybrid A vehicle that uses a small internal combustion engine to charge the batteries that power the electric drive motor.

Híbrido de serie Vehículo que utiliza un pequeño motor de combustión interna para cargar las baterías que alimentan el motor eléctrico.

Shift forks Mechanisms attached to shift rails that fit into the synchronizer hub for change of gears.

Horquillas de cambio de velocidades Mecanismos sujetados a las ranuras del anillo sincronizador del embrague cónico para realizar los cambios de velocidad.

Shift rails Rods placed within the transmission housing that are a part of the transmission gearshift linkage.

Barras de cambio de velocidades Varillas ubicadas dentro del alojamiento de transmisión que forman parte de la biela motriz del cambio de velocidades de la transmisión.

Side gears Gears that are meshed with the differential pinions and splined to the axle shafts (RWD) or drive shafts (FWD).

Engranajes laterales Engranajes que se engranan con los piñones del diferencial y son ranurados a los árboles motores en vehículos de tracción trasera o a los árboles de mando en vehículos de tracción delantera.

Silicone oil A very viscous oil with high-shearing resistance and volumetric change when its temperature increases. Silicone oil is typically used in a viscous coupling.

Aceite de silicona Un aceite muy viscoso con alta resistencia al cizallamiento y cambio volumétrico cuando aumenta su temperatura. El aceite de silicona se utiliza típicamente en un acoplamiento viscoso.

Slave cylinder Located at a lower part of the clutch housing. Receives fluid pressure from the master cylinder to engage or disengage the clutch.

Cilindro secundario Ubicado en la parte inferior del alojamiento del embrague. Recibe presión de fluido del cilindro primario para engranar o desengranar el embrague.

Sliding gear transmission A transmission in which gears are moved on their shafts to change gear ratios.

Transmisión por engranaje desplazable Transmisión en la cual los engranajes se mueven sobre sus árboles para cambiar la relación de los engranajes.

Slip joint In the powertrain, a variable length connection that permits the drive shaft to change its effective length.

Junta corrediza En el tren transmisor de potencia, una conexión de longitud variable que le permite al árbol de mando cambiar su longitud eficaz.

Solenoid An electromagnet with a movable core. The core is used to complete an electrical circuit or to cause a mechanical action.

Solenoide Un electroimán de núcleo móvil. El núcleo sirve para completar un circuito eléctrico o para causar una acción mecánica.

Solution Formed when a solid dissolves into a liquid, its particles break away from this structure and mix evenly in the liquid.

Solución Una solución se forma cuando se disuelve un sólido en un líquido y sus partículas se separan de esta estructura y se combinan de forma pareja con el liquido.

Solvent The liquid in a solution.

Solvente El líquido en una solución.

Specific gravity The weight of a volume of any liquid divided by the weight of an equal volume of water at equal temperature and pressure. The ratio of the weight of any liquid to the weight of water, which has a specific gravity of 1.000.

Gravedad específica El peso de un volumen de cualquier líquido dividido por el peso de un volumen igual de agua a igual temperatura y presión. El índice del peso de cualquier líquido respecto del peso del agua, que tiene una gravedad específica de 1,000.

Speed The distance an object travels in a set amount of time.

Velocidad La distancia que se mueve un objeto durante un plazo de tiempo fijo.

Spline A slot or groove cut in a shaft or bore; a splined shaft onto which a hub, wheel, gear, and so forth, with matching splines in its bore is assembled so that the two must turn together.

Lengüeta Hendidura o ranura que se hace en un árbol o un calibre; árbol ranurado sobre el que se montan un cubo, una rueda, un engranaje, etc., con lengüetas que hacen juego en su calibre, para que ambos giren juntos.

Sprocket A series of teeth on a wheel that engages links on a drive chain.

Piñón Serie de dientes de una rueda que engarzan los eslabones de una cadena de transmisión.

Spur gear Gears cut on a cylinder with teeth that are straight and parallel to the axis.

Engranaje recto Engranajes cortados en un cilindro que tienen dientes rectos y paralelos al eje.

Squeak A high-pitched noise of short duration.

Rechinamiento Sonido agudo de corta duración.

Squeal A continuous high-pitched noise.

Chirrido Sonido agudo continuo.

Standard shift A common name for a manual transmission.

Cambio de velocidades estándar Nombre común para la transmisión manual.

Stick-shift A common name for a manual transmission.

Palanca de marcha Nombre común para la transmisión manual.

Straddle-mounted pinion A differential drive pinion gear that uses two tapered roller bearings and a pilot bearing on the back of the pinion.

Piñón montado a ambos lados Engranaje del piñón de mando del diferencial que utiliza dos cilindros ahusados de cojinete y un cojinete piloto en la parte trasera del piñón.

Stud A threaded shaft that resembles a bolt without a head.

Espárrago Un eje roscado que se parece a un perno sin cabeza.

Sun gear The central gear in a planetary gear system around which the rest of the gears rotate. The innermost gear of the planetary gear set.

Engranaje principal Engranaje central en un sistema de engranaje planetario alrededor del cual giran los demás engranajes. Es el engranaje más interior del tren de engranaje planetario.

Synchromesh transmission Transmission gearing that aids the meshing of two gears or shift collars by matching their speed before engaging them.

Transmisión de engranaje sincronizado Engranaje transmisor que facilita el engrane de dos engranajes o collares de cambio de velocidades igualando la velocidad de estos antes de engranarlos.

Temper To change the physical characteristics of a metal by applying heat.

Templar Cambiar las características físicas de un metal aplicándole calor.

Tensile strength The amount of pressure per square inch an object can withstand just before breaking when being pulled apart.

Resistencia a la tracción La cantidad de presión por pulgada cuadrada que puede soportar un objeto antes de romperse cuando se lo estira.

Tension Effort that elongates or "stretches" a material.

Tensión Fuerza que alarga o estira un material.

Three-quarter floating axle An axle in which the axle housing carries the weight of the vehicle while the bearings support the wheels on the outer ends of the axle housing tubes.

Eje flotante de tres cuartos Puente trasero que soporta el peso del vehículo mientras los cojinetes soportan las ruedas en los extremos exteriores de los tubos del puente trasero.

Thrust bearing A bearing designed to resist or contain side or end motion as well as reduce friction.

Cojinete de empuje Un cojinete diseñado para resistir o contener el movimiento hacia los lados o los extremos así como también para reducir la fricción.

Torque A twisting motion, usually measured in ft.-lbs. (N•m).

Par de torsión Fuerza de torsión, normalmente medida en libras-pies (N•m).

Torque steer A self-induced steering condition in which the axles twist unevenly under engine torque. An action felt in the steering wheel as the result of increased torque.

Dirección de torsión Condición automática de dirección en la cual los ejes giran de manera desigual bajo el par de torsión del motor. Acción que se advierte en el volante de dirección como resultado de un aumento en el par de torsión.

Torque tube A fixed tube over the drive shaft on some cars. It helps locate the rear axle and takes torque reaction loads from the drive axle so the drive shaft will not sense them.

Tubo de eje cardán Tubo fijo sobre el árbol de mando en algunos automóviles. Ayuda a localizar el eje trasero y remueve las cargas de reacción de torsión del eje de mando para que no las reciba el árbol de mando.

Transaxle A type of construction in which the transmission and differential are combined in one unit.

Transeje Tipo de construcción en la que la transmisión y el diferencial se combinan en una sola unidad.

Transfer case An auxiliary transmission mounted behind the main transmission used to divide engine power and transfer it to both front and rear differentials, either full time or part time.

Caja de transferencia Transmisión secundaria montada detrás de la transmisión principal. Se utiliza para separar la energía del motor y transferirla a los diferenciales delantero y trasero, ya sea a tiempo completo o a tiempo parcial.

Transistor A three-terminal semiconductor device used for circuit amplification or switching.

Transistor Dispositivo semiconductor de tres terminales utilizado para amplificar o interrumpir circuitos.

Transmission The device in the powertrain that provides different gear ratios between the engine and drive wheels as well as reverse.

Transmisión Mecanismo en el tren transmisor de potencia que provee diferentes relaciones de engranajes tanto entre el motor y las ruedas motrices como en la marcha atrás.

Transmission case An aluminum or iron casting that encloses the manual transmission parts.

Caja de transmisión Pieza fundida en aluminio o en hierro que encierra las piezas de la transmisión manual.

Transmission control unit (TCU) A common name for the transmission or driveline control computer.

Unidad de control de la transmisión (TCU, en inglés) Un nombre común que se le da a la computadora de control de la transmisión o de la línea de conducción.

Tripod (also called tripot) A three-prong bearing that is the major component in tripod CV joints. It has three arms (or trunnions) with needle bearings and rollers that ride in the grooves or yokes of a tulip assembly.

Trípode Cojinete de tres puntas; componente principal en juntas trípode de velocidad constante. Tiene tres brazos (o muñequillas) con cojinetes de agujas y rodillos que van montados sobre las ranuras o los yugos de un conjunto tulipán.

Two-mode hybrid system A hybrid system that fits into a standard transmission housing and is basically two planetary gear sets coupled to two electric motors, which are electronically controlled. This combination results in a continuously variable transmission and motor/generators for hybrid operation. This arrangement has two distinct modes of hybrid drive operation: one mode for low-speed and low-load conditions and the other is used while cruising at highway speeds.

Sistema híbrido de dos modos Un sistema híbrido que se ajusta en la carcasa de una transmisión estándar y que está formado básicamente por dos juegos de engranajes planetarios acoplados a dos motores eléctricos, que se controlan electrónicamente. Esta combinación da como resultado una transmisión continuamente variable y motores/generadores para operación híbrida. Esta esquema tiene dos modos diferentes de operación de conducción híbrida: un modo para baja velocidad y condiciones de carga baja y el otro que se utiliza para desplazarse a velocidades de autopista.

Underdrive The condition caused by a gear driving another gear at a lower speed.

Transmisión baja La condición causada por un cambio que impulsa a otro cambio a una velocidad inferior.

Universal joint (U-joint) A mechanical device that transmits rotary motion from one shaft to another shaft at varying angles.

Junta universal Dispositivo mecánico que transmite movimiento giratorio de un árbol a otro a ángulos cambiantes.

Velocity The speed of an object in a particular direction.

Velocidad La rapidez de un objeto que va en una dirección específica.

Viscous coupling A coupling using alternating plates connected to two drive shafts or axles. The coupling is sealed and contains a silicone-based fluid that responds to speed differences between the alternating plates.

Acoplamiento viscoso Acoplamiento que usa placas alternantes conectadas a dos ejes o árboles de transmisión. El acoplamiento está sellado y contiene un fluido a base de silicona que reacciona a las diferencias de velocidad de las placas alternantes.

Voltage Electrical pressure that causes current to flow.

Voltaje Presión eléctrica que causa que fluya la corriente.

Voltage generating device A type of sensor that generates voltage in response to the movement of a shaft or other device.

Dispositivo generador de voltaje Tipo de sensor que genera voltaje en respuesta al movimiento de un eje u otro dispositivo.

Voltmeter The instrument used to measure electrical pressure or potential.

Voltímetro Instrumento que se utiliza para medir la presión eléctrica o la energía potencial.

Volts The unit of measurement for electrical pressure or EMF.

Voltio Unidad de medida de la presión eléctrica o EMF (la fuerza electromagnética).

Wave Any single swing of an object back and forth between the extremes of its travel through matter or space.

Onda Cualquier oscilación de un objeto hacia adelante y hacia atrás entre los extremos de su trayectoria de desplazamiento a través de la materia o el espacio.

Wavelength The distance between each compression of a sound or electrical wave.

Longitud de onda La distancia entre cada compresión de una onda de sonido o electricidad.

Weight A force on a mass by the gravitational force.

Peso Fuerza ejercida sobre una masa por la fuerza de gravedad.

Wet clutch A clutch in which the friction disc (or discs) is operated in a bath of oil.

Embrague de disco húmedo Embrague en el que el disco de fricción (o discos) funciona bañado en aceite.

Work What is accomplished when a force moves a certain mass a specific distance.

Tarabajo Lo que se logra cuando una fuerza mueve una masa particular por una distancia especifica.

Worm gear A gear with teeth that resemble a thread on a bolt. It is meshed with a gear that has teeth similar to a helical tooth except that it is dished to allow more contact.

Engranaje sinfín Engranaje con dientes parecidos a los filetes de tornillo en un perno. Se engrana con un engranaje cuyos dientes son parecidos a un diente helicoidal, pero se comba para permitir un mejor contacto.

INDEX

Note: Page numbers followed by "f" indicate material in a figure.